Oil Well Production Mechanism

Training Manual on Well Production Operations for Non-production Engineers (Oil and Gas Production Operations)

RIVER PUBLISHERS SERIES IN ENERGY SUSTAINABILITY AND EFFICIENCY

Series Editor

PEDRAM ASEF
Lecturer (Asst. Prof.) in Automotive Engineering,
University of Hertfordshire,
UK

The "River Publishers Series in Sustainability and Efficiency" is a series of comprehensive academic and professional books which focus on theory and applications in sustainable and efficient energy solutions. The books serve as a multi-disciplinary resource linking sustainable energy and society, fulfilling the rapidly growing worldwide interest in energy solutions. All fields of possible sustainable energy solutions and applications are addressed, not only from a technical point of view, but also from economic, social, political, and financial aspects. Books published in the series include research monographs, edited volumes, handbooks and textbooks. They provide professionals, researchers, educators, and advanced students in the field with an invaluable insight into the latest research and developments.

Topics covered in the series include, but are not limited to:

- Sustainable energy development and management;
- Alternate and renewable energies;
- Energy conservation;
- Energy efficiency;
- Carbon reduction;
- Environment.

For a list of other books in this series, visit www.riverpublishers.com

Oil Well Production Mechanism

Training Manual on Well Production Operations for Non-production Engineers

(Oil and Gas Production Operations)

Mohammed Ismail Iqbal

University of Technology and Sciences,
Nizwa, Oman

Vamsi Krishna Kudapa

UPES, Dehradun, India

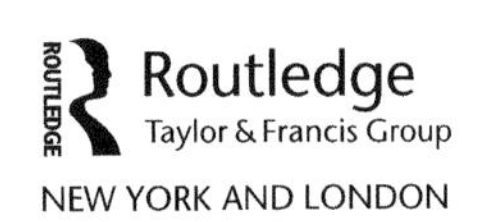

Published 2025 by River Publishers
River Publishers
Alsbjergvej 10, 9260 Gistrup, Denmark
www.riverpublishers.com

Distributed exclusively by Routledge
605 Third Avenue, New York, NY 10017, USA
4 Park Square, Milton Park, Abingdon, Oxon OX14 4RN

Oil Well Production Mechanism: Training Manual on Well Production Operations for Non-production Engineers (Oil and Gas Production Operations) / Mohammed Ismail Iqbal and Vamsi Krishna Kudapa.

Routledge is an imprint of the Taylor & Francis Group, an informa business

ISBN 978-87-7004-203-1 (hardback)
ISBN 978-87-7004-782-1 (paperback)
ISBN 978-87-7004-773-9 (online)
ISBN 978-8-770-04772-2 (ebook master)

Contents

Preface

The Training Manual on Production Operations for Non-Production Engineering is a comprehensive reference for understanding the intricacies of production operations in the oil and gas sector. Written by experienced industry specialists, it provides valuable insights into the technical, environmental, and economic elements of oil and gas exploration and production.

The book starts with an overview of the oil and gas sector, offering historical context and stressing its worldwide relevance and main actors. It then delves into basic reservoir ideas, including methods for finding and appraising prospective reserves. It also discusses well-completion, construction, and intervention procedures, emphasizing production optimization and the crucial phases required to commence and sustain production.

The handbook emphasizes responsible practices while also covering health, safety, and environmental aspects throughout drilling and production activities, as well as methods to reduce ecological consequences. Furthermore, it investigates emerging technologies such as hydraulic fracturing, directional drilling, and the expanding significance of digitalization, offering insights into future industry trends and difficulties.

This book serves as an essential resource for gaining a deep understanding of the oil and gas production processes for both academicians and industrialists.

List of Figures

List of Tables

List of Abbreviations

AGA	American Gas Association
AMJIG	Atlantic Margin Joint Industry Group
API	American Petroleum Institute
BHA	Bottom hole assembly
BHP	Brake horse power
BOP	Blow-out preventor
BP	British Petroleum
BPV	Backpressure valve
CCL	Casing collar locator
CEC	Cation exchange capacity
CF	Coarse filtration
CHAS	Combined habitat alignment system
CPM	Cross-polarized microscopy
CSR	Conceptual study report
CTC	Casing tubing cleanout
DO	Deoxygenation
DOT	Deoxygenation tower
DP	Dynamic positioning
DSC	Differential scanning calorimetry
DSV	Diving support vessel
E&D	Exploration and Development Directorate
ECP	External casing packer
EDTA	Ethylene diamine tetra acetic acid
EOR	Enhanced oil recovery
EOS	Equation of state
EPR	Emergency pipeline repair
ESD	Emergency shut-down system
ESP	Electrical submersible pump
FCM	First contact miscible
FDPSO	Floating drilling, production, storage and offloading
FF	Fine filtration
FOLT	Fiber optic light transmittance

FPF	Floating production facility
FPS	Floating production system
FPSO	Floating production storage and offloading
GGS	Group gathering station
GOR	Gas oil ratio
GVF	Gas volume fraction
HAS	Habitat alignment system
HCl	Hydrochloric acid
HEC	Hydroxyethylcellulose
HF	Hydrofluoric acid
HFEMW	High-frequency electromagnetic wave
HFZ	Hydrate formation zone
HPG	Hydroxypropyl guar gum
HWO	Hydraulic workover units
ICB	International competitive bidding
ID	Internal diameter
IFT	Interfacial tension
IMR	Inspection, maintenance, and repair
IOR	Improved oil recovery
IPR	Inflow performance relationship
IRM	Inspection, repair, and maintenance
IRR	Internal rate of return
IRS	Institute of reservoir studies
LACT	Lease automatic custody transfer
MCM	Multiple contact miscible
MCS	Master control system
MEG	Mono ethylen glycol
MER	Maximum efficient rate
MF	Multiplication factor
MINAS	Minimal National Standards
MIP	Main injection pump
MMP	Minimum miscibility pressure
MMS	Minerals management services
MMSCM	Million standard cubic meter
MMT	Million metric tons
MOPNG	Ministry of Petroleum and Natural Gas
MPY	Mills per year
MSV	Multi-purpose support vessel
NGHP	National gas hydrate program
NPSH	Net positive suction head

NPV	Net positive value
NPV	Net present value
NTU	Nephelometric turbidity unit
OOIP	Original oil in place
P3D	Pseudo-3D
PCP	Progressive cavity pump
PD	Positive displacement
PI	Productivity index
PLC	Programmable logic controller
PLEMS	Pipeline end manifolds (PLEMS)
POOH	Pull out of hole
PPD	Pour point depressants
PSI	Pressure sensing instrument
PVT	Pressure, volume, and temperature
RCP	Resin-coated proppant
REC	Reserve estimate committee
RIH	Run in hole
ROI	Return on investment
ROU	Right of User
ROV	Remotely operated underwater vehicle
SBM	Single buoy moorings
SCR	Steel catenary riser
SCSSV	Surface controlled sub-surface safety valve
SGMA	Self-generating mud acid
SMYS	Specified minimum yield strength
SPAR	Single point anchor reservoir
SPM	Single-point mooring
SPMC	Single-phase multi-chamber
SRP	Sucker rod pump
STO	Stock tank oil
SWLP	Seawater lift pump
TDS	Total dissolved solids
TIC	Tubing intake analysis
TLP	Tension leg platform
TPI	Tilted plate interceptor
TPTZ	2,4,6-Tripyridyl-s-triazine
TSO	Tip screen out
TSS	Total suspended solid
UMC	Underwater manifold center
VIT	Vacuum-insulated tubing

VLP	Vertical lift performance
WAG	Water alternate gas
WAT	Wax appearance temperature
WD	Water depth
WOR	Water/oil ratio

1

Production Operations

1.1 Introduction

1.1.1 History and occurrence of petroleum

The word petroleum is a combination of words "petra" meaning rock and "oleum" meaning oil. Hence, rightly many times we refer kerosene as rock oil. Since earliest times recorded by mankind, the petroleum is in use as a fluid having medicinal value and in religious ceremonies. However, with the discovery of large quantities underground, the use of petroleum spread slowly in the middle of nineteenth century. Further with the invention of internal combustion engines and turbines, which use petroleum fluids as the source of energy and invention of numerous useful chemicals derived from petroleum fluids, petroleum became one of the most important natural resources of modern civilization. Hence, the amount of petroleum being produced and used by a country has become an indicator of the economic strength of that country.

Petroleum deposits occur in a variety of ways and forms both as surface deposits and subsurface deposits. These deposits are found on all continents and regions, although the abundance of finding may vary very widely. Oil and gas seepages, tar asphalt, or bitumen are various forms of petroleum that occur at the surface of the ground. The subsurface occurrences can be classified as pools, fields, and provinces. Pool is a body of oil or gas or both found underground occurring as a separate reservoir and under a single-pressure system. Several such pools having a single geologic feature constitute a Field. For example, Gandhar and Ankleshwar fields having many major and minor pools. Petroleum province is a region, in which a number of oil and gas pools and field occur in a similar geologic region like Southeastern Oklahoma, Kansas, and Western Texas of US and Upper Assam and South Gujarat in India.

There is not a single way or method to find a petroleum reservoir that would enable easy finding of these deposits. It takes years together and crores

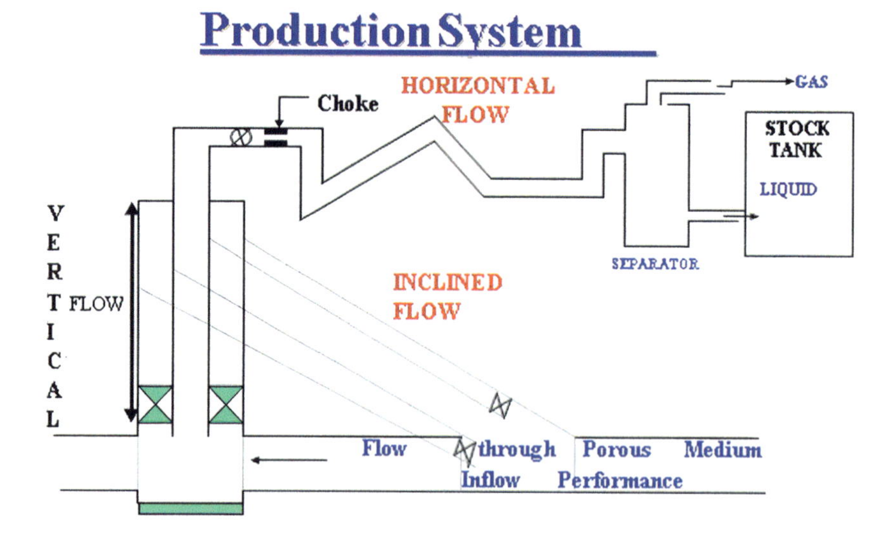

Figure 1.1 Depicts various types of production fluid flows.

of rupees to find an oil pool/field and develop it for commercial exploitation. Furthermore, petroleum being a non-reusable energy once used cannot be regenerated; for the continued development of the country, we need to keep on striving to find new oil fields on one hand while on the other hand we should recover economically as much as we can from the presently producing fields. Technically speaking, we need to improve the recovery factor (ratio of oil produced to original oil in place) of the existing fields. Production engineer plays a very important role in maximizing the recovery of the hydrocarbons from the petroleum reservoir through various production activities and by employing state-of-the-art technologies.

1.2 Production Operations

Production operations in an E & P company can be broadly classified into two groups: (a) exploitation and (b) processing. It is the responsibility of the production-engineering department to devise means to exploit the oil and gas economically and efficiently and to process the fluids extracted to make suitable for the use by different consumers downstream (Figure 1.1).

Following are the important production operations carried out during exploitation of oil and gas from the fields.

- Field development
- Pressure maintenance

- Well completion
- Well intervention
- Well stimulation
- Well analysis
- Artificial lifting
- Abandoning

1.2.1 Field development

Once the presence of hydrocarbons and its extent is established through the reservoir and well-testing analysis, the field is taken up for development. Reservoir engineers through various studies and simulations come out with several wells to be drilled and their approximate locations. The following parameters play a very important role in the development of the field:

- Type of reservoir
- Petroleum fluid characteristics
- Expected surface pressures and the quantities
- Type of pressure maintenance required
- Type of artificial lift method required
- Expected type of enhanced oil recovery methods

The type of reservoir like an active water-driven reservoir, depletion drive, or gas cap-driven reservoir decides upon the type of secondary means of maintaining reservoir pressure. For example, a reservoir may have an active bottom water drive in which case the reservoir pressure almost remains constant or there is a very little reduction in reservoir pressure, needing no pressure maintenance system, while if a reservoir is depletion type without any aquifer support, then we may have to build injection facility along with the establishing production facility. While if a reservoir has a large gas cap, we may plan to inject back extra gas into the gas-cap to maximize the oil recovery. Characteristics of petroleum fluids decide upon the facilities to be installed inside GGS to deal with flow assurance, separation of oil, gas, and water, etc. Like if a crude has strong emulsion or the crude is highly viscous, then we may have to design facilities for heating, injecting flow improvers and demulsifies, etc. As the field is put on production at a certain point of time, the wells will need an artificial lift to lift the fluids. Then suitable installations like high-pressure gas compressor in case of gas lift and other facilities like captive power generator may be required to be installed.

An oil field would essentially have fluid collecting centers called group gathering stations or gas collection systems, where oil, gas, and water are separated and pumped out to a central collecting center. A central bulk storage facility called central tank farm from where the bulk oil is dispatched to the refinery. In addition to oil handling system, gas handling and processing centers like LPG, C2-C3 extraction plants and compressor plants for the compression of gas and supplying to the consumers are also created.

1.2.2 Pressure maintenance

Whenever we put a well on production, the pressure inside the wellbore against formation gets reduced, causing an increase in volume of oil, gas, solution gas, and water and decrease in pore volume of rock, causing the flow of reservoir fluids into the wellbore. The fluids in the well travel through the tubing, utilizing the pressure energy and expansion of free gas and liberation and expansion of solution gas. As the fluid is produced during the course of time, the reservoir energy starts getting depleted, and hence to maximize the recovery from a field, pressure of the reservoir is maintained through water/gas injection into the formation.

1.2.2.1 Need for maintaining reservoir pressure

Need for pressure maintenance can be summed up as follows:

a) As we produce, the reservoir pressure declines due to withdrawal of fluids from the reservoir. Reservoir pressure decline adversely affects the oil production as the force that pushes the oil from the reservoir to the wellbore reduces.

b) Decline in reservoir pressure causes solution gas to be evolved out of oil. This evolved gas occupies more and more void space, resulting in increase in gas saturation. Once the gas saturation exceeds critical gas saturation, the gas starts moving toward the wellbore, which reduces the quantity of oil reaching the wellbore.

c) With the evolution of solution gas, the viscosity of oil in the reservoir increases, making it less mobile; hence, more oil is left behind.

d) In case of water drive reservoirs, if the reservoir fluid replacement by water is not sufficient enough to fill up the void space created by production, then there would be rapid reduction in reservoir pressures resulting in poor recovery. In such a case, we need to inject water at oil water contact to maintain the reservoir pressure.

e) In case of solution gas drive, the reservoir pressure would fall rapidly as the flow of fluids into the wellbore is caused due to the expansion of the oil and liberated gas and as there is no or very little water flowing into the reservoir. If we do not maintain reservoir pressure, the gas saturation in the reservoir would increase, thereby causing a rapid increase in GOR. This would result in the production of more and more gas and much of the oil would be left behind in the reservoir. In order to maximize the recovery from such a reservoir, we need to maintain the reservoir pressure above the bubble point pressure as far as possible and fill up the voidage created by production.

   The best example of pressure maintenance by water injection is GS 12A sand. In this, water injection is being carried out by injecting water down dip while the production is being taken from the wells in the up dip.

Therefore, it is required to maintain reservoir pressure above certain values by injecting water, gas, or any other miscible fluid from surface into the reservoir. As a thumb rule, it is required to maintain the reservoir pressure above bubble point pressure to prevent segregation of solution gas from oil within the reservoir in case of depletion-driven reservoirs. While in case of reservoirs with a combination drive or gas cap gas drive, there is need of maintaining higher reservoir pressures for longer time so that as far as possible, the release of solution gas within the reservoir is put under control. Another purpose of pressure maintenance is to replace natural displacing forces like gas with more efficient displacing force like water.

Sometimes nitrogen/natural gas is also used as injectant. But however water being abundantly available and cheap with very few handling problems, water injection is the most widely used method for maintaining the reservoir pressures.

1.2.3 Well completion

Well is the most important component of oil and fields. Well is the gateway to the reservoir through which we produce the reservoir fluids and also the only conduit through which we communicate with the reservoir, to know the reservoir health and we carry out various operations through wells to maximize the recovery. A well completion should be designed such that we are able to carry out all future operations effectively and safely.

A well mainly constitutes a set of tubular and sealing elements. Tubulars that are used to isolate the surrounding rocks are called casing pipe. The number of different sizes of casing pipes used in a well depends upon the depth of the well and the presence of thief zones, fracture gradients, and the density of mud needed to drill. Generally, in ONGC 3 casing policy is used. The casing is cemented with the outer casing or the rock to achieve the zonal isolation and ensure the integrity of the well so that the high-pressure oil and gas do not migrate into upper low pressure zones causing underground blow-outs. At the top, the casings are held within the outer casing through casing hangers and sealing elements. The outer most casing is known as outer casing while the innermost casing, which holds tubing, is called production casing. The diameter of the production casing depends both on the depth and the deliverability of the wells. For wells needing a tubing size less than 3 ½ inch, 5 ½ or 7 inch casing is used, while for larger tubing sizes, we need 9 ½ inch production casing as in the case of Mumbai High Asset.

The major factors influencing well completion are:

a) Reservoir parameters

b) Well service type

c) Fluid parameters expected to be handled

d) Environmental aspects and government regulations.

1.2.3.1 Completion methods

Completion methods can be categorized depending upon the type of production casing completion, tubing casing configurations, and tubing and packer configurations.

1.2.3.1.1 Production casing completions

A well can be completed as open hole or cased and perforated or can be liner completed. In case of open hole, the production casing is run in up to the top of the formation and cemented. This facilitates fully opening the face of the formation for production. It is suitable for tight zones like basement completions. The main disadvantage is of lesser control over production and multiple difficulties to complete multiple zones. A few wells of Borholla field are completed as bare-foot completion. Cased and perforated wells are preferred over the bare foot completion owing to the flexibility of completion of different zones spaced over different point of times and can be completed for commingled production. With the use of packer, it gives many variations in completions to suit the particular need of a field/well. In ONGC almost all the wells are cased and perforated. A well could be liner completed with liners to deal with a sludgy formation or with a perforated liner to deal with unconsolidated formations. The basic advantage is of lesser cost as the liner need not be run from the surface and can be hung inside a regular casing just above the formation.

1.2.3.1.2 Different tubing and casing configurations
A well could be completed as tubing-less or with a tubing hanging from the tubing hanger. A tubing-less completion is useful for producing multiple reservoirs from a single well. This kind of completion is economical as tubings for each reservoir can be run in large hole and cemented and perforated and produced to the surface. However, completion with hanging tubing is the most commonly used completion in ONGC because of easy adaptability for future artificial lift requirements and easy well interventions.

1.2.3.1.3 Different tubing and packer configurations
The use of packers along with tubing gives flexibility of producing different zones separately through different tunings like in dual completed wells and/or to produce the different zones through a single tubing using isolation packers and producing in commingled fashion.

1.2.4 Well intervention

When a well becomes sick and stops producing due to mechanical and/or reservoir/completion problems, a production engineer analyzes the probable causes of sickness of the well and a workover plan is made and the well is repaired by deploying a workover rig, which is essentially a small capacity drilling rig. However, in offshore, the drilling rig carries out the workover operations also. Once a rig is deployed, the well is subdued with suitable gravity workover fluid and the wellhead is removed. After installing a blow-out preventor, the tubings are pulled out and the planned repair jobs are carried out.

Sometimes, a well is worked out in live condition by using a specialized unit called snubbing unit. This unit is designed to carry out workover operations in a well in live condition without killing a well. This helps in minimizing the formation damage due to the use of workover fluids and also enables a repair job on a well, which cannot be subdued due to various reasons.

1.2.5 Well stimulation

1.2.5.1 Objective
Restoration and/or enhancement of well's productivity or injectivity are the prime objective of well stimulation. Restoration is necessitated

when formation damage (measured in terms of skin factor) is suspected. Enhancement techniques are generally employed in low productivity/injectivity conditions. These objectives can be explained through Darcy's law for fluid flow in a well

$$q = \frac{7.08kh(\text{Pr- } Pwf)}{\mu_O B_O \left[\ln(r_e / r_w) - 3/4 + S\right]}.$$

The presence of skin factor, *S*, diminishes the productivity of the well as is evident from the above equation. Stimulation methods (like matrix acidization) for restoration of well's productivity aims to reduce this value of skin factor.

In a low permeability condition, the well's potential is inherently low; a suitable stimulation technique (like fracturing) aims to reduce the "r_e/r_w" factor in the above equation. To sum up, there are two main objectives of well stimulation, viz:

- Restoration of well's productivity/injectivity by removing formation damage.
- Enhancement of well's productivity/injectivity by improving fluid flow path in the reservoir.

1.2.5.2 Types of well stimulation

Following are the main types of stimulation treatment used in oilfields to repair the damages and to enhance the performance of a well:

- Matrix acidization
- Hydraulic fracturing
- Acid fracturing
- Solvent treatment
- Scale treatment

1.2.5.2.1 Matrix acidization

Matrix acidizing is defined as the injection of acid into the formation porosity (intergranular, vugular, fracture) at a pressure, which is below the pressure at which a fracture can be opened. The goal of a matrix acidizing treatment is to achieve, more or less, radial acid penetration into the formation without disturbing the rock mechanics. It is done to remove formation damage, which is either natural or induced.

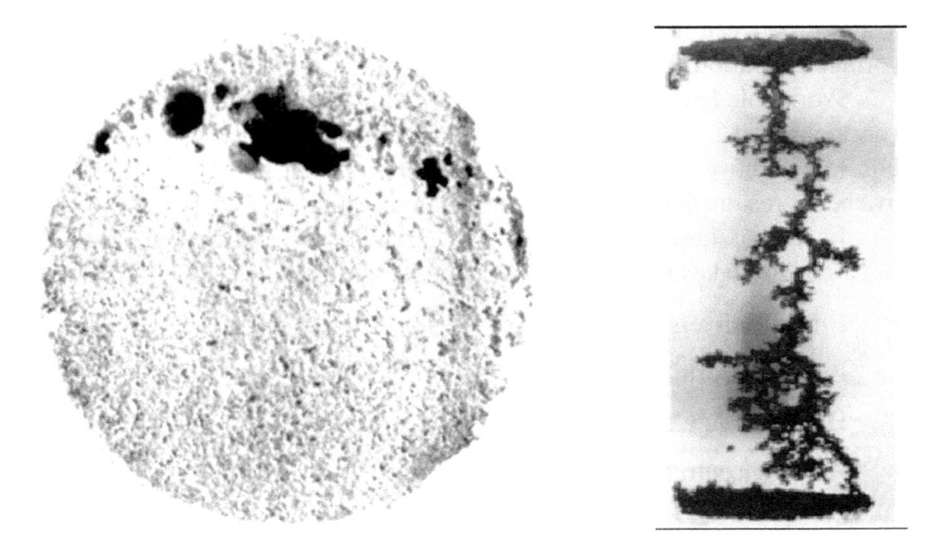

Figure 1.2 Carbonate core with large wormholes.

Matrix acidization in sandstone: Matrix acidizing in sandstone achieves the natural true permeability of the formation by removing clay around the wellbore. The acid is exposed to a large area in the vicinity of the wellbore; it is effective only for a short distance in the reservoir. Retarded hydrofluoric acid systems can reach a longer distance. The required volume of acid is determined in the lab with representative core plugs. Alternatively, acid volumes are also obtained through computer simulators. Mud acid treatment is the most commonly carried out acidization treatment in sandstone reservoirs. It is basically a mixture of hydrofluoric acid and hydrochloric acid. Hydrofluoric acid reacts with carbonates, silicates, and siliceous materials.

Matrix acidization in limestone: It removes damage in the vicinity of the wellbore and also improves near-wellbore permeability by enlarging the pore throats and flow channels. HCl is injected into the formation. A major problem in carbonate acidizing is that HCl usually reacts too fast. The main factors that affect HCl reaction rate on limestone and dolomite are area/volume ratio and diffusion. HCl reaction time is indirectly proportional to the carbonate surface area in contact with the acid. Diffusion is the time required for spent HCl on the formation surface to be replaced by live acid in solution.

When HCl is injected into a carbonate reservoir under matrix conditions, acid preferentially flows into the areas of highest permeability. Acid reaction in the high-permeability regions causes the development of large, highly conductive flow channels called wormholes (Figure 1.2). High reaction rates, as observed between all concentrations of HCl and carbonates, tend to favor wormhole creation.

Wormhole length normally is controlled by the fluid-loss rate from the wormhole to the formation matrix. It can be substantially increased with fluid-loss additives. Emulsified acids, because of their high viscosity, often give better results than plain HCl to contain the fluid loss.

In view of these major areas of concern in matrix acidizing, application for carbonates takes into consideration the following:

- Effective diversion
- Limiting worm holing and excessive fluid leak-off
- High- and low-temperature applications
- Acid concentration

1.2.5.2.2 Acid fracturing

Acid fracturing is injection of acid into the formation at a pressure high enough to fracture the formation or open existing fractures. Its benefit is that a highly conductive flow channel remains open after treatment. The length of conductive fracture depends on the rate of acid reaction and rate of fluid loss from the fracture to the formation.

In an acid fracturing treatment, acid, or a fluid used as a pad before the acid, is injected down the well casing or tubing at rates higher than the reservoir will accept. This produces a buildup in wellbore pressure until it exceeds the compressive earth stresses and tensile rock strength. At this pressure, the formation fails, allowing a crack (fracture) to be formed. The fracture so formed is propagated by continued fluid injection.

Fracture width can be maximized by:

- Using pad fluid with high viscosity
- Injecting fluid at high rate
- Injecting a large volume of fluid
- Reducing the rate of fluid loss to the formation by adding fluid-loss additives

The design of an acid fracturing treatment to stimulate production involves the following five steps:

i. Determination of formation rock and fluid properties
ii. Selection of variable parameters, including the fracturing fluid to be used as a pad, injection rate, etc.

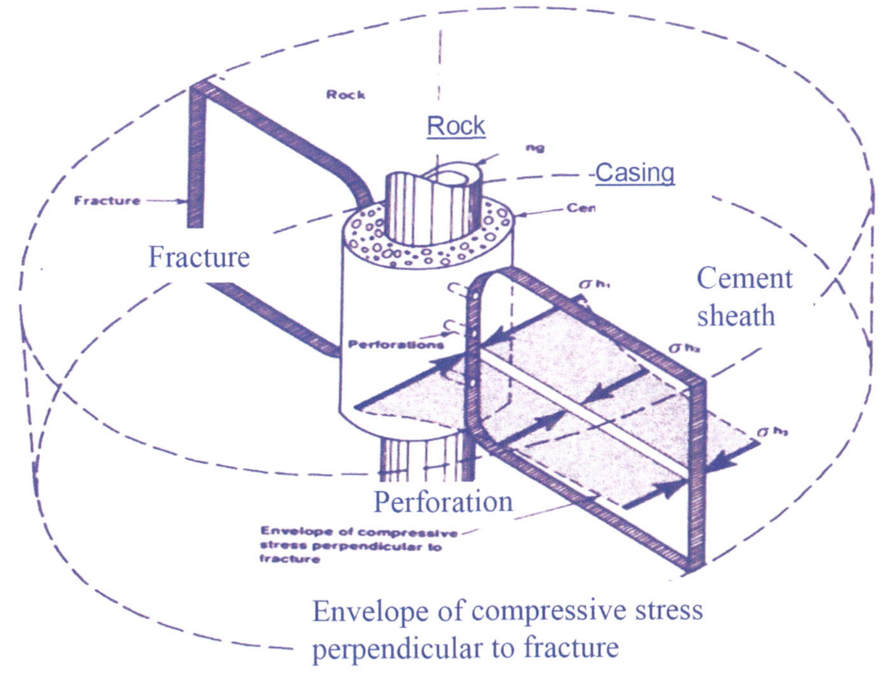

Figure 1.3 Propagation of fracture in outward direction.

iii. Predicting the fracture geometry and the acid penetration distance for the fracturing fluid and acid of interest

iv. Predicting the fracture conductivity and expected stimulation ratio for pad and acid volumes of interest

v. Selecting the most economic treatment

1.2.5.2.3 Hydraulic fracturing

Hydraulic fracturing greatly differs from matrix stimulation. It can be regarded as one of the most complex procedures performed on a well. It involves rock mechanics. Fluids at high rates and pressures are injected into the formation either through producers or injectors and the induced pressure is sufficient to overcome native stresses in the formation and create a fracture in the porous medium (Figure 1.3). Hydraulic fracturing is one of the primary engineering tools for improving well productivity, which is achieved by:

- creating a conductive channel through near wellbore damage and thus bypassing the damaged zone, and
- extending the channel to a greater depth into the reservoir to further increase productivity.

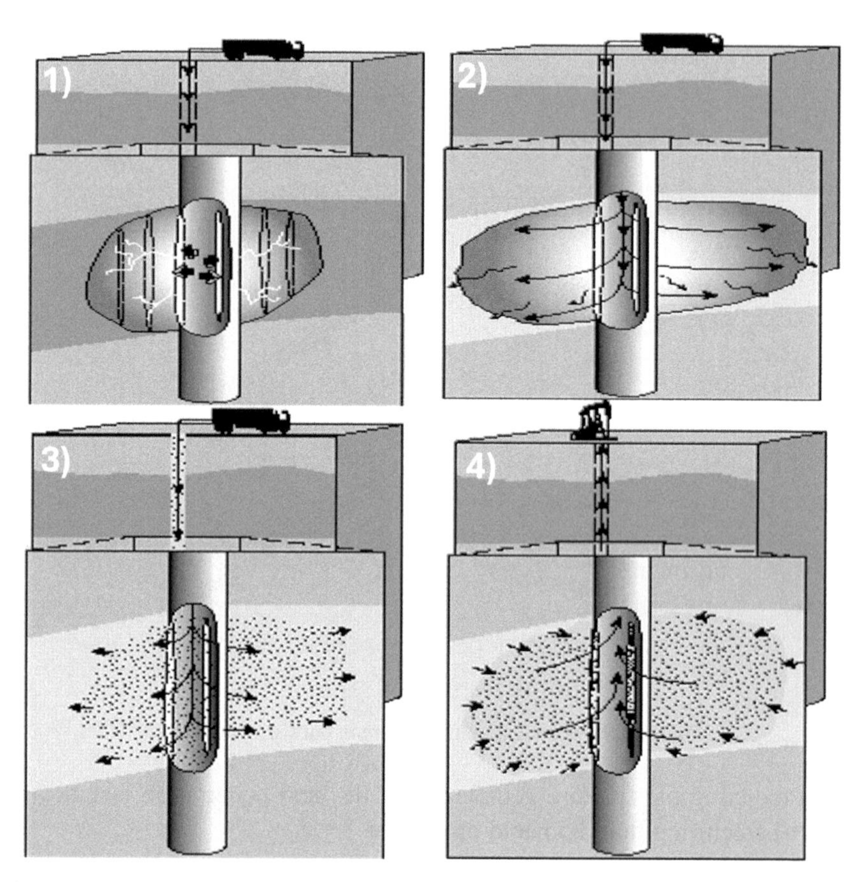

Figure 1.4 (1) Injection pressure breaking the wellbore. (2) Propagation of fractures. (3) Proppants in the fracture. (4) Improved flow into wellbore.

If the fluid is pumped into a well faster than the fluid can escape into the formation, consequently, pressure rises, and at some point, rock cannot sustain the pressure and it breaks. Because most wells are vertical and the smallest stress is the minimum horizontal stress, the initial splitting (or breakdown) results in a vertical, planar parting in the earth.

Once the fractures are created, proppants are placed inside the fractures so that the fractures generated do not close on the release of the fracturing fluid pressure (Figure 1.4).

1.2.6 Well analysis and remedial measures

Depending on the economics of a particular situation, a sick well may be related within specific limits to low oil or gas production, high gas–oil ratio,

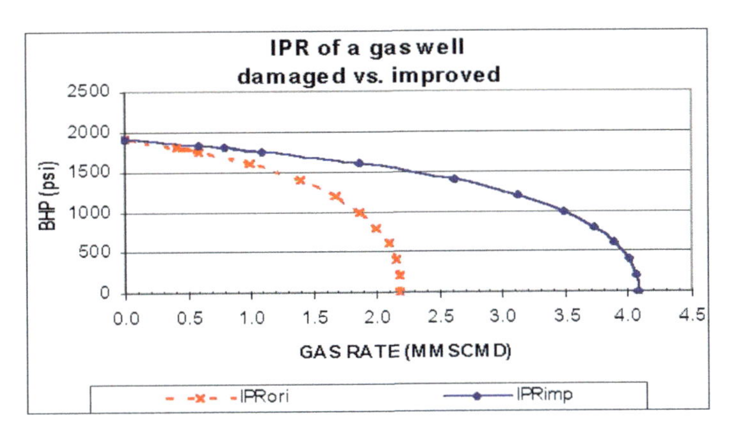

Figure 1.5 Inflow performance curve (IPR) of gas well from a normal well and well with more skin damage.

high water cut, or mechanical problems. Problems in injection or disposal wells may be related to high injection pressures and low injection volumes, or any mechanical problems. Often, a quantifying limit, e.g., a minimum oil production rate, is used as a basis to distinguish sick wells. Problem well analysis may be handled on a reservoir basis, an area basis, or an individual well basis.

A comprehensive sick well analysis is always considered an ideal approach before the decision is taken to bring a workover rig at the well location for carrying out the sequence of jobs required. A periodic sick well analysis not only helps in increasing oil production by identifying the production problem and taking the requisite corrective measure but also helps in a better resource planning (e.g., annual planning for services of workover rig, MSV, etc.), which could be an important aspect in the management of offshore fields.

Most oil and gas wells may become sick due to *wellbore plugging*, *perforation restrictions*, or *formation damage*. Therefore, removing or bypassing these inflow restrictions will increase production. Such damage may be indicated by production well tests, pressure buildup and drawdown tests, comparison with offset wells, and analysis of production history. In suspect well, production logs may be needed to define which zones are producing (or not producing) and the volumes of oil and/or water contributing from each.

The following figure illustrates the difference in pressure drawdown in a normal well as compared with one having serious skin damage (Figure 1.5).

Inflow restrictions due to wellbore plugging, paraffin or asphaltene plugging, emulsion blocks, condensate blocks, water blocks, plugging due to fines of the formation, and water and gas channeling are some of the reasons for a well to become sick. Well analysis helps to identify the probable reasons for a well becoming sick so that suitable remedial actions can be taken.

1.2.7 Methodology of sick well analysis

1.2.7.1 Apparent well problem

- Apparent problem identified based on well production history, pressure and production logs.
- Similar problems existing in nearby wells of the same reservoir/field/layer are identified.

1.2.7.2 Analysis of well history

- Drilling history/methodology to be examined to check for mud composition, fluid loss and formation damage.
- Evaluation of primary cement job by CBL-VDL-USIT logs, for checking primary cementation.
- Study of available initial completion data, which includes:
- Casing policy, downhole equipment, perforation history, layer that is completed
- Details of well stimulation and results
- Well test data.
- Workover history.
- Analysis of production history, which includes test results of oil, water, and gas rate.

1.2.7.3 Reservoir considerations

- Pay zone thickness, permeability, porosity, water saturation, and relative permeability.
- Maps like structure contour, iso-pay, and iso-porosity to understand depositional history of reservoir rock.
- The operating reservoir drive mechanism.
- Forecasted future reservoir behavior.

1.2.7.4 Comparison of performance with nearby wells

- Structural position, fluid contacts, and completion interval of each well considered.
- Performances of nearby producers and injectors are compared.

1.2.7.5 Special checks/surveys

- Check for wellbore communication.
- For pumping wells, fluid level with echo-meter survey.
- For gas lift wells, P-T survey and two-pen pressure recorder charts.
- Recent and initial water sample analysis for comparison.
- Bottomhole sample analysis for paraffin, asphaltene, or scale.
- Corrosion problems and casing leak history of the well.

1.2.8 Artificial lifting

The purpose of artificial lift is to create a sustained and steady low pressure or reduced pressure in the wellbore against the sand face, so as to allow the well fluid to come into the wellbore continuously as per the productivity index of the well and the difference between the average reservoir pressure and the wellbore pressure at the sand face. Thus, a steady stream of production would result. In other words, maintaining a required and steady low pressure against the sand face, which is called steady flowing bottomhole pressure, is the fundamental basis for the design of any artificial lift system.

1.2.8.1 Artificial lift methods

As per the statistics available, the first oil well was exploited commercially in October 1859. Colonel Edwin Drake used a downhole pump to produce a mixture of oil and water from a depth of 10 feet from the surface. It can thus be said that this was the first use of artificial lift to produce oil commercially. Today, that is after 143 years, downhole pumps are still the most predominant mode of lift of producing oil, gas, and water mixture. Out of the total number of artificial lift wells, the pump-operated wells are more than 80%.

For more than a century, engineers operating in E&P industry have been constantly looking to evolving and employing of various methods of artificial lift to produce oil wells in the most economic manner with the maximum efficient rate (MER) of production from the respective oil wells. As a result, various types of artificial lift are in operation in oil wells worldwide.

The commonly used artificial lift methods can be broadly categorized into two groups, namely those that use pumps and those that use high-pressure gas.

1.2.8.2 Artificial lift methods – pumps

Essentially, all pumps raise the pressure in a fluid by converting mechanical work into potential energy, which is called pump pressure. Fluid enters the pump at a given pressure, which is called suction pressure. The pump by its mechanical action delivers energy to the incoming fluid and as a result, fluid leaves the pump at a higher pressure, which is called discharge pressure. Therefore, the gain in pump pressure is determined as the difference between the discharge and suction pressure. However, the gain in pump pressure or gain in potential energy represents only a fraction of the total external work used to drive a pump that accounts the efficiency of the pump. So, while selecting the pump for lifting of oil from a well, its capability suiting to the individual's requirement, cost, and efficiency are always taken into consideration.

Pumps are broadly divided into two groups, namely *positive displacement pumps* and *dynamic displacement pumps* according to the methods used to transform driving forces into pressure.

Positive displacement pumps develop pressure by moving a piston or cam to reduce the volume of a compression chamber. The reduction of the volume of the compression chamber raises the pressure of the liquid in the chamber and that is why, in a positive displacement pump, the pumping rate is independent of pump discharge pressure. It is because of this reason, safety arrangements like relief valve, safety valve, or overload cut out is provided at the discharge side to avoid the pump parts/body from getting damaged or burst in case the pump discharge pressure by chance exceeds its rated pressure.

The most common type of positive displacement pumps for artificial lift in oil wells is sucker rod pumps (SRP).

1.2.8.2.1 Sucker rod pump

It is a plunger-type reciprocating pump, where the plunger creates pumping action by reciprocating inside the barrel. Its prime mover and operating mechanisms are installed on the surface at the wellhead. The rotating motion of the prime mover is converted to reciprocating motion with the various link mechanisms on the surface. The reciprocating motion is communicated to the sucker rod pump by a string of sucker rods. During the down stroke of the

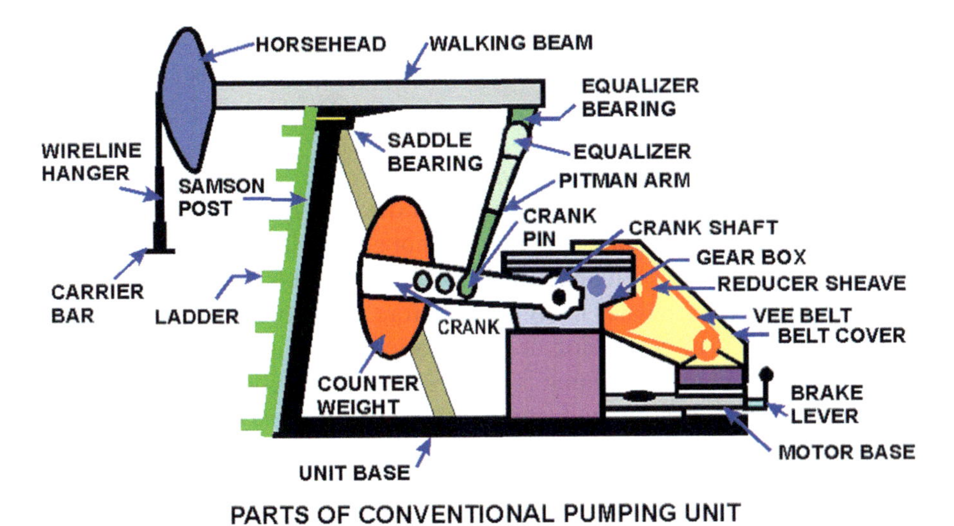

Figure 1.6 Pictorial representation of sucker rod pump (SRP) with various components labelling.

pump plunger; the well fluid enters the pump above its plunger and during the upstroke of the plunger, the pump delivers the well fluid to the surface (Figure 1.6).

SRP is a very old technique in the oil industry for lifting of crude oil from the wells and in fact it is the most widely used mode of artificial lift system in the present-day scenario. As per published data, approximately 80% to 90% of artificial lift wells have been operating on SRP. An SRP operated by conventional beam pumping unit is more versatile and more common among other types of surface operating mechanisms. SRP is generally preferred for producing very low to moderately producing wells. All the SRP operating wells are located onshore, since this type of pump is difficult to be installed on offshore wells mainly because of the constraint of space and load on the offshore platforms. Progressive cavity pump is another positive displacement pump commonly used for lifting high viscous crude.

Dynamic displacement pumps develop pressure by a sequence of acceleration and deceleration of the pumped liquid. Dynamic energy as supplied from an external source is required to accelerate the liquid and to build up required levels of kinetic energy in this process. Thereafter, the liquid is decelerated, and thereby the kinetic energy is transformed into potential energy. This energy transformation results in pump pressure. In dynamic displacement pumps, the pumping rate is low at higher discharge pressure and high at lower discharge pressure. The most common types of dynamic displacement pump being used for oil well's production are electrical submersible pump (ESP) and hydraulic jet pump.

1.2.8.2.2 Electrical submersible pump

ESP is a multistage turbine type of pump for developing high heads. In a turbine or regenerative pump, the liquid does not get discharged freely from the tip of the impeller but is circulated back to a lower point on the impeller diameter. The liquid re-circulates many times before it finally leaves the impeller. As a result of this re-circulation or regeneration, pump develops high heads. Each stage of a multistage pump is just a small turbine consisting of a rotating impeller and a stationary diffuser. The fluid enters the eye of the impeller, where it is accelerated with the help of the rotating impeller and gains velocity, i.e., it develops kinetic energy. Thereafter, the fluid gets discharged into stationary diffuser, where it decelerates; consequently, kinetic energy is transformed into potential energy, which corresponds to pressure increase. ESP is basically a multistage centrifugal pump. Its stages are arranged in series as a pancake, where the discharge end of one stage is the suction to its adjacent overlying stage. The total pressure developed across the pump is the sum of the pressures from the individual stages. Its diameter is limited by the diameter of the well. The pump, motor, etc., are interconnected with a long shaft and the whole assembly is submerged in the well fluid (Figure 1.7). The pump is usually installed below the tubing and electricity is supplied through a special heavy-duty armored cable. It is generally preferred for a very high rate of production.

Other pumping systems like hydraulic jet pump and hydraulic piston pump are also in use but are not as popular as the other above-mentioned pumping systems.

1.2.8.2.3 Artificial lift methods – gas lift

The use of high-pressure gas for artificial lift, commonly called "gas lift," is the other important artificial lift method. The invention of bellows-type gas lift valve and mandrel, which houses gas lift valve, has made this type of artificial lift technology possible for lifting well fluid to the surface.

1.2.8.2.3.1 Continuous gas lift

A steady stream of high-pressure injection gas is conveyed down the well deep enough through one conduit. The injection gas then enters the other conduit, usually the tubing, through a deeply installed gas lift valve and beneath a column of flowing well fluid. The injection of the gas into the well fluid

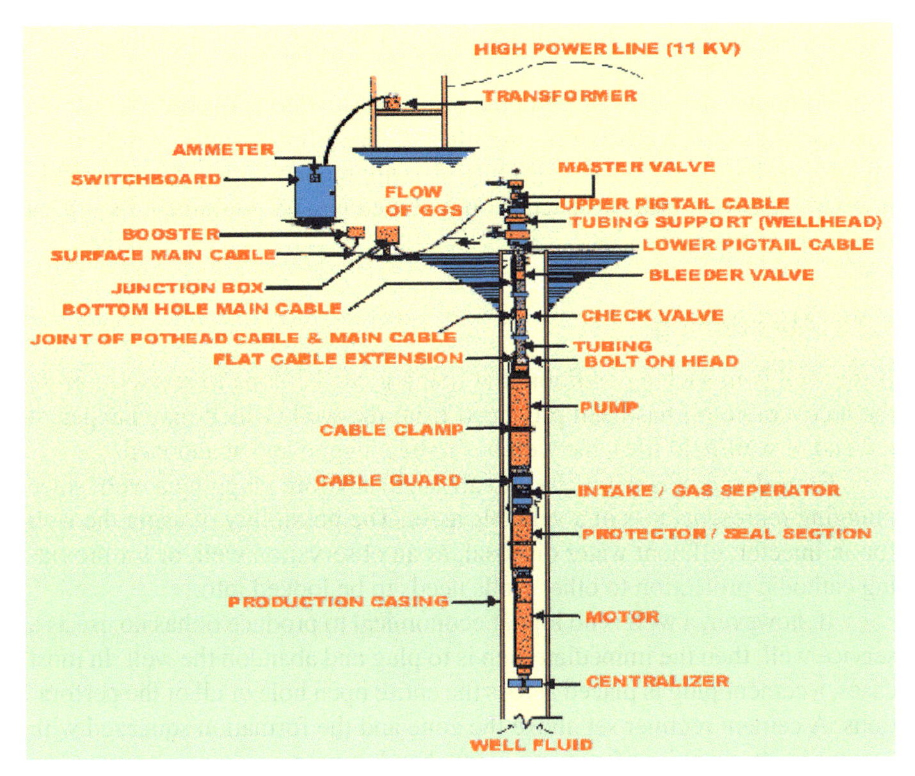

Figure 1.7 Pictorial representation of electrical submersible pumps (ESP) with various components labeling.

column causes aeration of the well fluid from the point of gas injection to the surface, causing the well fluid column to become lighter. As a result, the required flowing bottomhole pressure is obtained at the sand face sufficient enough to produce fluid from the well at the desired rate. This form of gas lift is termed as continuous gas lift because gas injection is done continuously. This type of lift is generally installed in moderate to high producing wells.

1.2.8.2.3.2 Intermittent gas lift

Intermittent gas lift is another common form of gas lift. In this type of lift, high-pressure gas of sufficient volume is injected beneath a well fluid column in the other conduit, usually tubing, through a gas lift valve located deep in the well. The injection period is very short and gas injection is done at regular intervals. The sudden injection of gas at a very high rate forms a gas slug, which acts almost similar like a piston that pushes the accumulated well fluid slug to the surface. This form of gas lift is known as intermittent gas lift because gas injection is done intermittently. This type of lift is applicable for very low producing wells.

Although the share of pumping wells worldwide is over 80%, still the number of gas lift wells is very significant. Gas lift is usually preferred for offshore wells, mainly because the gas-lift equipment is largely located downhole (in offshore, platform space is limited) and gas-lift equipment requires a minimum of maintenance.

1.2.9 Abandoning

If the well is in such a condition that nothing can be done to revive it, or all the recoverable oil has been produced from the well (which may happen at the end of well/field life), the well has to be plugged and abandoned.

Consideration is to be given to all options before plugging a well, since plugging represents loss of a valuable asset. The possibility of using the well for an injector, effluent water disposal, as an observation well, or for providing cathodic protection to other wells needs to be looked into.

If, however, a well is no longer economical to produce or has no use as a service well, then the immediate step is to plug and abandon the well. In most cases, a cement plug is placed across the entire open hole or all of the perforations. A cement retainer set above the zone and the formation squeezed with cement is often preferred as permanent abandonment.

Cement plugs should also be placed across any zones where the casing has been severed or is in danger of external corrosion. It is necessary to protect the freshwater sands near the surface.

2

Basic Reservoir Concepts

2.1 Basic Reservoir Concepts

A "reservoir" is defined as an accumulation of oil and/or gas in a porous and permeable rock.

Reservoir engineering is that segment of petroleum engineering which is concerned primarily with reservoirs. Over the years, reservoir engineering evolved gradually as a separate function of petroleum engineering as it became apparent that maximum recovery could be achieved only by controlling reservoir behavior as a whole.

During the 1940s, reservoir engineering made remarkable advances as a result of growing demand and because of the relatively large increases in ultimate recovery, which could be obtained by utilizing the principles of reservoir engineering. The present state of technology in the petroleum engineering demands that all petroleum engineers in the industry have a good understanding of the principles governing reservoir behavior. Thus, the production engineers and drilling engineer must be well grounded in the fundamentals of the reservoir engineering, since all these functions are closely related.

The principal function of a reservoir engineer is to predict the future behavior of a petroleum reservoir under the various producing mechanisms, which are, or may become, available. The economics of various operating plans is an integral part of any reservoir engineering study. A study of the recovery to be expected from various operating plans along with an economic analysis of these plans will determine the need for pressure maintenance, secondary recovery, cyclic, or other operations. From his studies, the reservoir engineer must recommend an operating plan, which will yield maximum recovery of oil.

2.1.1 Reservoir dynamic behavior

The reservoir and well behavior under dynamic conditions are key parameters in determining what fraction of hydrocarbons initially in place will be produced to surface over the lifetime of the field, at what rates they will be produced, and which unwanted fluids such as water are also produced. The reservoir and well performance are linked to the surface development plan and cannot be considered in isolation; different subsurface development plans will demand different surface facilities. The prediction of reservoir and well behavior is therefore a crucial component of field development planning, as well as playing a major role in reservoir management during production.

2.1.2 The driving force for production

Reservoir fluid and the rock matrix are contained under high temperatures and pressures; they are compressed relative to their densities at standard temperature and pressure. Any reduction in pressure on fluids or rock will result in an increase in the volume, according to the definition of compressibility.

Gas has much higher compressibility than oil or water and therefore expands by a relatively large amount for a given pressure drop. As underground fluids are withdrawn, any free gas present expands readily to replace the voidage, with only a small drop in the reservoir pressure. If only oil and water were present in the reservoir system, a much greater reduction in reservoir pressure would be experienced for the same amount of production.

The expansion of the reservoir fluids, which is a function of their volumes and compressibility, acts as a source of driving energy, which can act to support primary production from the reservoir. Primary production means using the natural energy stored in the reservoir as a driving mechanism for production. Secondary recovery would imply adding some energy to the reservoir by injecting fluids such as water or gas, to help to support the reservoir pressure as production takes place.

One additional contribution to driving energy is by pore compaction. As the pore fluid pressure reduces due to production, the grain to grain stress increases, which leads to the rock grains crushing closer together, thereby reducing the remaining pore volume and effectively adding to the drive energy. The effect is usually small but can lead to reservoir compaction and surface subsidence in cases where the pore fluid pressure is dropped considerably and the rock grains are loosely consolidated.

2.2 Reservoir Drive Mechanism

Oil recovery operations *traditionally* have been subdivided into three stages: primary, secondary, and tertiary. Primary production, the initial production stage, resulted from the displacement energy naturally existing in a reservoir.

Secondary recovery, the second stage of operations, usually was implemented after primary production declined. Traditional secondary recovery processes are water flooding, pressure maintenance, and gas injection. Tertiary recovery is the third stage of production, which was obtained after water flooding. Tertiary processes used miscible gases, chemicals, and/or thermal energy to displace additional oil after the secondary recovery process becomes uneconomical.

Oil recovery is *now* classified as primary, secondary, and EOR processes.

Primary recovery: Natural energy displacement, solution gas drive, and natural water drive, fluid and rock expansion and gravity drainage.

Secondary recovery: The augmentation of natural energy through injection of water or gas to displace oil toward producing wells.

EOR: Injection of gases or liquid chemical and/or the use of thermal energy.

IOR: Which includes EOR but also encompasses a broader range of activities; e.g., reservoir characterization, improved reservoir management, and infill drilling.

2.2.1 Primary recovery

Results from the use of natural energy present in a reservoir as the main source of energy for the displacement of oil to producing wells. These natural energy sources are solution gas drive, natural water drive, fluid and rock expansion, and gravity drainage. Three sets of initial conditions can be distinguished and reservoir and production behavior may be characterized in each case (Table 2.1):

2.2.1.1 Solution gas drive

It occurs in a reservoir that contains no initial gas cap or underlying active aquifer to support the pressure and therefore oil is produced by the driving

Table 2.1 Represents various initial conditions for different drive mechanisms.

Drive mechanism	Initial conditions
Solution gas drive	Under-saturated oil
Gas cap drive	Saturated oil with gas cap
Water drive	Saturated or under-saturated oil

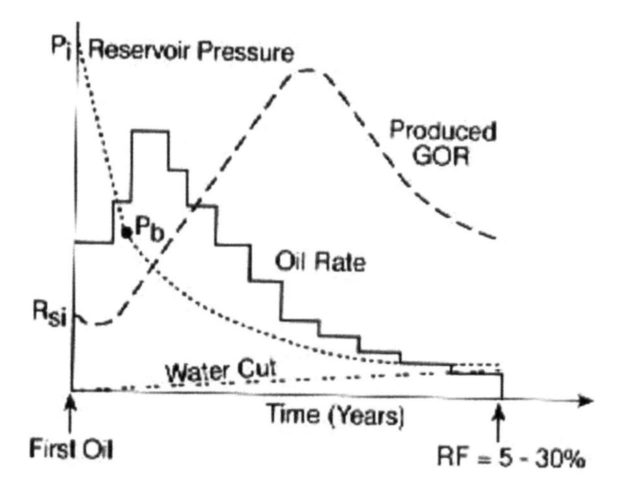

Figure 2.1 Depicts the well production profile during the solution gas drive mechanism.

force due to the expansion of oil and connate water plus any compaction drive. The contribution to drive energy from compaction and connate water is small, so the oil compressibility initially dominates the drive energy. Because the oil compressibility itself is low, pressure drops rapidly as production takes place, until the pressure reaches the bubble point.

Once the bubble point is reached, solution gas starts to become liberated from the oil, and since liberated gas has a high compressibility, the rate of decline of pressure per unit of production slows down. Once the liberated gas has overcome a critical gas saturation in the pores, below which it is immobile in the reservoir, it can either migrate to the crest of the reservoir under the influence of buoyancy forces or move toward the producing wells under the influence of the thermodynamic forces caused by the low pressure created at the producing well (Figure 2.1). To make use of the high compressibility of the gas, it is preferable that gas forms a secondary gas cap and contributes to the drive energy. This can be encouraged by reducing the pressure sink at the producing wells and by locating the producing wells away from the crest of the field.

The typical recovery factor from a reservoir developed by solution gas drive is in the range of 5%–30%, depending largely on the absolute reservoir pressure, the solution GOR of the crude, the abandonment conditions, and the reservoir dip. The upper end of this range may be achieved by a high dip reservoir, with a high GOR, light crude, and a high initial reservoir pressure.

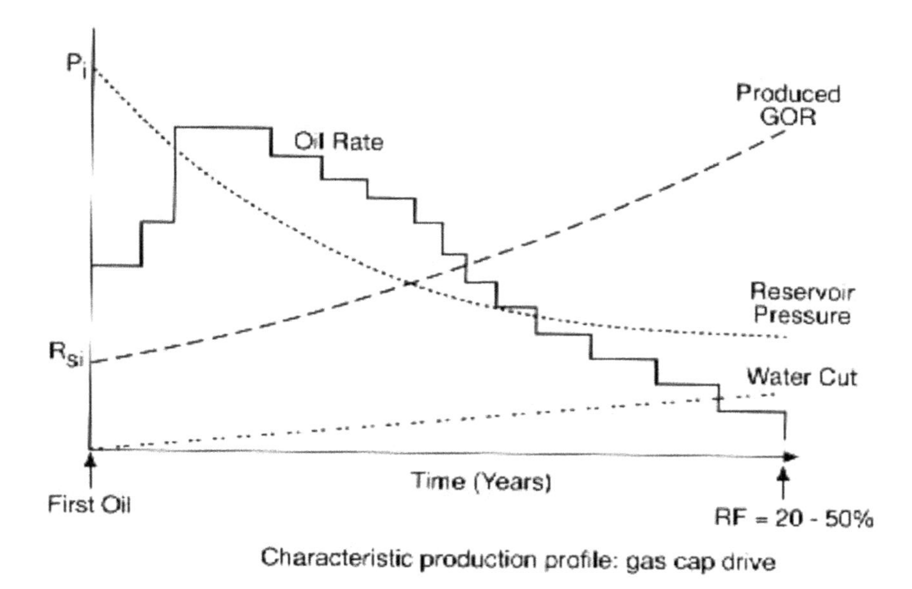

Figure 2.2 Depicts the well production profile during the gas cap drive mechanism.

2.2.1.2 Gas cap drive

Requires as an initial gas cap. The high compressibility of gas provides drive energy for production, and larger the gas cap, the more energy is available. The well positioning follows the same reasoning as the solution gas drive; the objective being to locate the producing wells and their perforations as far as away from the gas cap as possible, but not so close to the OWC to allow significant water production via coning. Compared to solution gas drive case, the typical production profile for gas cap drive shows a much slower decline in reservoir pressure, due to the energy provided by the highly compressible gas cap, resulting in a more prolonged plateau and a slower decline (Figure 2.2). The producing GOR increases as the expanding gas cap approaches to producing wells, and gas is coned or cusped into the producers. Typical recovery factors for gas cap drive are in the range of 20%–60%.

2.2.1.3 Water drive

Occurs when the underlying aquifer is both large and the water is able to flow into the oil column, i.e., it has a communication path and sufficient permeability. If these conditions are satisfied, then once production from the oil column creates a pressure drop, the aquifer responds by expanding, and water moves into the oil column to fill the void created by production. Since water compressibility is low, the volume of water must be large to make the process effective, hence the need for the large connected aquifer.

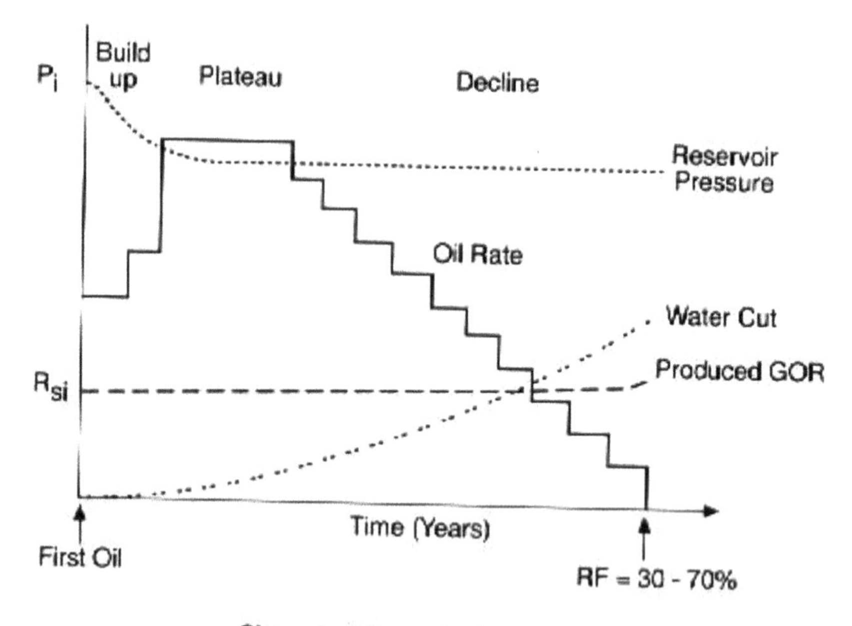

Figure 2.3 Depicts the well production profile during the water drive mechanism.

The aquifer response may maintain the reservoir pressure close to initial pressure, providing a long plateau period and slow decline of oil production. The producing GOR may remain approximately at the solution GOR if the reservoir pressure is maintained above the bubble point. The outstanding feature of the production profile is the large increase in the water cut over the life of the field, which is usually the main reason for abandonment (Figure 2.3). The recovery factor is in the range of 30%–70%, depending on the strength of the natural aquifer, or the efficiency with which injected water sweeps the oil.

2.2.1.4 Combination drive

It More than one of these drive mechanisms may occur simultaneously; the most common combination being gas cap drive and natural aquifer drive.

2.3 Enhanced Oil Recovery

EOR results principally from the injection of gases or liquid chemicals and/or the use of thermal energy. Hydrocarbon gases, CO_2, nitrogen, and flue gases are among the gases used in EOR processes. Several liquid chemicals are commonly used, including polymers, surfactants, and hydrocarbon solvents. Thermal processes typically consist of the use of steam or hot water, or rely on the *in situ* generation of thermal energy through oil combustion in the reservoir rock.

EOR processes involve the injection of fluid or fluids of some type into a reservoir. The injected fluid or injection processes supplement the natural energy present in the reservoir to displace oil to producing wells. Injected fluid interacts with reservoir rock/oil system to create conditions favorable for oil recovery. Examples: results in lower IFTs, oil swelling, oil viscosity reductions, wettability modification, or favorable phase behavior. The interactions are attributable to physical and chemical mechanisms and to the injection or production of thermal energy.

General classification and description of EOR processes:

- Mobility control
- Chemical
- Miscible
- Thermal
- Microbial EOR

2.3.1 Mobility control

It is a generic term describing any process where an attempt is made to alter the relative rates at which injected and displaced fluids are moved through a reservoir. The objective of mobility control is to improve the volumetric sweep efficiency of a displacement process. Mobility control is usually discussed in terms of the mobility ratio, M, and a displacement process is to have mobility control if M is less than or equal to 1.0. Volumetric sweep efficiency generally increases as M is reduced.

Because it is often not feasible to change the properties of the displaced fluid when it is oil or the permeabilities of the rock to the displaced fluids, most mobility control processes of current interest involve the addition of chemicals to the injected fluid. These chemicals increase the apparent viscosity of the injected fluid and/or reduce the effective permeability of rock to the injected fluid. The chemicals are primarily polymers when the injected fluid is water and surfactants that form foams when injected fluid is a gas. In some cases, mobility control is attained by WAG (water alternate gas) injection (Figure 2.4).

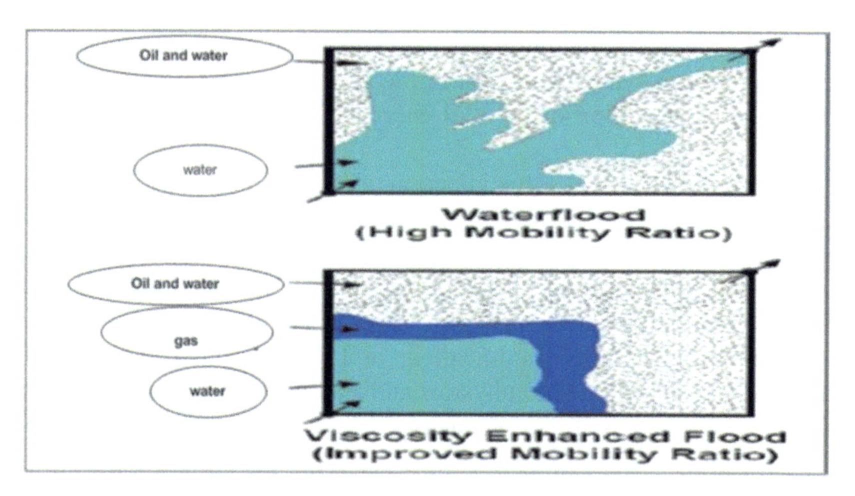

Figure 2.4 Mobility control process during WAG injection.

2.3.2 Miscible displacement processes

Miscible displacement processes are defined as processes where the effectiveness of the displacement results primarily from miscibility between oil in place and the injected fluid. Displacement fluids, such as hydrocarbon solvents, $CO_{2,}$ flue gas, and nitrogen are considered (Figure 2.5). The displacement processes treated here are classified as FCM (first contact miscible) and MCM (multiple contact miscible), i.e., on the basis on which miscibility is developed. In a specified fluid and reservoir system, MMP (minimum miscibility pressure) is an important parameter for these processes.

2.3.3 Chemical flooding

The process treated in depth, called the micellar/polymer process, is based on the injection of a chemical system that contains surface-active agents, i.e., surfactants. The process improves recovery efficiency primarily using displacing fluid that has a low interfacial tension (IFT) with the displaced crude oil. In the process, injection of micellar solution usually is followed by injection of an aqueous solution to which polymer has been added to maintain mobility control (Figure 2.6).

2.3.4 Thermal recovery processes

They rely on the use of thermal energy in some form both to increase the reservoir temperature, thereby reducing oil viscosity, and to displace oil to

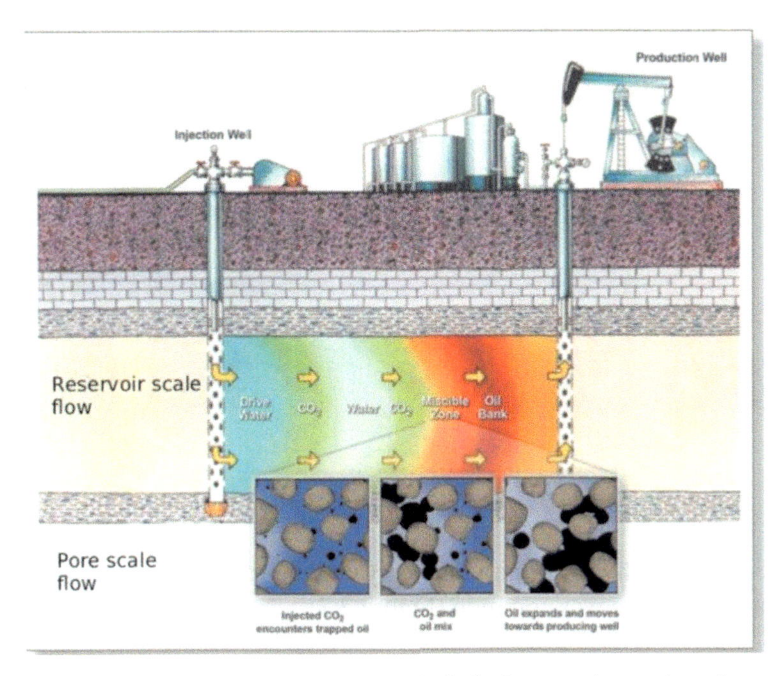

Figure 2.5 A diagram depicting the process of oil displacement by a solvent in a vertical cross-section of an oil reservoir, where the two substances are capable of mixing completely.

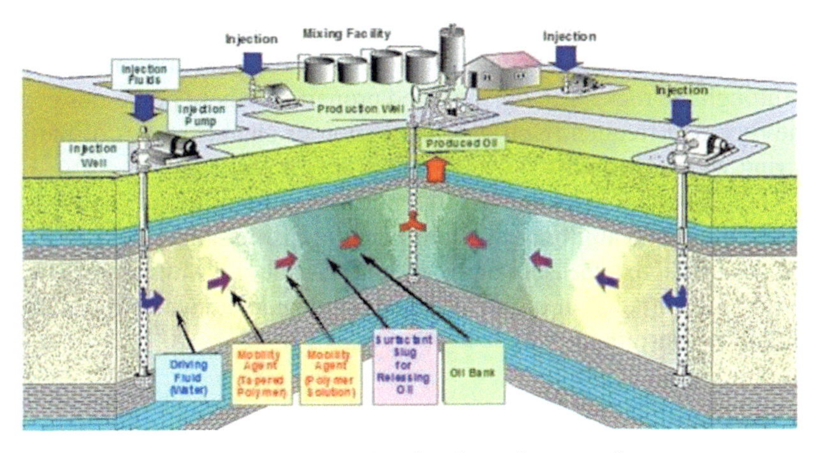

Figure 2.6 Utilizing chemical flooding to improve oil recovery.

produce well. The processes can be subdivided into cyclic steam stimulation, steam-drive, and *in situ* combustion. The motivation for developing thermal recovery processes was the existence of major reservoirs all over the world that were known to obtain billions of barrels of heavy oil and tar sand that could not be produced with conventional methods. In many reservoirs, the oil

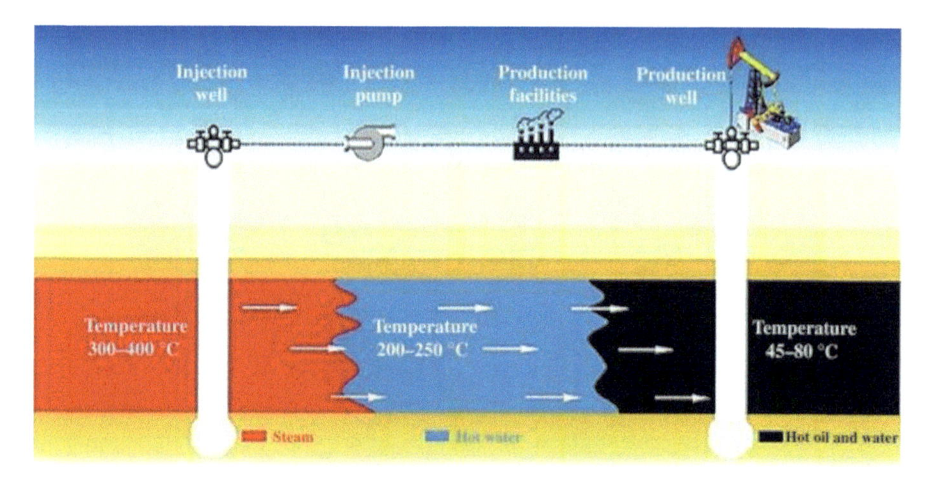

Figure 2.7 Thermal recovery process via steam injection.

viscosity is so high that primary recovery on the order of a few percent of the original oil in place was common (Figure 2.7).

2.4 Fundamental Properties of Fluid-Permeated Rocks

Naturally occurring rocks are in general permeated with fluid, water, oil, gas, or a combination of these fluids. The reservoir engineer is concerned with the quantities of fluids contained within the rock, the transmissibility of fluid through the rocks, and other related properties. These properties depend on the rock and frequently on the distribution or characteristics of the fluid occurring within the rock.

The properties discussed are the porosity, a measure of void space in the rock; the permeability, a measure of the transmissibility of a rock; and the fluid saturation, a measure of gross fluid distribution within a rock. These properties constitute a set of fundamental parameters by which rocks can be quantitatively described.

2.4.1 Porosity

Porosity is a measure of the space in a reservoir rock, which is not occupied by the solid framework of the rock. It is defined as the fraction of the total bulk volume not occupied by solids.

$$\phi = \frac{\text{Total Volume of Void Space}}{\text{Total Bulk Volume}} * 100.$$

As the sediments accumulated and the rocks were being formed during past geologic time, some of the void spaces, which developed, became isolated from the other void spaces by excessive cementation. Thus, many of the void spaces will be interconnected, while some of the void space remains isolated. This leads to two distinct types of porosity: (1) absolute porosity and (2) effective porosity.

$$\phi_{eff} = \frac{\text{Interconnected Pore Volume}}{\text{Total Bulk Volume}} * 100.$$

Reservoir porosity can be measured directly from core samples or indirectly using logs. However, as core coverage is rarely complete, logging is the most common method employed, and the results are compared against measured core porosities where core material is available.

2.4.2 Fluid saturation

More than one fluid normally occupies the pore space of the reservoir. From the history of the formation of the petroleum reservoirs, it is noted that the pore of the rocks were initially filled with water, since most of the petroleum-bearing formations are believed to be of marine origin. The oil/gas then moved into the reservoir displacing the water to some minimum residual saturation. Thus, when a reservoir is discovered, there may be oil, water, and gas distributed in some manner throughout the reservoir. The term fluid saturation is used to define the extent of occupancy of the pore spaces by any particular fluid. Fluid saturation is defined as that fraction or percent of the total pore space occupied by a particular fluid.

$$So = \frac{\text{Oil Volume}}{\text{Total Pore Volume}} *100.$$

Nearly all reservoirs are water bearing prior to hydrocarbon charge. As hydrocarbons migrate into a trap, they displace the water from the reservoir but not completely. Water remains trapped in small pore throats and pore spaces. The most common method for determining hydrocarbon saturation is by logging with a resistivity tool.

The fluids in most of the reservoirs are believed to have reached a state of equilibrium, and therefore will have become separated according to their density. There will be connate water distributed throughout the oil and gas zones, the irreducible water. The forces retaining the water in the oil and gas zones are referred to as capillary forces because they are important only in pore spaces of capillary size. The connate water saturation is a very

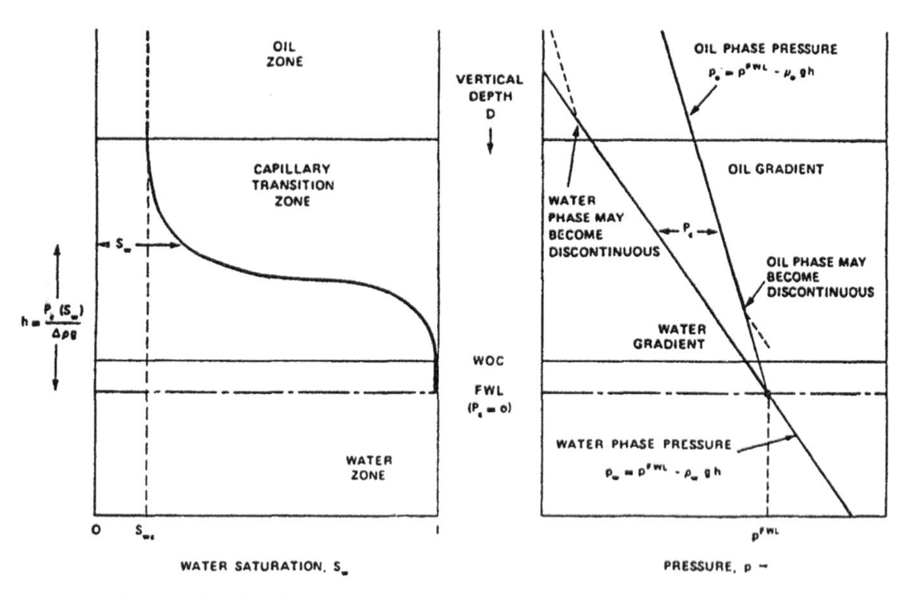

Figure 2.8 Capillary pressure in a water-wet reservoir's transition zone.

important factor, as it determines the fraction of pore space, which can be filled with oil.

$$Pc = h * \Delta\rho * g = \frac{2\gamma \, Cos \, \theta}{r}$$

where

Pc = capillary pressure
h = height above the free liquid surface
$\Delta\rho$ = difference in density
g = acceleration due to gravity
γ = interfacial tension between the fluids
θ = contact angle
r = radius of capillary

Capillary pressure is related to the height above the free water level. The capillary pressure data was converted to a plot of h vs. Sw. Changes in pore size and changes in reservoir fluid densities will alter the shape of the capillary pressure curve and the thickness of the transition zone (Figure 2.8). A reservoir rock system with small pore sizes will have a longer transition zone than a reservoir rock system composed of large pore sizes.

2.4.3 Fluid flow in reservoirs

The basic work on flow through porous material was published in 1856 by Darcy, who was investigating the flow of water through sand filters for water purification. Later investigations found that other fluids flowing in a porous media also can be modeled with Darcy equation as long as they do not react with the porous media. Reacting normally means any type of interaction that changes the size or shape of the flow channels or changes the surface forces between the rock and fluid.

$$q = kA\frac{(h_1 - h_2)}{L}.$$

In the above equation, A is the cross-sectional area and h_1 and h_2 are heights above the standard datum of the water in manometers at the input and output faces, respectively. L is the length of sand pack, and k is a constant proportionality found to be the characteristics of the porous media.

In reservoir engineering, it is neccssary to modify the equation to reflect differing fluid viscosity, dip angle for flow, and various flow geometries. Also note that flow must be laminar.

Darcy equation for linear horizontal system is

$$q = \frac{-A\ k\ (\Delta P)}{\mu\ L}$$

where:

k = permeabilities, darcies
q = outlet flow rate, cc/sec
μ = fluid viscosity, cp
L = system length, cm
A = cross-sectional area, cm^2
ΔP = pressure differential across porous medium, atm

2.4.4 Flow geometry

Several different flow geometries have been considered in reservoir fluid flow. The three most common are: linear, radial, and 5-spot.

2.4.4.1 Horizontal steady-state single-phase flow of fluids

The basic equation is

$$\text{Linear flow: } q = -A\frac{k}{\mu}\frac{dp}{dL}$$

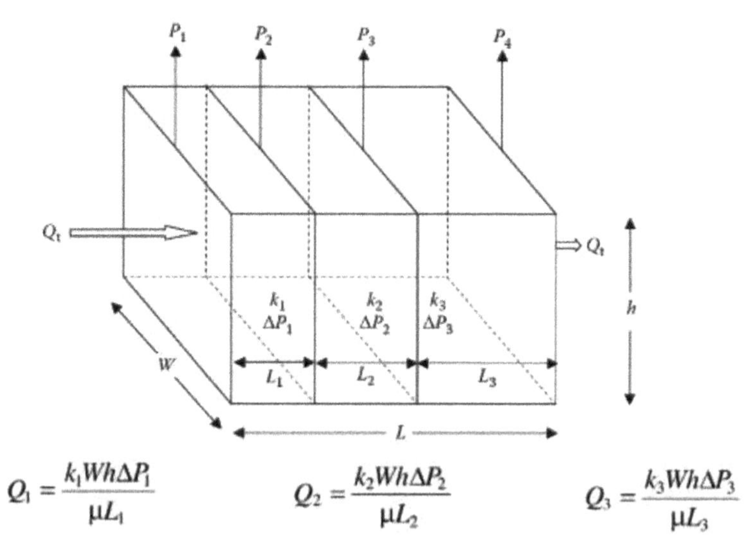

Figure 2.9 Fluid flow through the linear bed in series.

$$\text{Radial flow: } q = 2\pi \frac{kh}{\mu} \frac{\Delta P}{\ln\ (re/rw)}$$

2.4.4.2 Linear bed in series

Occasionally, calculations are necessary where linear beds in series are considered present (Figure 2.9). Note that pressure drops are additive.

$$k_{avg} = \frac{\Sigma\ L_i}{\Sigma\ (L_i\ /\ k_i)}$$

2.4.4.3 Linear bed in parallel

Fluid flow in the reservoir through parallel strata having different permeabilities. The total flow is the sum of the individual flow in each zone (Figure 2.10).

$$k_{avg} = \frac{\Sigma\ k_i h_i}{h_i}.$$

2.4.4.4 Radial bed in series

Depositionally, it is hard to imagine a radial bed in series occurring in an actual reservoir (Figure 2.11). However, this condition is needed due to the alteration of reservoir properties that can occur near well bores during drilling, production, and stimulation operations.

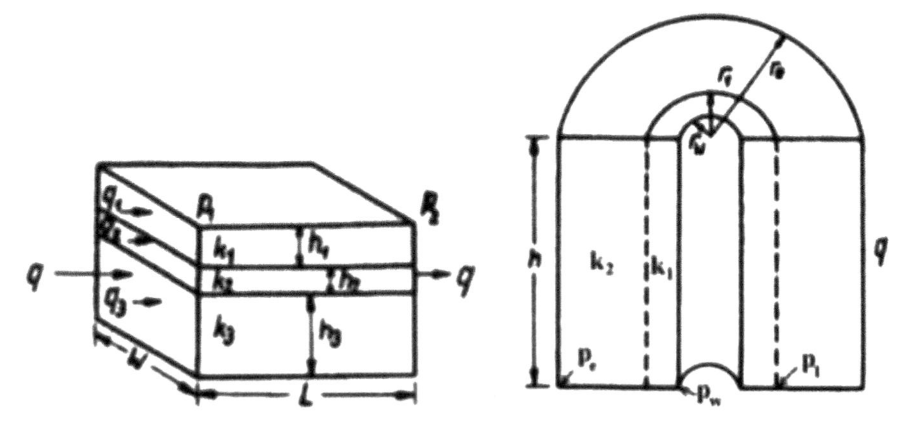

Figure 2.10 Fluid flow through the linear bed in parallel.

Figure 2.11 Fluid flow through the radial bed in series.

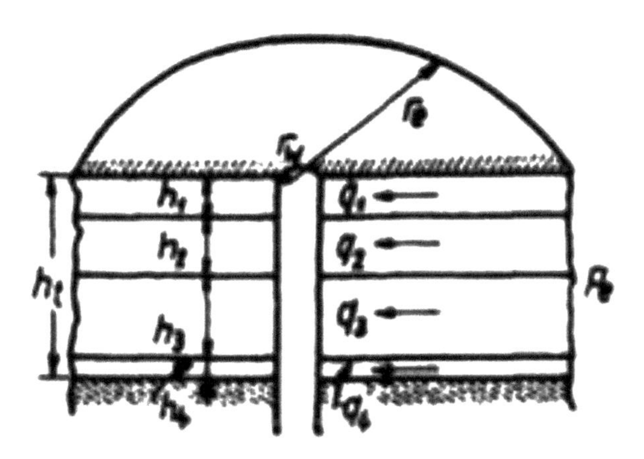

Figure 2.12 Fluid flow through the radial beds in parallel.

$$k_{avg} = \frac{k_a \, k_e \, \ln \, (re/rw)}{k_a \, \ln \, (re/ra) \, + \, k_e \, \ln \, (ra/rw)}$$

2.4.4.5 Radial bed in parallel

Most sedimentary reservoirs are composed of strata of different properties. Where calculation of the producing rates is desired, it is useful to determine the average permeability or that permeability which will allow the system to be treated as a single radial bed with total thickness (Figure 2.12).

$$k_{avg} = \frac{k_1 h_1 \, + \, k_2 h_2 \, + \ldots + \, k_n h_n}{h_t}$$

2.5 Well-bore Damage and Improvement Effects

The popular method for representing well-bore condition by steady-state pressure drop at well face in addition to normal transient pressure drop in the reservoir. The additional pressure drop called “skin effect” occurs in an infinitesimally thin skin zone. In flow equations, the degree of damage (or improvement) is expressed in terms of a “skin factor” s, which is positive for damage and negative for improvement. It can vary from about –5 for a hydraulically fractured well to +∞ for a well that is too badly damaged to produce.

$$\Delta p_s = \frac{141.2\, qB\mu}{kh} s.$$

Since damaged zone thickness is infinitesimal, the entire pressure drop caused by the skin occurs at the well face. The thin skin approximation results in a pressure gradient reversal for well-bore improvement ($s < 0$). Although this situation is physically unrealistic, the skin factor concept is valuable as a measure of well-being improvement.

If the skin is viewed as a zone of finite thickness with permeability k_s, then

$$s = \left(\frac{k}{k_s} - 1\right) \ln \left(\frac{r_s}{r_w}\right).$$

The flow efficiency (also called the condition ratio) indicates the approximate fraction of a wells undamaged producing capacity.

$$F_{low}\ E_{fficiency} = \frac{J_{actual}}{J_{ideal}} = \frac{P_{ws} - P_{wf} - \Delta P_s}{P_{ws} - P_{wf}}.$$

The damage ratio and damage factor are also relative indicators of well-bore condition. The inverse of flow efficiency is the damage ratio. By subtracting the flow efficiency from 1, we obtain the damage factor.

2.5.1 Effective permeability

At least two fluids are present in most petroleum reservoirs and in many cases, three different fluids may be present and flowing simultaneously. Therefore, the concept of absolute permeability must be modified to describe the flow conditions when more than one fluid is present in the reservoir. Effective permeability is defined as the permeability to a fluid when the saturation of that fluid is less than 100%. Effective permeability can vary from zero, when the saturation of the measured phase is zero, to the value of the absolute permeability, when the saturation of measured phase is equal to 100%.

One of the phenomena of effective permeabilities is that the sum of effective permeabilities is always less than the absolute permeability.

2.5.2 Relative permeability

Relative permeability is defined as a ratio of effective permeability to absolute permeability. Relative permeability is a very useful term since it shows how much the permeability of a particular phase has been reduced by the presence of another phase.

$$kr = \frac{ke}{k} * 100.$$

The wetting properties of a reservoir rock have a marked effect on the relative permeability characteristics of the rock. The tendency of a liquid to spread over the surface of a solid is an indication of the wetting characteristics of the liquid for the solid. The tendency of a liquid to spread over a solid surface can be expressed more conveniently and more precise nature by measuring the angle of contact at solid liquid surface. It is always measured through the liquid to the solid, and is called the contact angle. Complete wetting would be evidenced by a zero contact angle, and complete non-wetting would be evidenced by a contact angle of 180°. The wetting phase fluid will preferentially cover the entire solid surface of reservoir rock and will be held in the smaller pore spaces of the rock because of the action of capillarity. On the other hand, the non-wetting phase will tend to be expelled from contact with the surface of rock. Thus, at small saturations, the non-wetting phase will tend to collect in the larger pore openings of the reservoir rock.

The distribution of the reservoir fluids according to their wetting characteristics results in characteristic wetting and non-wetting phase relative permeability. Another important phenomenon associated with fluid flow through porous media is the concept of residual oil saturations. When one immiscible fluid is displacing another, it is impossible to reduce the saturation of displaced fluid to zero. At some small saturation, which is presumed to be the saturation at which the displaced phase ceases to be continuous, the flow of displaced fluid will cease. This is an important concept as it determines the maximum recovery from the reservoir. Conversely, a fluid must develop a certain minimum saturation before the phase will begin to flow, which is called the equilibrium saturation.

Theoretically, the equilibrium saturation and irreducible minimum saturation should be exactly equal for any fluid; however, they are not identical.

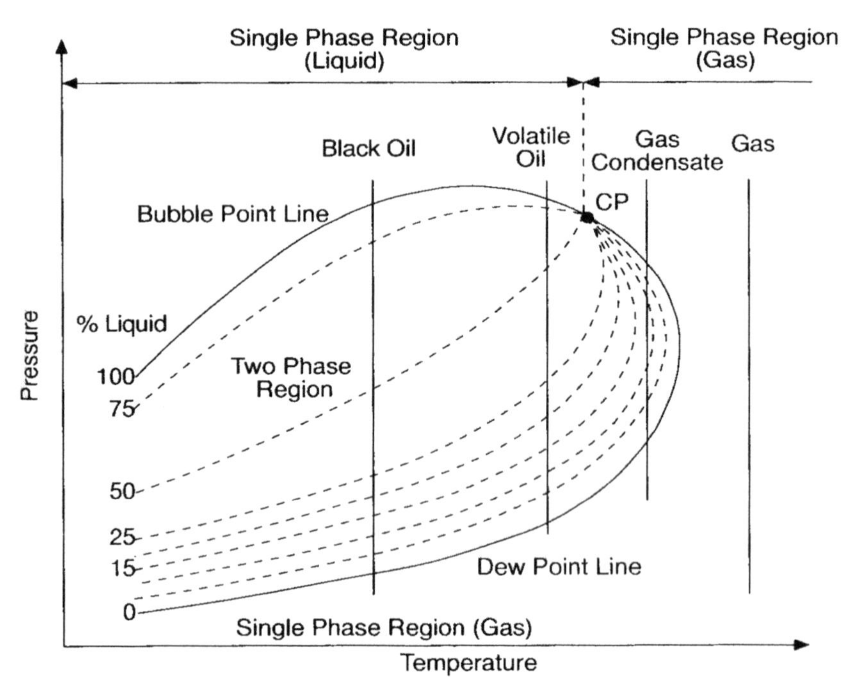

Figure 2.13 Pressure Temperature phase envelopes for main hydrocarbon types.

Equilibrium saturation is measured in the direction of increasing saturation, while irreducible minimum saturation is measured in the direction of reducing saturation.

2.6 Phase behavior

As the conditions of pressure and temperature vary, the phases in which hydrocarbon exists and the composition of the phases may change. It is necessary to understand the initial conditions of the fluids to be able to calculate surface volumes represented by subsurface hydrocarbons. It is also necessary to be able to predict phase changes as the temperature and pressure vary both in a reservoir and as the fluids pass through the surface facilities so that the appropriate subsurface and surface development plans can be made.

Phase behavior describes the phase or phases in which a mass in fluid exists at given conditions of pressure, volume, and temperature (PVT) (Figure 2.13). Typical PVT analysis in laboratory consists of sample validation, a compositional analysis of the individual, and recombined samples, measurements of oil and gas density and viscosity over a range of temperatures, and determination of basic PVT parameters Bo, Rs, and Bg.

3

Well Completion and Well Construction

3.1 Background

The well is the only communication channel with the reservoir in the development of oil and gas fields. It influences its own cost, operation, production/recovery, and economics of exploitation of reservoir as a whole. Therefore, every care and effort is directed to achieve the optimum well design.

Petroleum formed in the rock (known as source rock) has migrated to new rock (known as reservoir rock) under the influence of subsurface of pressure differential. The accumulation of petroleum is due to its trapping in the porous and permeable rocks, overlain and underlain by non-porous rocks. Geologists and geophysicists carry out geological prospection for oil and gas by analyzing the data obtained from gravity, magnetic, and seismic surveys. The drilling and casing program is the outcome of a study run by the drilling department. This study will consider any geological data such as loss of circulation, geo-pressured or under-compacted zones, etc., and will define the diameter and the casing shoe depth for the casing strings (conductor pipe, surface casing, intermediate casing, and production casing). The completion specialist will be involved as a party in the definition of the production casing especially when development wells are concerned. He will give his specific requirements (production casing diameter, pressure, and temperature stress on casing due to anticipated operations such as formation treatment, etc.).

Oil and gas production from a well is resumed after completing the drilled well. Usually, more than one casing pipe is required. A surface casing is essentially lowered to protect the groundwater from possible contamination by hydrocarbons. Intermediate casing is required for protecting the hole from caving, etc., during drilling, whereas production casing (oil string) is lowered and cemented from the surface to the target depth to serve as a conduit. The well is completed with tubing, pipe, wellhead assembly including a Christmas tree, and flowlines connected to an oil gas separator situated at the oil collecting station (OCS). Oils and gas from different wells are collected at OCS and separated. Any water produced is separated. Emulsified water is treated with chemicals, electrical methods, or heat.

An oil well may be naturally flowing if the formation pressure is high. When the natural flow becomes difficult, then it is required to lift the oil from the bottom hole to the surface artificially. The most commonly used artificial techniques today are sucker rod pumping, progressing cavity pump (or screw pump), gas lift, electrical submersible pumping, hydraulic pumping, and jet pumping. In addition to oil and gas wells, some wells are drilled for injection of fluids for pressure maintenance, enhanced oil recovery (EOR), or improved oil recovery (IOR). Water flooding is the most used method of EOR, although other methods such as polymer, surfactant caustic, CO_2, gas, steam flooding, and *in-situ* combustion (thermal) methods are also being used.

3.2 Types of Completions

According to the casing program, there are basically three methods for completing a well, which are usually known as:

- Open hole completion where the production casing is set on top of the pay zone.
- Perforated completion where the producing interval is covered by the production casing.
- Liner completion where the production casing is set on top of the pay zone and is followed by a liner.

The completion engineer should aim for safety, operability, and simplicity, within the requirements dictated by the optimum (cost vs. recovery) development of the field.

Major factors that influence designs are as follows.

3.2.1 Open hole completion

In the open hole type of completion, casing is set only at the top of or slightly into the completion interval. Neither casing nor cement is set opposite the production formation to restrict its flow (Figure 3.1).

Advantages of open hole completion:

- Casing set at the top of the pay allows for special drilling techniques, which minimizes formation damage

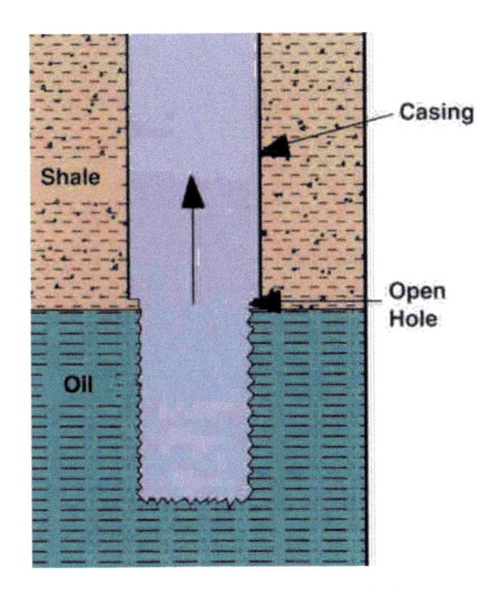

Figure 3.1 Open hole completion.

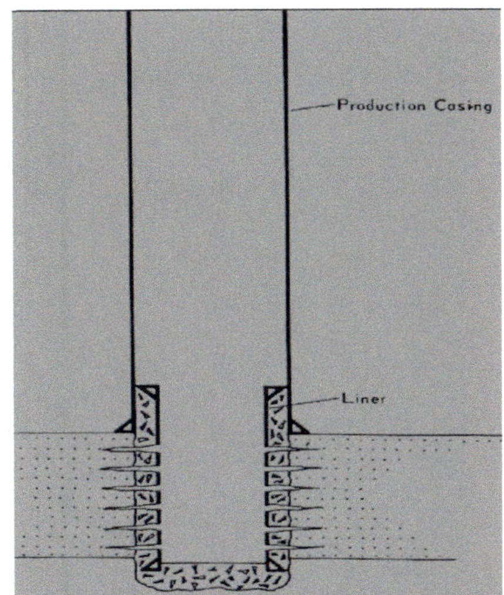

Figure 3.2 Liner completion.

- Full hole diameter is available to flow
- No perforation is required
- Hole is easily deepened and converted to a liner completion
- High productivity is maintained when gravel is packed for sand control

Disadvantages of open hole completion:

- No way to regulate fluid flow from or into well bore
- Cannot control gas or water production effectively
- Difficult to selectively stimulate producing intervals
- Well bore may require periodic cleanout

3.2.2 Liner completion

In this type of completion, casing is set above the producing zone (refer Figure 3.2). In the pay section, either an uncemented screen and liner assembly is lowered or a liner assembly is cemented in place. In the latter, the liner is required to be perforated.

Advantages:

- Formation damage is minimized

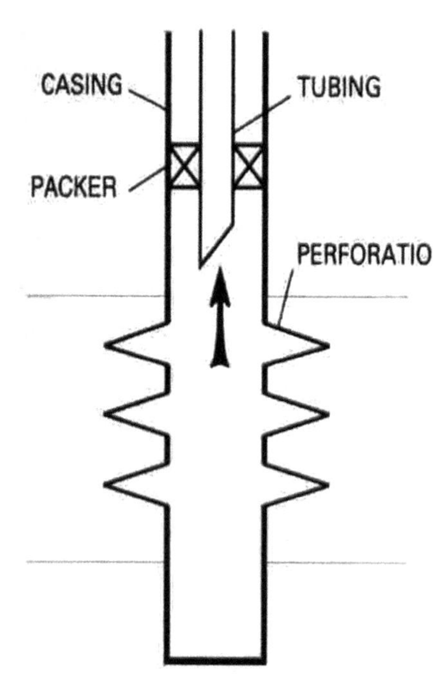

Figure 3.3 Conventional completion.

- Selective stimulation is possible in cemented liner
- Perforation expense is avoided in screen liner
- Cleanout problem is avoided in screen liner

Disadvantages:

- Diameter across the pay is minimized
- Good quality cementation is difficult in cemented liner

3.2.3 Conventional completion

In this type of completion, casing is set, cemented through pay zone, and then perforated. A packer is set above the producing zone in the string (Figure 3.3).

Advantages:

- The tubing controls the internal corrosion of the casing because produced fluid flows through it and do not contact the casing

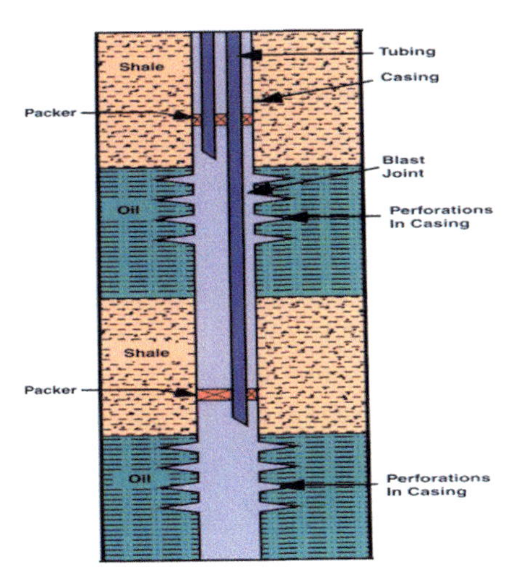

Figure 3.4 Dual completion.

Disadvantages:

- Tubing restricts the flow of produced fluid
- The completion is more expensive because of the cost of packer, tubing, and auxiliary equipment

3.2.4 Multiple completions

Each production zone is produced through its own tubing string and is isolated from the other zones by packers.

A dual completion (Figure 3.4) operation usually begins with both zones being perforated with a wireline casing perforator. Two packers are set above each production interval. The lower or longest tubing string is run with sealing elements spaced to seal both packers. The second string is run with one sealing element to fit the upper packer.

Advantages:

- Possible to produce from/inject into more than one production/injection zone through a single well, thereby reducing overall development costs.
- Selective zone well treatment is possible.
- Use of natural energy from one zone to artificially produce another zone is possible.

Disadvantages:

- Large number of equipment downhole creates a problem.
- Expensive and more complicated completion and workover technique.
- Possible loss of production in zone due to mechanical problems and formation damage during workover.

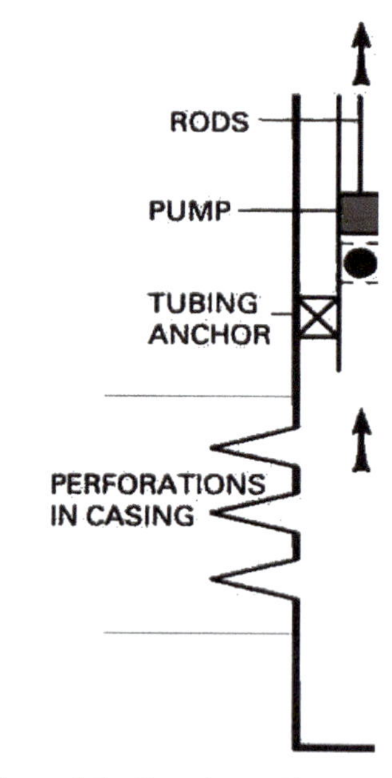

Figure 3.5 Pumping well completion.

3.2.5 Pumping well completion

Casing has been set through the pay zone and perforated. Tubing and a tubing anchor (a mechanical hold-down device that uses slip to grip the casing) are run. The anchor minimizes the tubing string movement as the pump reciprocates. The downhole pump may be run as part of the tubing string or run on rods and set in the tubing (Figure 3.5).

Advantages:

- Useful where reservoirs drive pressure is low.
- It allows large amount of water cut oil to be produced, which might otherwise be uneconomical to get out of the reservoir.

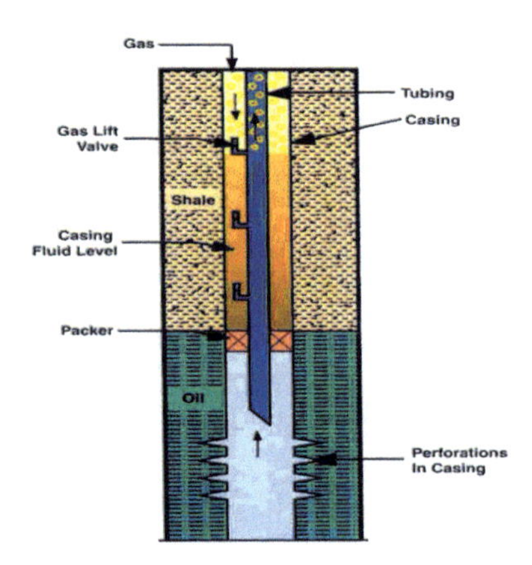

Figure 3.6 Gas lift completion.

3.2.6 Gas lift completion

The casing is set through the pay zone and then perforated. A packer is set above the producing zone with the tubing, which has the required number of gas lift valves. A readily available natural gas is injected down the casing through the gas lift valves and into the tubing. It may be injected at various intervals. This gas is used to lift the reservoir fluid to the surface (Figure 3.6).

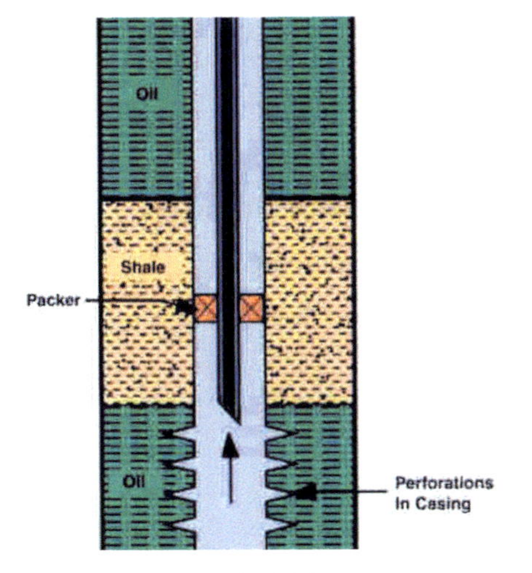

Figure 3.7 Corrosive high-pressure completion.

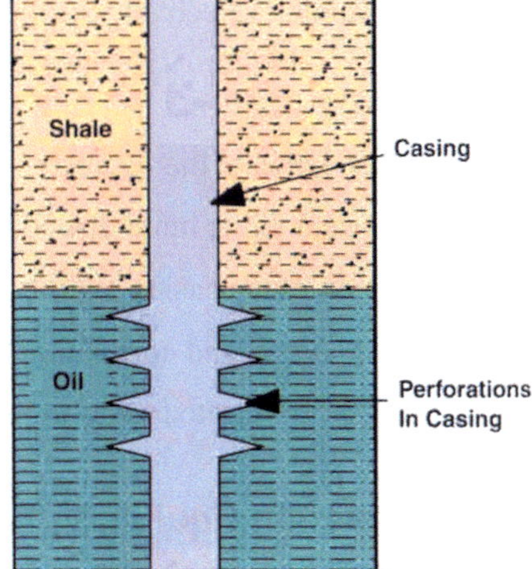

Figure 3.8 Tubing-less completion.

3.2.7 Corrosive high-pressure well completions

Casing is set through the production zone and perforated and a packer is then set above the pay zone. A small diameter string is run through the production tubing, which provides a convenient way to kill the well, where pressures are high and corrosion may have weakened the production tubing. This concentric tubing string also allows for easy treating. If necessary, a small amount of chemical can be pumped down continuously for the corrosion control in the production tubing (Figure 3.7).

3.2.8 Tubing-less cased hole completion

In tubing-less cased hole completion, casing is set into or through the producing formation and cemented (refer Figure 3.8). The casing is then perforated to provide communication between the well bore and formation. This type of completion is usually followed in slim-holes in isolated areas to keep the completion cost low. All remedial operations are carried out with coiled tubing.

Advantages:

- A tubing-less cased hole completion creates very little flow restriction and may be deepened easily.
- Well cleanout is performed more easily.

- Salt-water production can be controlled more easily and formation can be stimulated selectively.

Disadvantages:

- Cement opposite the formation may cause reduced production.
- Gravel packing is more difficult through perforations.
- Corrosion of the casing exposed to the produced fluid can be a problem.
- Log evaluation and correlation are essential.
- Pressures are limited to casing strength.

3.3 Completion Equipment

3.3.1 Wellhead equipment

- Wellhead equipment are attached to the top of the various casing strings used in a well to support the tubular strings, hang them, provide seals between strings, and control production from the well.
- These equipment are covered by American Petroleum Institute (API) Specification-6A.

3.3.2 Wellhead assembly

A typical wellhead assembly is shown in Figure 3.9.

Lowermost casing head:

- The lowermost casing head is a unit or housing attached for supporting the other strings of pipe, and sealing the annular space between the two strings of casing.
- It is composed of a casing hanger bowl to receive the casing hanger for attaching blow-out preventors (BOPs), or other intermediate casing heads or tubing heads and a lower connection (a female or male thread or a slip-on socket for welding).

Lowermost casing hanger:

- It seats in a bowl of a lowermost casing head or an intermediate casing head to suspend the next smaller casing securely and provide a seal between the suspended casing and the casing bowl.

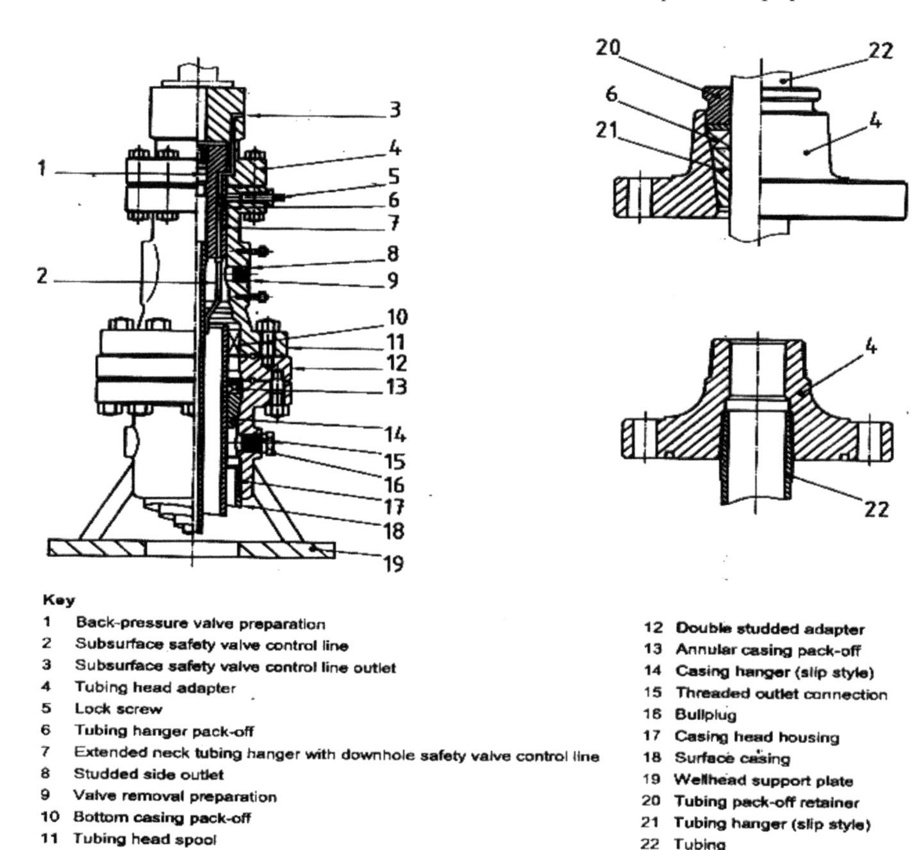

Figure 3.9 A typical wellhead assembly.

- It usually consists of a set of slips and a sealing mechanism. It is latched around the casing and dropped through the BOPs to the casing bowl.

Intermediate casing heads:

- An intermediate casing head is a spool-type unit or housing attached to the top flange of the underlying casing head, to provide a means of supporting the next smaller casing string and sealing the annular space between the two casing strings.
- It is composed of:

  1. A lower flange (counterbored with a recess to accommodate a removable bit guide, or a bit guide and a secondary-seal assembly)

2. One or two side outlets
3. A top flange with an internal casing hanger bowl

Intermediate casing hangers:

- These are identical in every respect to casing hangers used in lowermost casing heads and are used to suspend the next smaller casing string in the intermediate casing head.

Tubing head:

- It is a spool-type unit or housing attached to the top flange of the uppermost casing head to provide a support for the tubing string and to seal annular space between the tubing string and the production casing string.
- It also provides access to the casing/tubing annulus through side outlets (threaded, studded, or extended flanged).
- It is composed of a lower flange (or could have a threaded bottom that screws directly on the production casing string), one or two side outlets, and a top flange with an internal hanger bowl.
- On the double-flanged type, in the lower flange a recess is provided to accommodate a bit guide or a bit guide and a secondary-seal assembly. Lock screws normally are included in the top flange to hold the tubing hanger in place and/or to compress the tubing hanger seal, which seals the annular space between the tubing and the casing (Figure 3.10).
- In selecting a tubing head, the following factors should be considered to always maintain a positive control over the well:
  - The lower flange must be of the proper size and working pressure to fit the uppermost flange on the casing head below or the crossover flange attached to the casing head flange, if one is used.
  - The bit guide, or a bit guide and a secondary-seal assembly must be sized to fit the production casing string.
  - The side outlets must be of the proper design, size, and working pressure.
  - The working pressure of the unit must be equal to or greater than the anticipated shut-in surface pressure.
  - The top flange must be sized to receive the required tubing hanger, and of the correct working pressure to fit the adapter flange on the

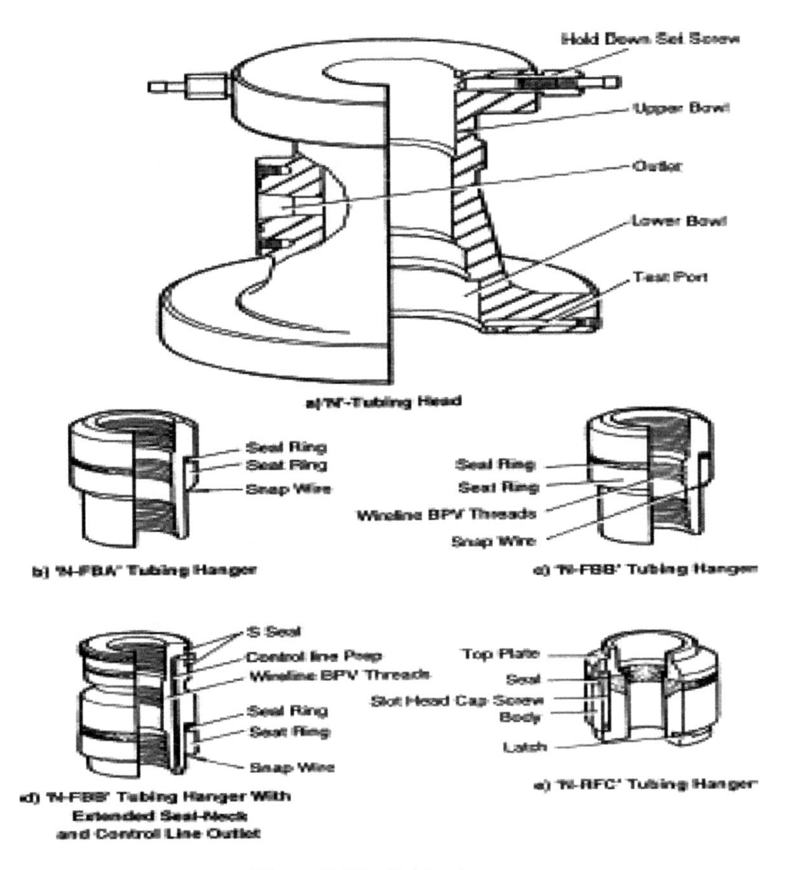

Figure 3.10 Tubing hanger.

Christmas tree assembly. Lock screws should also be included in the top flange.

- ○ The tubing head should have a full opening to provide full-sized access to the production casing string below and be adaptable to future remedial operations as well as to artificial lift.

- It is used to provide a seal between the tubing and the tubing head, or to support the tubing and to seal between the tubing and tubing head.
- Several types of tubing hangers are available and each has a particular application. More commonly used types are wrap-around, polished-joint, ball-weevil, and stripper rubber.
- The most popular is the wrap-around type. It is composed of two hinged halves, which include a resilient sealing element between two

steel mandrels or plates. The hanger can be latched around the tubing, dropped into the tubing-head bowl, and secured in place by the tubing-head lock screws.

- The lock screws force the top steel mandrel or plate down to compress the sealing element and form a seal between the tubing and tubing head.
- Full tubing weight can be temporarily supported on the tubing hanger, but permanent support is provided by threading the top tubing thread into the adapter flange on top of the tubing head. The hanger then acts as a seal only.
- In selecting a tubing hanger, it should be ensured that the hanger will provide an adequate seal between the tubing and tubing head for the particular well conditions (metal-to-metal seals are desired in most cases) and that it is of standard size, suitable for lowering through full opening drilling equipment.

Adapter:
It is used to connect two flanges of different dimensions or connect a flange to a threaded end.

Crossover flange:
A crossover flange is an intermediate flange used to connect flanges of different working pressures. These are usually available in two types:

- A double-studded crossover flange is studded and grooved on one side for one working pressure, and studded and grooved on the other side for the next higher-working-pressure relating. The flange must also include a seal around the inner string of the pipe to prevent pressure from the higher-working-pressure side reaching the lower-working-pressure side. The seal may be of the resilient type, plastic packed type, or welded type.
- Another type of crossover flange includes a restricted O-ring groove in the topside of the flange to fit a corresponding restricted-ring groove in the mating head. The restricted-ring groove and the seal between the flange and the inner casing string act to restrict the pressure to a smaller area, thereby allowing a higher pressure rating.

3.4 X-Mas Tree

A Christmas tree is usually the first device encountered by a workover crew (Figure 3.11). It is an assembly of valves, spools, flanges, and connections

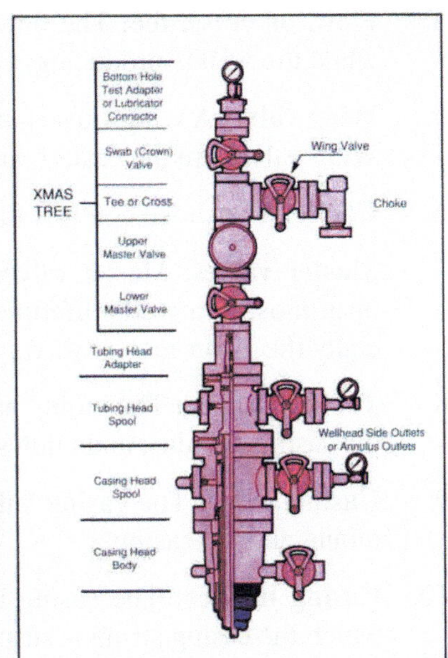

Figure 3.11 X-mas tree and its components.

that control the flow of fluids from the well. Because a Christmas tree controls the flow of fluids from a well, crew members must be careful not to damage it when they move in and rig up. Carelessness could prove fatal to personnel and could destroy the rig.

Many types of Christmas tree are available. Some, like those on pumping wells, may be simple and consist mainly of a stuffing box. On the other hand, complex trees with numerous master and wing valves may be required on deep, high-pressure gas wells. Each well is unique and requires a specific type of tree. In spite of the wide variety of trees available, they share certain basic components:

1. **Pressure gauges:** Pressure gauges monitor tubing pressure and casing, or annular pressure.

2. **Gauge flange, or cap:** The gauge flange seals the top of the tree and has a fitting for a pressure gauge. When the gauge flange is removed, the tubing becomes accessible and the bottom-hole test or lubricator equipment can be installed.

3. **Crown, or swab, valve:** The crown valve shuts off pressure and allows access to the well for wireline, coil tubing, or other workover units to be rigged up.

4. **Flow, or cross, tee:** The flow tee allows tools to be run into the well while the well is producing.

5. **Wing valve:** A wing valve shuts in the well for most routine operations. Wing valves are the easiest valve to replace on the tree.

6. **Choke:** The choke controls the amount of flow from the well.

7. **Master valves:** Master valves are the main shut-off valves. They are open most of a well's lifetime and are used as little as possible, especially the lower master valve, to avoid wear or damage to them.

8. **Tubing hanger:** The tubing hanger supports the tubing string, seals off the casing annulus, and allows flow to the Christmas tree.

9. **Casing valve:** The casing valve gives access to the area between the tubing and the casing.

10. **Casing hanger:** The casing hanger is a slip-and-seal assembly from which the casing string is suspended.

11. **Casing:** Casing is a string of pipe that keeps the well bore from caving in and prevents communication from one zone to another.

12. **Tubing:** Tubing is a string of pipe through which produced fluids flow.

13. **Backpressure valve (BPV) (Figures 3.12 and 3.13):** This valve can be installed in the tubing hanger. Two types of BPV are available. One type is one-way check valve; it allows fluid to be circulated down the tubing but closes against well pressure when required. Another type is a two-way check valve; it holds pressure from both directions. BPVs allow the X-mas tree to be removed or replaced without having to kill the well. Once the tree is removed, other devices, such as BOPs, can be installed. Also, the tree can be pressure tested and the master valves repaired or replaced.

3.4.1 Installing X-mas tree

The exact procedures for installing a tree will vary depending upon factors such as tree rating and type, well conditions, etc. The following is a general procedure. It assumes that the packer fluid has been conditioned.

1. Space out of hole to install the surface controlled sub-surface safety valve (SCSSV). After the SCSSV has been made up, attach the control line and test the working pressure. Maintain pressure to the control line and space out and run in the hole with tubing and control line.

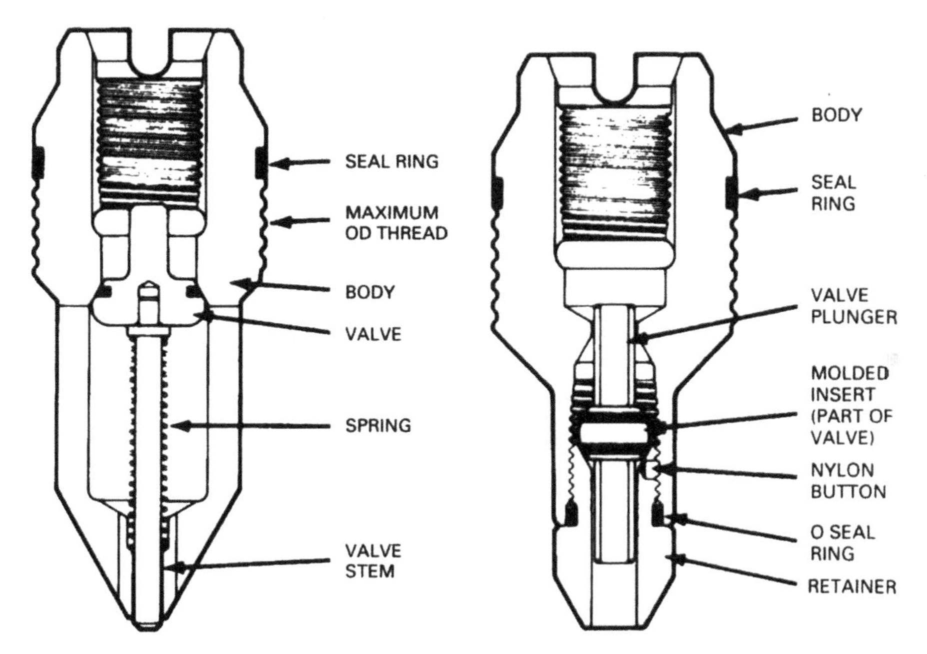

Figure 3.12 Type-H backpressure valve.

Figure 3.13 Type-H check valve.

Attach control line to tubing with banding material or plastic tie wraps. It is advisable to utilize a minimum of one control line protector per joint.

2. Make sure that all tubing hanger lock-down bolts are fully backed out.
3. Install tubing hanger and landing joint. The landing joint (or joints) should have a full open safety valve (in the opened position) installed at the top.
4. Bleed pressure off the control line and connect it to the bottom and top of the tubing hanger. Test the control line integrity and maintain pressure.
5. Drain the BOP stack at the tubing spool.
6. Keep the tubing hanger centered (to avoid damage to the seals) and lower it into BOP stack.
7. String seal assembly into packer and land tubing hanger.
8. Run in all tubing hanger lock-down bolts and torque properly. Pressure test the casing, seal assembly, and tubing hanger to the required pressure through the tubing spool.

9. Bleed pressure off the control line to close the SCSSV.
10. Remove landing joints and set backpressure valve (BPV) in tubing hanger.
11. Nipple down BOPs.
12. Clean and inspect the seal surfaces on the tubing hanger neck. Install the top seal ring.
13. Clean and inspect the bottom seal of the tubing hanger bonnet. Install the main run of the tree.
14. Tighten all studs properly and evenly to energize seals and ring gasket. Re-tighten and check torque on tubing hanger lock-down bolts. Pressure test the tubing bonnet.
15. Nipple up the remaining tree valves. Install a blanking plug or backpressure valve. Test tree (hydrostatically) to required pressure. Bleed off the pressure.
16. Pull blanking plug or backpressure valve. Connect the emergency shut-down system (ESD); pressure up on the tree to equalize and open SCSSV, and activate the emergency shut-down system on the tree with remote on the rig.
17. Rig up and test flowlines to test the heater, separator, and tank if required.
18. If perforating through tubing, displace tubing with completion fluid. Perforate.
19. Test well.
20. Close SCSSV and test by bleeding off the pressure. Bleed half of the tubing pressure off (above SCSSV) and observe for leaks.
21. Set the BPV and test by bleeding off the remaining tubing pressure. If ready to produce, remove the BPV; if not, then secure tree.

3.4.2 Removal of X-mas tree and tubing hanger

Prior to entering the well for workover operations, the well must be killed and all associated surface and subsurface safety devices accounted for. The exact procedures for removing a tree will, of necessity, vary. Many factors such as tree rating and type, well conditions, etc., may alter the set or standard procedures. The following is a general procedure.

1. Prior to the kill procedure, note the shut-in tubing pressure (SCSSV open). Check the production and surface casing for pressure. Hold the SCSSV open with the control line pressure.
2. If pressure is found on the casing, it may have resulted from the thermal expansion of the packer fluid. To determine if this is the case, bleed a small volume of fluid from the casing. If the pressure bleeds off or falls to 0 psi (bar), it is probably the thermal expansion causing the pressure. If gas is bled, there may be a packer seal, a tubing connection leak, or a hole in the tubing. If there is a rapid build-up of pressure, it may be an indication of a serious leak.
3. If the casing pressure does bleed to 0 psi, then rig up the pump and top off the casing with kill weight fluid. Pressure test the casing, if procedures dictate. If a bullhead kill is to be performed, then it is advisable to pressure the casing with several hundred psi of pressure (200–500 psi will suffice in most cases). This pressure should be monitored throughout the kill procedure.
4. If the bullhead kill technique is to be used, rig up on the tubing and the bullhead kill weight fluid. This should be of sufficient volume to displace the capacity of tubing and well bore to perforations. If there is no communication between the tubing and casing, shut down the pump, and observe whether the well is dead.
5. If communication is noted between the tubing and casing, a standard or reverse circulating kill method may have to be utilized. This may require the tubing to be perforated above the packer, or a sliding sleeve opened with the wireline.
6. Make sure that all tubing and casing pressures have been bled off.
7. Release pressure from the SCSSV control line, rig up the wellhead lubricator on the tree, and set the backpressure valve in the tubing hanger. Note: It is normally desirable to keep the SCSSV open if any wireline work is required. Some SCSSVs can be locked out with the wireline. This should be done after the well has been killed and before removing the tree. If the SCSSV cannot be mechanically locked out, a valve should be installed on top of the tubing hanger and pressured up with a hand-pump to hold the SCSSV open. The valve is then closed and the hand-pump disconnected.
8. Remove the Christmas tree. Inspect and lubricate the tubing hanger lift threads. These may be corroded and notable to support the string weight.

9. Install and test BOPs. To test the blind rams, the backpressure valve must be pulled and a two-way check valve or blanking plug installed. Test blind rams and then install a landing joint and test bonnet seals; flange breaks, pipe rams, annular preventer, and choke manifold.

Exercise caution when using the dry rod to release and pull backpressure valves or blanking plugs. The stack should be filled with water and the backpressure valve slowly released. Never use excessive force. A backpressure valve is installed (normally) by one person using a 24″ or 18″ pipe wrench. The same wrench should also be used to remove it. Avoid overtightening.

An alternate procedure may be used to avoid using a dry rod as follows:

1. Install and test the blind rams.
2. Flange up on the blind ram BOP with proper connection back to the lubricator.
3. Using the polished rod, retrieve the backpressure valve or blanking plug.
4. Close the blind rams (they have been tested).
5. Nipple up the remaining BOP stack, open blind rams, and install the landing joint.
6. Test the remaining BOPs. The well can be monitored through a side outlet below the blind rams.
7. After testing BOPs, remove the landing joint and retrieve the two-way check valve or tubing plug.
8. Back out the tubing hanger hold-down pins.
9. Pick up to pull out of seals (or release packer) and remove tubing hanger.

3.5 Casing

3.5.1 Types of casing

(i) Stove pipe/marine conductor:

It is run to:

- Prevent wash out of soft soil near surface (unconsolidated formation).

- Provide a circulating system for drilling mud and to ensure stability to the ground surface upon which the rig is operated.

These pipes do not carry any wellhead equipment and can be driven into the ground with a pile driver.

(ii) Conductor pipe:
It is run to some shallower depth to protect surface unconsolidated formation and to seal off shallow water depth to provide protection against shallow gas flow. The size ranges from 36″ to 14 3/4″.

(iii) Surface casing:
This casing is set on a compact rock such as hard limestone. This eliminates the caving of weak formations at shallow depths and safeguards formations from fracturing at higher hydrostatic pressures of mud. This also protects against shallow blow-out. Its size ranges from 20″ to 14 3/4″ in diameter.

(iv) Intermediate casing:
This is set in a transition zone below or above an over-pressured zone or it is run to seal off severe fluid loss zones or to protect against problematic formations such as salt section or caving shale.

The size ranges from 9 5/8″ to 14 3/4″ in diameter.

(v) Production casing:
This is the last casing run to the target zone. This permits a single or selective production in multizone completion.

It is generally of 7″ or 5 ½″ or 5″. In offshore, 9 5/8“ is common.

3.5.2 Production tubing

This is the major conduit for the well's produced fluids. It also protects the casing from pressure and corrosion. Its size may run from several inches to a fraction of an inch. The most common sizes are 2–7/8″ OD (73.02 mm) and 2–3/8″ OD (60.32 mm). Tubing normally runs from the wellhead to the production zone. Tubing is classified by size (OD, ID, tool joint OD, ID), weight (lbs/ft), and grades such as J-55 and N-80. Tubing may be constructed of exotic materials to withstand the pressures, velocities, and corrosivities of the well's produced fluid and environment. Internal coating may be applied for corrosion protection. There are many types of connections to couple or screw joints of tubing together. Care and handling of tubing is similar to that of casing (Figure 3.14).

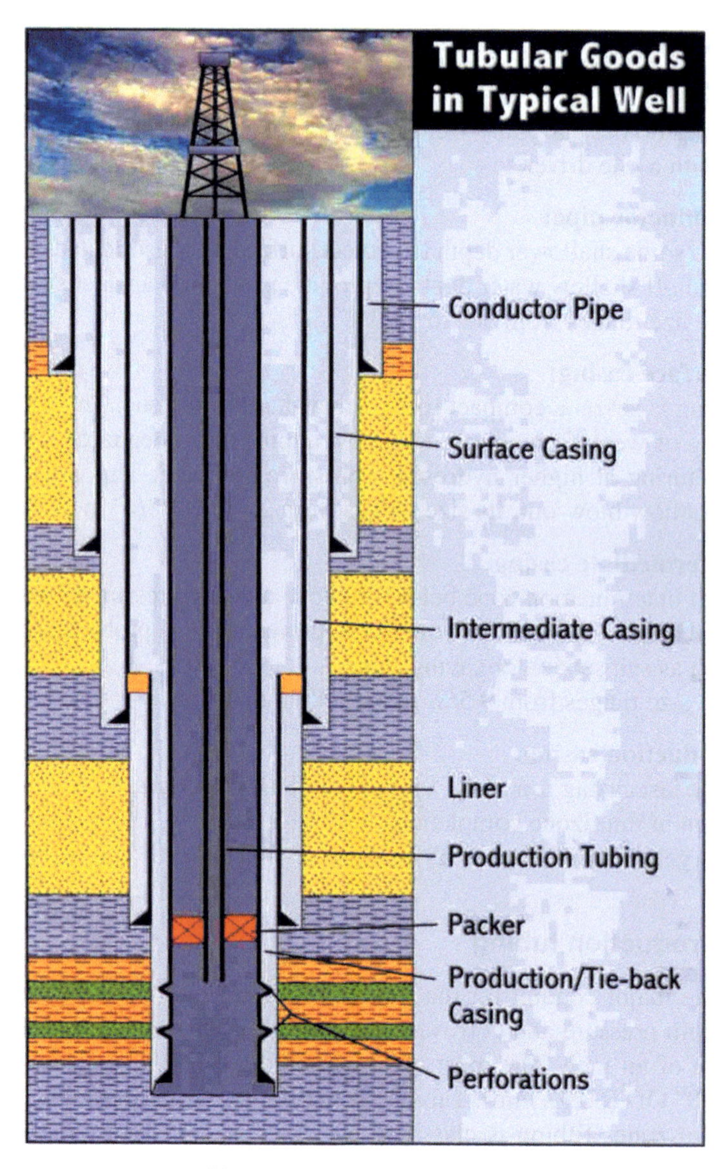

Figure 3.14 Casing and tubing.

3.6 Horizontal and Multilateral Well Completion

- Horizontal and multilateral wells move the well bore closer to oil and gas in place.
- Multilateral wells allow multiple producing well bores drilled radially from a single section of a "parent" well bore.

- Both parent and lateral extensions produce oil and gas.
- Typical horizontal and multilateral applications include
  - Improving productivity from thin reservoirs.
  - Draining multiple, closely spaced target zones with horizontal exposure of each zone.
  - Improving recovery in tight, low permeability zones.
  - Preventing water and/or gas coning.
  - Controlling sand production through lower draw down at the sand face.
  - Improving the usability of slot constrained platform structures.
  - Improving water flood and enhance oil recovery efficiency.
  - Intersecting vertical fractures.

3.6.1 Horizontal well completion

- Horizontal wells enhance reservoir contact and thereby enhance well productivity.
- A horizontal well is drilled parallel to the reservoir-bedding plan.
- While analyzing a horizontal well performance geometric configuration of reservoir, bedding zones is important.
- The type of completion affects the horizontal well performance.
- Certain types of completions are possible only with certain types of drilling techniques.
- Horizontal wells may be completed as (Figure 3.15)
  - Open hole
  - With slotted liners
  - Liners with external casing packers (ECPs)
  - Cemented and perforated liners

3.6.2 Open hole completion

- Early horizontal wells completed open hole.
- Inexpensive but limited to competent rock formations.

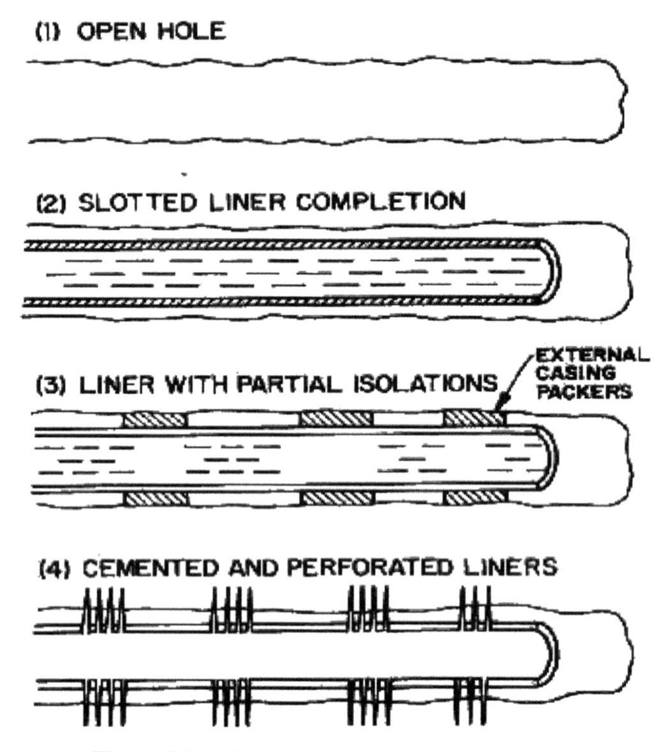

Figure 3.15 Types of horizontal well completion.

- Difficult to stimulate open hole wells and control either injection or production along the well length.

3.6.3 Slotted liner completion

- Guard against hole collapse.
- Liner provides a convenient path to insert downhole tools and CT.
- Three types of liners are available:
  - Perforated liners
  - Slotted liners
  - Pre-packed liners, which provide limited sand control
- In unconsolidated formations, wire wrapped slotted liners are used to control sand.
- In this completion, effective well stimulation is difficult.

3.6.4 Liners with partial isolation

- External casing packers (ECPs) are installed outside the slotted liner.
- ECPs divide a long horizontal well bore into several small segments.
- Provides limited zonal isolation.
- Wells can be selectively stimulated and produced.

3.6.5 Cemented and perforated liners

- It is possible to cement and perforate medium- and long-radius wells.
- Cementation of short-radius horizontal wells is not economical.
- Segregation of cement at bottom and water near top results in bad cementation.

3.7 Multilateral Well Completion

- The following three configurations of multilaterals are possible:
  - Commingled production.
  - Commingled production with individual branches that can be shut off by gates and re-entered easily.
  - Individual production tubing tied back to the surface.
- Multilateral well completions are driven by the following criteria:
  - Pressure, temperature, and chemical properties of produced fluids and gas.
  - Zonal segregation and isolation requirements.
  - Lateral well bore rock properties and stability.
  - Regulatory requirements for zone production and management.
  - Workover/re-entry options and methods.
  - Sand and water production potential.
  - Abandonment requirements.
  - Equipment performance records and risk analysis.

- Three types of multilateral well designs are available:
  - Open hole multilateral.
  - Limited isolation/access multilateral.
  - Complete multilateral wells.

3.7.1 Open hole multilateral well completion

- In open hole multilateral wells, the first primary well bore is drilled to a depth above the producing interval and lateral kick-off points.
- Surface or intermediate casing is set and cemented.
- Laterals are drilled from the parent bore hole.
- Most open holes involve short horizontal sections (~1200 ft).
- Drainage objective for such a completion is to get the highest initial production to drain the reservoir as rapidly as possible.
- Therefore, reservoirs permit commingled production from each lateral.
- Lateral-bore may plug; so for open hole completion to be stable, non-sloughing, heterogeneous hard rock formation should be targeted.
- Re-entry access to any specific lateral bore is difficult.
- Typically, a single string of production tubing with a PKR set near the base of the parent casing.
- Loading nipples for plugging and safety valve can also be added.
- Open hole multilaterals pose minimal risk.
- Open hole multilateral wells present no more manageable risk implications than other types of open hole completions.

3.7.2 Limited isolation/access multilateral systems

- This type of completion (Figures 3.16, 3.17, and 3.18) is used where zonal isolation and re-entry access is required.
- Undesirable production thus can be shut off from the producing bore.
- A limited isolation/access system does not allow casing to be set in lateral bores and mechanically reconnected to the parent bore.

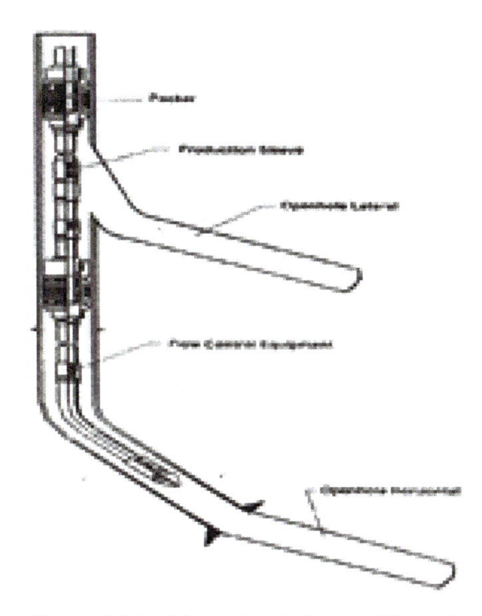

Figure 3.16 Limited-isolation multilateral completion.

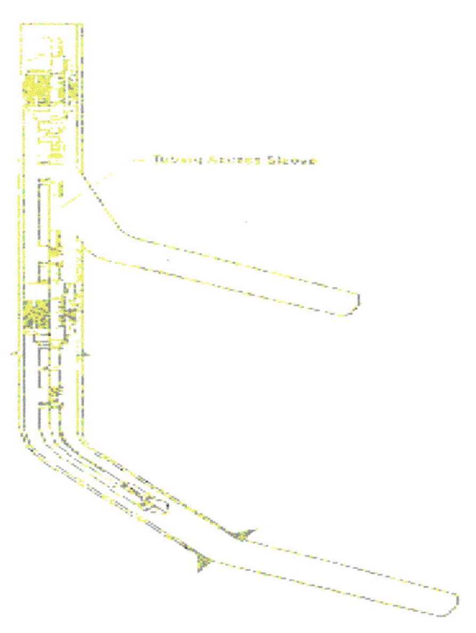

Figure 3.17 Limited-isolated (commingled)

- This system does not allow re-entry into the lateral bore hole during the drilling and initial completion phase.
- A stable, non-sloughing, hard-rock formation is the best candidate.
- Slotted liners can be hung on open hole inflatable PKR anchors.
- If sand control problems are anticipated, pre-packed screens can be installed in a similar manner.
- Failure in the completions tends to occur at unprotected, open hole sections adjacent to the parent-bore casings (the heel of the lateral).

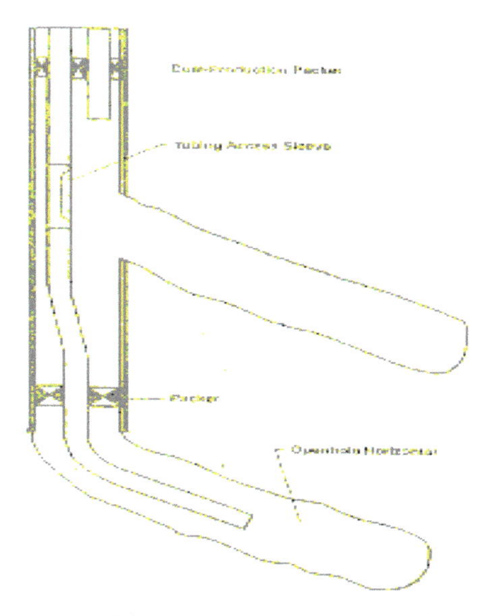

Figure 3.18 Limited-isolated completion.

- Once borehole is cased and logged, a retrievable whip stock is installed at the required lateral exit point.
- The wedge-shaped diverter (whip stock) is oriented for both depth and azimuth with the sloped face of the tool aimed at the direction of the planned lateral bore path.
- Pilot mills, window mills, or water melon mills are run against the face of whip stock to cut the section of the casing opposite to diverter face.
- Then a full gauge window opening is cut and sized in the casing.
- After completion of drilling of the lateral section, liner, screen, or other tools may be installed again using a whip stock as diverter into the open hole lateral.
- Additional lateral can be created in a similar manner.
- This type of completion offers more production management options than open hole laterals.
- Typically, production packers are installed in primary well bore casing above and below each later exit point.
- Single or dual strings of production tubing could also be installed to commingle or segregate production as required.
- Sliding sleeves are sometimes installed at lateral exit prints to provide on/off production control from each lateral bore.

3.7.3 Complete multilateral system

- A complete multilateral system (Figures 3.19 and 3.20) consists of two to five laterals from one new or existing well bore.
- Suits for deepwater or subsea environments.
- In a complete multilateral system, the lateral well bore is cased back to the primary bore exit and a liner casing string is mechanically connected to the primary bore casing.
- The lateral to main well bore junction must be hydraulically sealed.
- This system provides additional reservoir management options and increases the choice of target zone.

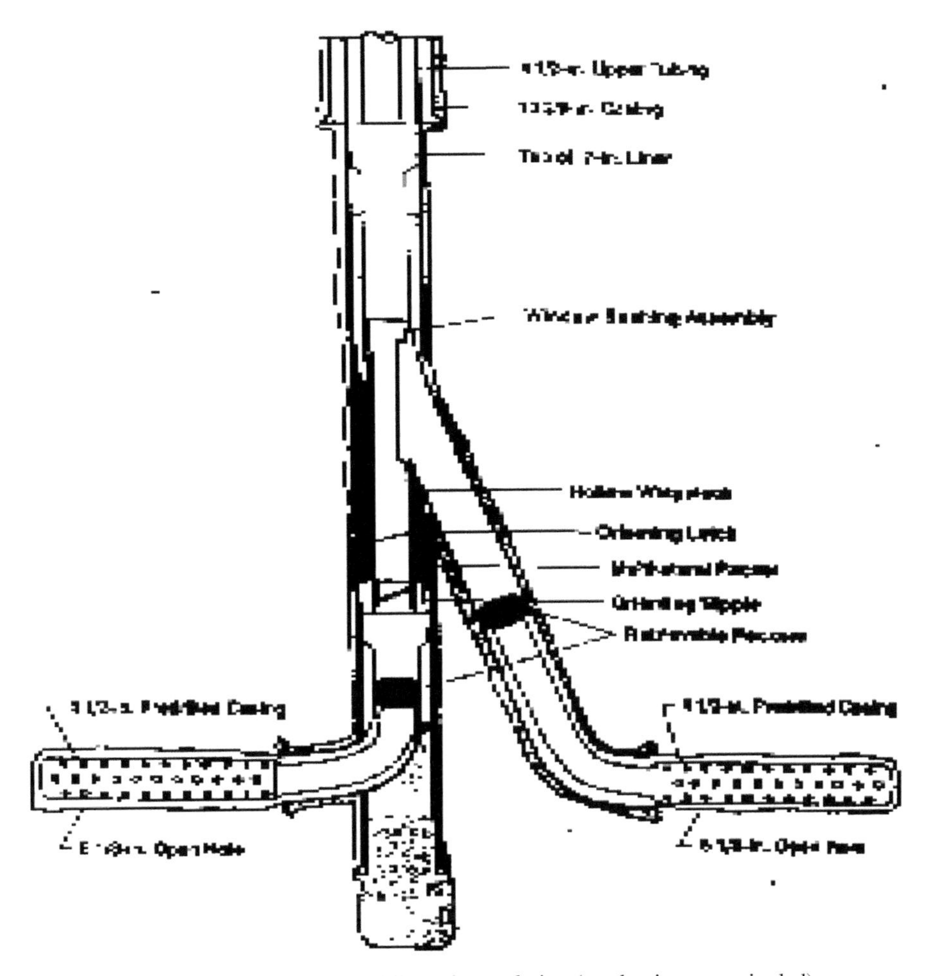

Figure 3.19 Advanced multilateral completion (production commingled).

- First the primary bore is drilled; then the primary production casing is set and cemented across all the anticipated lateral bore exit points.
- A combination of assemblies having a whip stock packer anchor, a hollow whip stack filled with a composite plugging material.
- A starter mill is run to the depth and oriented to azimuth and set hydraulically.
- A starter mill is sheared loose from anchored whip stock and the first section of the exit window is rotary milled.

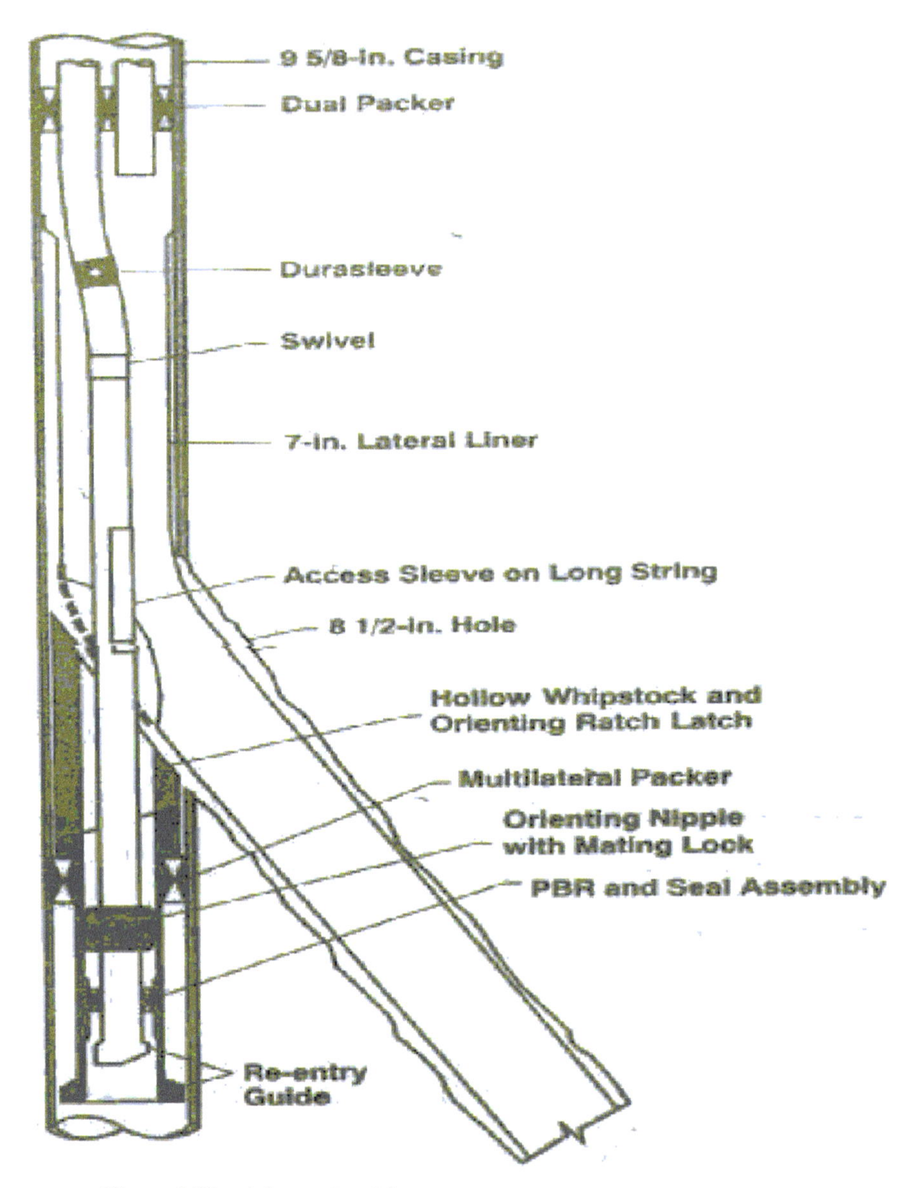

Figure 3.20 Advanced multilateral completion (production commingled).

- Additional window and watermelon mills are run to open and exit portal to a full gauge dimension.
- Lateral bore hole is drilled and then a liner casing is run to depth and hydraulically sealed or cemented.

- Cemented liner perforated and completed with production packers, sliding sleeves, sand control tools or other completion tools.
- Additional open hole drilling out the lower end of the lateral liner may also occur to meet completion objectives requiring open hole pre-packed screens or slotted liners.
- Completion options range from single string, commingled flow designs to multi-string, segregated production installations.
- Each lateral well bore can be re-entered through the production tubing with compatible tools or with a rig, if the tool required is longer than the tubing ID.

4

Well Intervention

4.1 Well Intervention

Many servicing operations can be conducted by workover rigs; however, live well intervention is preferred as killing a well risks fluid invasion of the formation, thereby causing potential formation damage. The primary objective of well intervention operations is the management of wells to provide optimum well production. This is achieved by conducting live well remedial operations, obtaining downhole reservoir data, or preparation of the well for a dead well workover (if a problem cannot be solved by live well servicing). Occasionally, gathering of downhole reservoir data is a secondary objective only opportunistically taken when an intervention is planned for other reasons.

4.1.1 Reasons for well interventions

There are many reasons for remedial live well intervention well operations, most notably to:

- remove obstructions to flow such as tubing blockage with sand, wax, or asphalt;
- eliminate excessive water or gas production;
- repair mechanical failure;
- improve production through well stimulation, re-completions, or multiple completions on low productivity wells;
- enhance production by conducting well stimulation such as hydraulic fractures on high productivity wells;
- increase production by bringing other additional potentially productive zones on stream;

- maintain control of oil, water, and gas in various zones or layers in stratified reservoirs;
- side-tracking passed severely damaged formations.

In what follows, these main well problems are discussed in detail.

4.1.2 Tubing blockage

Tubing blockage is generally caused by sand, wax, and asphalt production or scale build-up. It can usually be remedied with a well clean-out operation. Some of these can be prevented, or at least alleviated, by treating the formation with regular chemical inhibition treatments, pumped into the formation from surface.

Tubing blockage is one of the most commonly experienced production problems and which is remedied by clean-out operations conducted normally by snubbing or coiled tubing (C/T) intervention although dead well workover may also be considered. Asphalt can also be removed similarly by pumping solvents rather than hot oil.

Some well clean-outs may be accomplished with wireline methods using tools such as gauge cutters, which can remove wax from tubing walls and bailing to remove sand or other blockages, provided the amount to be removed is relatively small.

4.2 Control of Excessive Water or Gas Production

4.2.1 Control of water production

There are different reasons for water problems. Firstly, fingering of water in stratified or layered reservoirs where the water production is essentially from one zone. Secondly, advancing water level due to oil depletion. Thirdly, water coning in reservoirs where there is appreciable vertical permeability. Once a rock becomes more saturated with water, the relative permeability to water increases in regard to that of the other fluids. This leads to a self-aggravating cycle of increasing water flow and increasing relative permeability to water. Water production can usually be controlled by several differing methods depending upon the specific well design and well conditions:

- Sand placement in the sump.
- Setting a through tubing bridge plug.
- Cement squeezing.

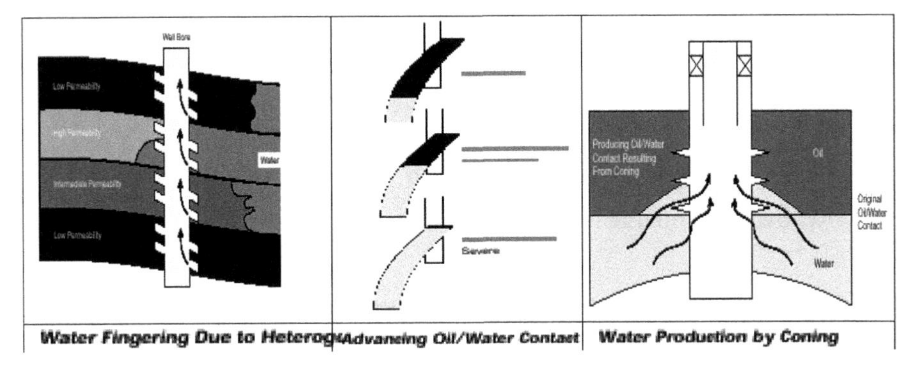

Figure 4.1 Various reasons for excessive gas production.

- Chemical treatment to produce a gel block.

4.2.2 Control of gas production

The most common reason for excessive gas production is the growth of the gas cap as oil is produced (Figure 4.1). A gas/oil contact will gradually move downwards causing an increase in the production of gas. The common method of remedying excessive gas coning is to squeeze the gas-producing zone and deepen the well by re-perforating (converse to water coning). In a layered reservoir, gas-producing zones can also usually be effectively squeezed off with cement.

4.2.3 Mechanical failure

Well service operations to repair mechanical completion failures are still relatively common in old wells; however, in new wells, less servicing is required due to the increasing reliability of modern completion equipment. Probably the most common reason for remedial mechanical operations today is tubing failure due to erosion or corrosion. Some completion failures can be repaired by wireline or C/T methods, but, in some circumstances, a full workover program to pull the tubing is necessary.

Typical failures are:

- Downhole safety valve mechanical failure or leak.
- Downhole safety valve leak.
- Casing, packer, or tubing leak.
- Casing collapse.

- Tubing collapse.
- Cement failure.
- Gas lift failure or inefficiency.
- ESP or hydraulic pump failure.
- Recover fish unable to be recovered by other methods.

4.2.4 Stimulation of low productivity wells

There are many reasons why a well may have low productivity, for instance:

- Formation damage.
- Low permeability.
- Pressure depletion.
- Liquid hold up in a gas well.
- Gas slip in an oil well.
- Excessive water or gas production.
- Sand or other fill or debris.
- Mechanical failure.
- Artificial lift failure.

The remedial methods of the above problem can be solved by stimulating the formation with different fluid and techniques. For instance, reservoir problems such as formation damage and low permeability can sometimes be improved by stimulation operations such as acidization or hydraulic fracturing.

4.2.4.1 Partially depleted reservoirs

Similar to low permeability wells, in a depleted oil reservoir, an effective artificial lift system can be installed to increase production. If a well was originally planned and designed for gas lift and completed with gas lift mandrels in the string, then the gas lift valves are simply installed by wireline intervention. However, if a re-completion is needed, a full dead well workover would be necessary. In high-angle wells, gas lift valves can be installed with coiled tubing methods.

4.2.4.2 Sand control

There are normally two solutions to control unconsolidated sand, and these are: to gravel pack or install a pre-packed screen although resins are occasionally used. The drawback of having to implement such sand control measures is that they reduce productivity typically by 10%–15%. The installation of a gravel pack entails a full workover and re-completion although new snubbing methods with HWO unit have now been developed. For a successful gravel pack, it must be clean and the second requirement is that the gravel is correctly sized in relationship to the formation sand to prevent further ingress or cause a blind-off.

4.3 Well Intervention Services

Well interventions are servicing operations conducted through the Xmas tree (through-tree) on live wells. These are carried out by the following methods:

- Wireline (both electric line and slickline).
- Coiled tubing.
- Snubbing.

Well service operations or workovers on dead wells where the Xmas tree is replaced by well control equipment, are carried out by:

- Drilling rigs.
- Workover rigs.
- Hydraulic workover units.

During workovers, it is probable that well interventions with wireline and/or coiled tubing are required as part of the work program to prepare the well for tree removal or establish production post workover. Many offshore installations have drilling rigs onboard used for the drilling phase of a development.

On subsea wells, normally the only means of conducting a well intervention is to use a semi-submersible vessel (drilling unit, DSV, or specialized well servicing unit) from which a workover riser can be deployed. However, if the work program can be conducted solely with wireline, this can be successfully carried out by subsea wireline systems deployed from well servicing vessels. These vessels also have the capability to carry out subsea tree change-outs once appropriate barriers have been installed by wireline. Well control equipment used on well interventions in live wells is specific to the particular service being used for the intervention.

4.4 Snubbing/Hydraulic Workover Units (HWO)

The snubbing/HWO unit is a well service unit utilized for both snubbing and dead well servicing. Snubbing technology has long been used in live well operations to snub out or strip in drill pipe and tubing in order to reestablish well control. With key modifications in equipment and procedures, this technology can be extended to other applications that enhance well production and reduce well drilling, well completion, or well workover costs.

They are also very useful when working in confined spaces and with small-diameter (skinny) pipe where a drilling rig's instrumentation is generally not sensitive enough.

An HWO unit would only be used before C/T on a snubbing job where:

- There is insufficient space above the wellhead or deck space.
- When rotational torque required on the pipe is greater than that available from downhole motors.
- Where pressures exceed the rating of C/T pipe, i.e., circa 5000 psi.

4.4.1 Snubbing applications include

4.4.1.1 New well completion

It is inclusive of perforation and subsequent stimulation followed by installation of the final production configuration. These completions commonly use snubbing and generally consist of tubing end plug, one joint of tubing, a profile nipple, and tubing to surface. The procedure consists of perforating when underbalanced and then snubbing tubing into the well. Such completions are simple, fast, and cost efficient.

Programming considerations:

a. Snubbing always requires a tubing hanger and, if possible, one that incorporates lag screws to lock it in place.

b. A tubing end plug eliminates the requirement for wireline service.

c. In dry wells, a small differential between annular pressure and tubing pressure is required to drop the end plug. If the well is wet or full of sand after the fracture, different methods may be necessary.

1. **Changing out the bottom hole assembly (BHA)**:
   This is the removal of existing well completion followed by installation of a modified BHA, smaller or larger tubing string.

2. **Well re-completion**:
   This is the abandonment of the existing zone followed by re-completion of an alternate interval, inclusive of perforation, stimulation, and re-installation of the production string.

Figure 4.2 Conventional snubbing unit.

Other snubbing applications (in completions):

1. **Air/N_2 drilling:**
   - Snubbing is often used when drilling dry, as with air or N_2, because it allows the well to encounter virgin gas with the option to proceed further or shut in the well to remove the drill string.
2. **Lubricating:**
   - Lubricating is the melding of wireline operation with snubbing and is used for the handling of long BHAs and fishing tools.
   - The use of a lubricator with the snubbing unit eliminates the need to "spool up" the snubbing unit with riser spools.
3. **Workovers/milling and fishing:**
   - On zones that take large amounts of kill fluid, snubbing proves economical as it eliminates trucking and fluid cost.
   - In fluid-sensitive zones, snubbing can save production.
   - On water injection/disposal wells, snubbing may sometimes be the only safe and economical way to workover (Figure 4.2).

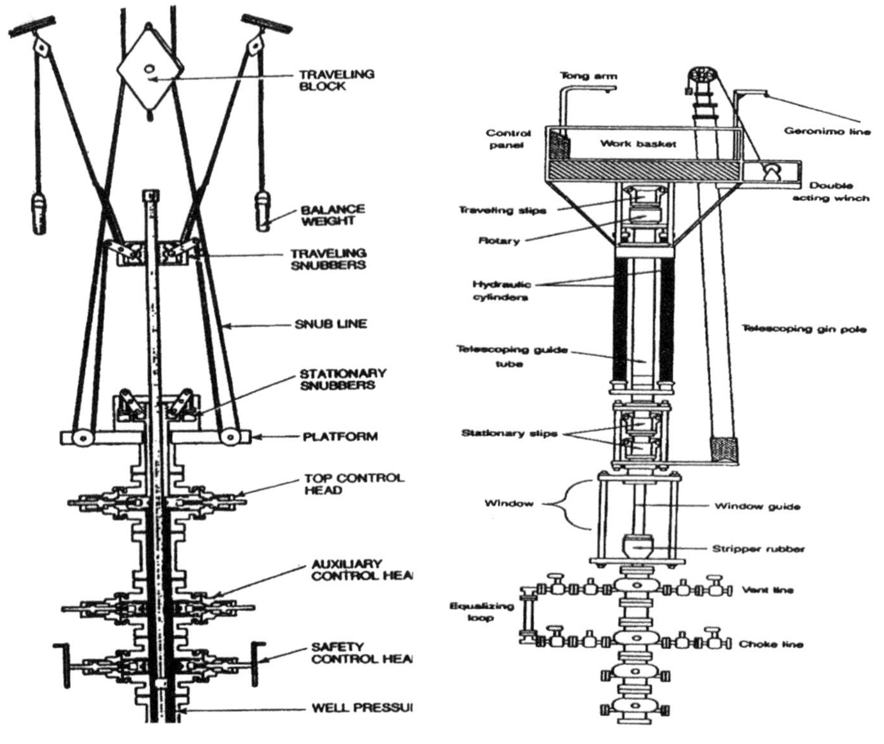

Figure 4.3 Mechanical snubber.

Figure 4.4 Hydraulic snubbing unit.

HWO units are supplied in a range of lifting capacities (lbs. in thousands), 60, 90, 120, 200, 250, 400, and 600 K. Snubbing capacity is half of this rating.

4.4.1.2 Hydraulic jack assembly

The jack assembly consists of one or more hydraulic cylinders that travel in a vertical direction to move pipe in or out of the hole.

Guide tube:

This is simply a tube, which prevents the buckling of the pipe under snubbing forces. It should be sized to be just larger than the tubing to be run or pulled to constrain lateral movement (Figure 4.3).

Splined tube:

Some units have a splined tube, which passes rotational torque force generated by the rotary table through to the bottom plate and hence to the wellhead. If a splined tube is not used, the forces are transmitted through the hydraulic cylinders possibly reducing the operating life (Figure 4.4).

Travelling slips:
The travelling slips, or snubbers, are attached to the upper end of the jack and grip the pipe to push it into or pull it from the hole. There are two sets, one for snubbing and one for lifting.

Stationary slips:
The stationary slips hold the pipe while the traveling slips are released for the next stroke.

Access window:
The access window (work window) is installed at the base of the jack between the stationary slips and the stripper and is the access for stripper rubber change-out or for installing tools in the string. It must also help guide the pipe like the guide tube.

Power swivel:
The power swivel (or rotary head) is used for rotating the pipe for drilling or milling operations.

Hanger flange:
A hanger flange (also known as a tubing hanger assembly) is a pressure containing component sometimes used in the blowout preventer stack to hold the pipe and tool string in both the "light" and "heavy" directions.

Power tongs:
Power tongs are used to make up and break out the pipe connections. They are located in the work basket and controlled hydraulically from the control panel.

Work basket:
The work basket is the work platform of an HWO unit and is located at the top of the hydraulic jack and on which the operator and assistant perform the manual functions including the picking up, laying down, stabbing, making up, or breaking out of the pipe joints.

Control panel:
The control panel is mounted in the work basket and is usually in two sections, one for the operator's use and one for his assistant.

Power pack:
The power pack and its accessories consist of a diesel engine and hydraulic pumps.

Hose package:
The hose package transports the hydraulic fluid to and from the various functions, some of which are high up on the unit and are therefore of considerable length.

BOP system:
The BOP configuration is dependent upon whether the HWO unit is being used as a rig on a well, which has been killed, or in the snubbing mode rigged up above the Xmas tree. If on the former, the BOP configuration will be like that in a drilling situation and may be covered by the operator's well control policies and procedures. If on a snubbing job, the configuration is quite different being rigged up above the Xmas tree.

Strippers:
The strippers control well pressure when snubbing or any time surface well pressure is encountered.

Circulating system:
Pumps, chiksans, Kelly hose, and a circulating swivel are the main components of the circulating system.

Types of snubbing unit:

- Short stroke snubbing unit
- Long stroke snubbing unit
- Rig-assisted snubbing unit

Wireline unit:
Wireline is the oldest and most common type of well servicing method. It is extremely efficient, economic, and relatively easy to rig up and deploy. Electric line services provide essential information about the reservoir and the completion and performs many services, typically: logging – depth determination, cement bonding, sonic, nuclear, temperature, pressure, spinner, caliper, density, dipmeter, profile, and so on.

- Calipering.
- Downhole sampling.
- Perforating.
- Setting bridge plugs, packers, and cement retainers.

This is achieved by communicating with the tools through the conductor cable.

Mechanical wireline, also known as slickline (as the line has a smooth OD), is used to conduct mainly mechanical operations such as:

- Installing flow controls.
- Installing gas lift valves.
- Depth finding.
- Plugging.
- Bailing.
- Paraffin cutting.
- Tubing gauging.
- Setting bridge plugs.
- Fault finding.
- Fishing.
- Logging – through-tubing BHP gauges or the latest electronic solid state logging tools such as spinners, CCLs, etc.

The slickline unit can also be rigged up with braided line for heavy duty wireline operations such as running heavy, large tools, or performing heavier duty fishing operations. A more recent development in wireline services is the heavy duty wireline unit used mainly for fishing jobs where regular fishing methods have failed. These units, in conjunction with heavy duty tooling, are so powerful that they can destroy normal wireline tools and devices, if desired. Its greatest limitation, due to using gravity as its motive force, is in working in high-angle or horizontal wells with inclination angles higher than 70°.

There are two types of wireline units – the electric line or logging unit and the mechanical or slickline unit. Both types of unit are constructed similarly in that they have:

- Power pack.
- Operator's/engineer's cabin.
- Winch, including a wireline drum or reel.
- Spooling or measuring head.
- Weight indicator and pulleys.

Wireline advantages over conventional w/o rig:

- The foremost advantage is economics.
- Takes less time of rig-up, rig-down, and operations time for special tasks.
- Performs tasks much faster.
- As support services to w/o rig to set packer, plugs or installs valves.

Disadvantages of wireline:

- Limited wireline strength.
- Not being able to rotate.
- Quite limited use in: high-angled and horizontal wells because they rely on gravity to lower the tool
- Not effective even with the use of heat shield electric instruments in a well having temperature of 240°C and above.

4.4.2 Power pack

The power pack is normally a diesel-driven hydraulic unit and provides hydraulic power through supply and return hoses to the winch. Power packs are normally fireproofed and certified for division 1, zone 2 hazardous areas.

Operator's/engineer's cabin:
The cabin is an integral part of the winch unit situated directly behind the drum for direct observation and monitoring of the wireline spooling. It contains the winch and possibly the power pack operating controls.

Winch:
The winch consists of the wireline reel driven by a hydraulic motor controlled from the console in the cabin, all of which is mounted in the unit frame. Hydraulic power is supplied from the power pack.

Spooling head:
The spooling or measuring head controls the winding of the wire off and onto the reel and also measures the length of the wire spooled off the drum. The depth measurement is given on an odometer via a cable drive and a precisely machined measuring wheel (one for each size wire).

Weight indicator and hay pulley:
The weight indicator can be mounted on the hay pulley or be an integral part of the spooling head. If mounted at the hay pulley, the weight sensor is a load cell placed between the hay pulley and the tie down chain.

4.5 Types of Wirelines

4.5.1 Electric line

Cable used on electric line units can be monoconductor, coaxial, or multiconductor braided line and supplied for various service conditions. Each particular type has a range of sizes and specific uses according to the required service or tool being run. Careful handling of electric lines is essential, especially with the smaller sizes and when rigging up, to prevent line damage and penetration of the core insulation leading to subsequent loss of signal.

4.5.2 Slickline

Slickline is a high strength monofilament steel line and is available in common sizes of 0.082, 0.092, 0.108, and 0.125 ins. These are also supplied for various service conditions. Being slick, the OD of the wire is easy to seal around using a simple packing device called a stuffing box, whereas the cable requires a grease seal arrangement.

4.5.3 Braided line

Braided wireline used for heavier duty wireline operations is supplied in 3/16 and 7/32 ins. sizes.

4.5.4 Wireline lubricators and accessories

The wireline lubricator when assembled acts like pressure vessel on top of the Xmas tree into which the wireline tools are "lubricated." It consists of:

- Wellhead adapter.
- Wireline BOPs or wireline valve.
- Lower lubricator section(s).
- Upper lubricator section(s).
- Stuffing box or grease head.
- Line wiper.

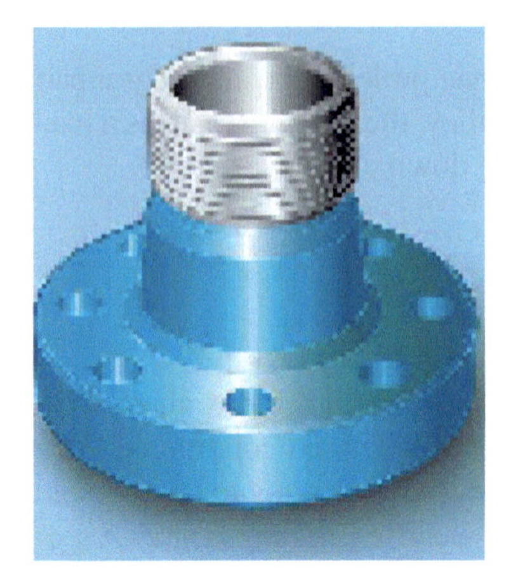

Figure 4.5 Wellhead adapter.

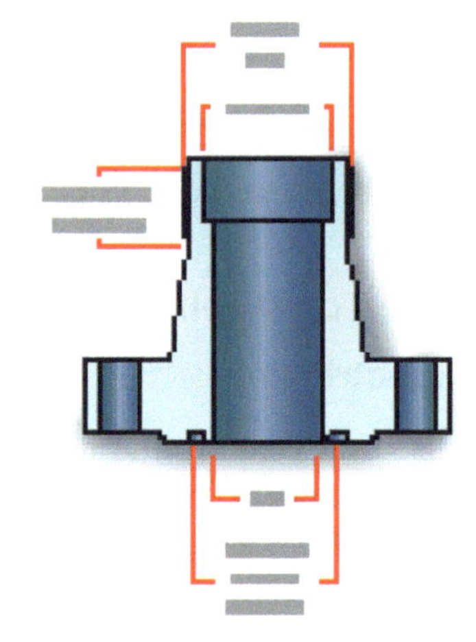

Figure 4.6 Wellhead adapter (cross section).

It is extremely important that a wireline lubricator pressure rating meets the maximum anticipated surface well pressure. Lubricators must be designed, not only to withstand the stress caused by the internal pressure but also from stresses caused by the jar action or high pulling forces. To install the tools, the lubricator must first be isolated from well pressure at the Xmas tree, usually the swab valve, and all pressure bled off through the bleed-off valve. The lubricator is then broken out at the connection immediately above the BOPs, and the tools, after attaching to the tool string, are pulled up into the lubricator bore and the lubricator re-installed. The lubricator should then be pressure tested before opening the tree and running in the hole.

4.6 Wellhead Adapter

This is basically a crossover to mate the BOP to the tree cap and is usually a quick type connection named a "quick union." In some cases, the adapter may be from a quick union to a tree flange. The wellhead adapter flange consists of a quick union machined on a standard API flange. The wellhead adapter flanges are available in various bore sizes from 2 1/16″ to 7 1/16″ and working pressure up to 15,000 psi (Figure 4.5 and 4.6).

4.7 Wireline BOPs

Wireline BOPs (sometimes referred to as wireline valves) are installed immediately above the wellhead adapter or on top of a wellhead riser. In some situations for ease of operation and safety, a BOP may be placed both above the tree and on top of a riser.

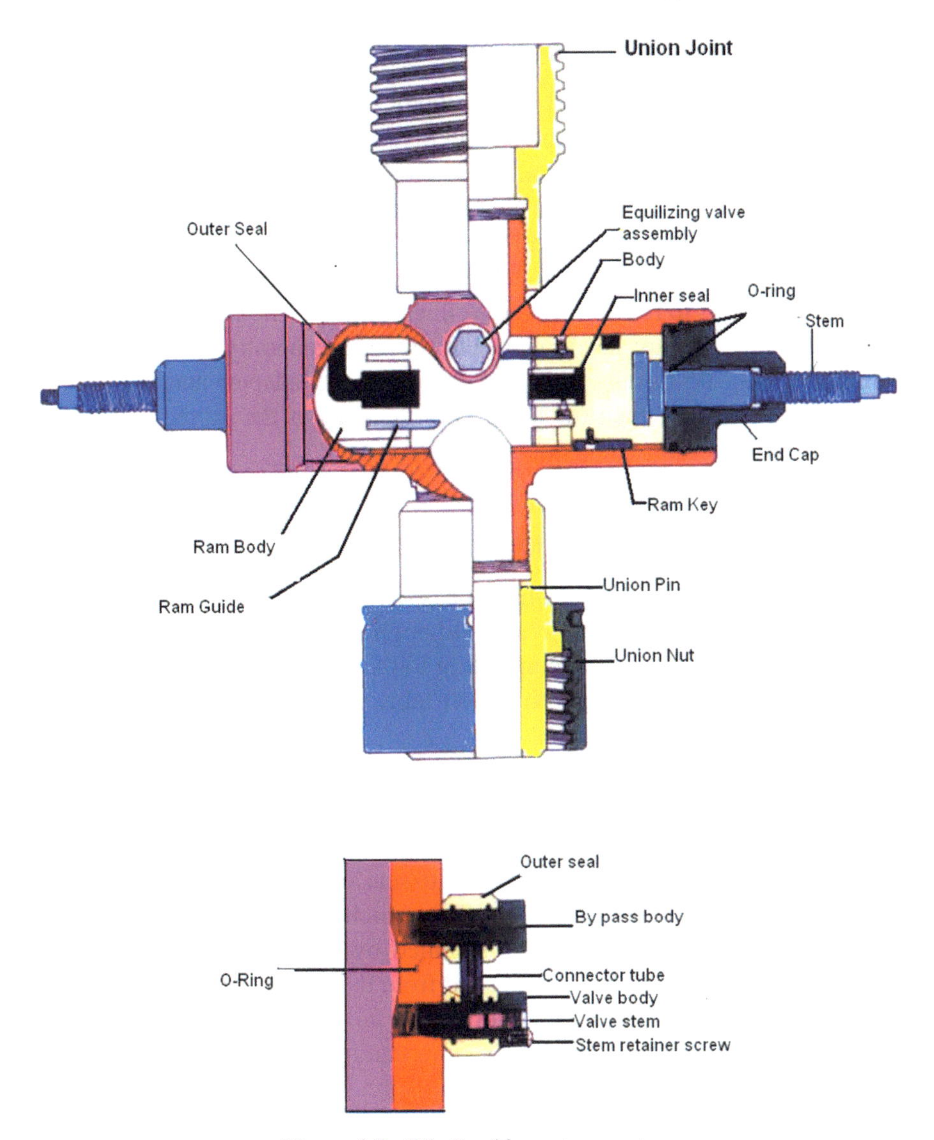

Figure 4.7 Wireline blowout preventer

On slickline operations in low-pressure wells, a single BOP is installed dressed with slickline rams to close and seal around the wire. On high-pressure wells, a dual BOP is used, the lower rams dressed for slickline and the uppers with blind. The injection point is used to pump grease if there is leakage past the rams. When running cable, a dual BOP is used with both rams dressed for the particular cable size and a grease injection point also available between the rams (Figure 4.7).

4.8 Lower lubricator sections

These are sections of thick wall tubes usually between 8 and 10 ft. long with quick union connections at each end and made up in a total length to accommodate the longest tool to be run. They are installed immediately above the BOPs and usually have the same bore size as the Xmas tree.

The section above the BOPs must have two bleed-off valves (contingency for one being plugged by debris or hydrates). Riser sections, used in offshore platforms to reach from the wellhead deck to another working deck, are similar to lubricator sections except they are generally much longer in length and may be installed between the wellhead adapter and the BOPs. They may also be of even thicker section to support the increased weight being carried.

4.8.1 Upper lubricator sections

These accommodate the tool string that has a smaller OD than the tool strings that are normally 1, 11/2, and 2 ins., although larger sizes are available for heavy duty work. The section connecting to the lower lubricator will have a connection to mate with that of the lower lubricator sections (or vice versa).

4.8.2 Stuffing box or grease head

The stuffing box or grease head terminates the top of the lubricator. The stuffing box contains packing that is squeezed to seal around the line. The packing is squeezed by an adjustable packing nut, which is hand adjusted although most stuffing boxes are now being supplied by remote hydraulic actuated packing nuts so that they can be adjusted from the deck eliminating the need for personnel to be lifted up to the top of the lubricator and, hence, is safer. The stuffing box also incorporates a sheave that turns the wire through 180°, from the outside of the lubricator into the bore. The grease head is used on braided line, electric line, or plain cable. It seals around cable by grease being pumped, at a higher pressure than that inside the lubricator, into the small annulus space between a set of flow tubes and the cable filling the cable interstices. The grease, being at higher pressure, tends to flow downwards into the lubricator and also upwards out of the tubes.

The upward flow is forced out through a return line for disposal by activating a cable pack off above the tubes. Downward flow is only constrained by the differential pressure applied between the grease and the lubricator pressure. Adjustments must be made to maintain the optimum conditions

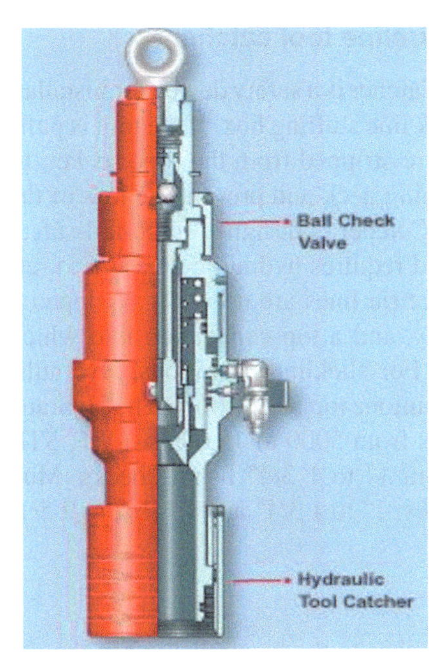

Figure 4.8 Hydraulic wireline tool catcher.

between grease lost to the hole, amount of gas entrained in the grease returns, and differential pressure.

4.8.3 Hydraulic tool trap with external indicator

The hydraulic tool trap with external indicator is installed between the wireline valve and the lubricator string. It prevents the loss of wireline tools downhole in the event of the wire being pulled off the rope socket, by retaining the dropping tool on a flapper. The flapper has an opening larger than the wireline but smaller than the tool string.

The flapper is operated by a hydraulic actuator, through an external handle, connected to the flapper with a low-torque pressure-balanced shaft. The external handle doubles as a tool passage indicator. The tool string may be pulled freely upwards through the tool trap.

The flapper will move upwards, together with the external handle, indicating tool passage. As soon as the string bottom has cleared the flapper, the flapper will drop back into the trap position, pushed by a torsion spring. The flapper can be lifted remotely when running a tool in the well by activating the piston in the hydraulic actuator. Pumping back the piston to its original position allows the flapper to return to its trap position (Figure 4.8).

4.8.4 Hydraulic wireline tool catcher

The Hydraulic Tool Catcher is a safety device for installation below the grease injection head or slick line stuffing box. If the tool is pulled into the top of the lubricator and the wire stripped from the rope socket, the Tool Catcher will engage the tool's fishing neck and prevent the loss of the tool string into the well bore. The Tool Catcher is designed to be fail-safe: it is permanently in the catch position and requires hydraulic pressure to release. The Hydraulic Tool Catchers for electric lines are normally equipped with an integral ball check valve assembly, and a top connection into which a grease head can be directly screwed. For slickline operations, Hydraulic Tool Catchers are available with quick unions top and bottom. The Hydraulic Tool Catchers are available with ratings from 5000 to 15,000 psi WP, STD, and H2S Service. Collet sizes vary from 1″ to 1 3/4″ fishing necks. Multi-catch options are available in two ranges: 1″ to 1 3/4″ and 1 3/16″ to 2 5/16″.

4.8.5 Line wiper

This is a tool that attaches to the hay pulley when the wire is being pulled to remove all contaminants from the wire before it is spooled.

4.9 Wireline Tools (Figure 4.9)

- Rope socket
- Knuckle joint
- Sinker bars
- Go devil
- Jars
- Wireline spear/grab/retriever
- Overshot
- Wireline pulling tool
- Installation equipment for side pocket gas lift valve
- Shifting/lockout tools

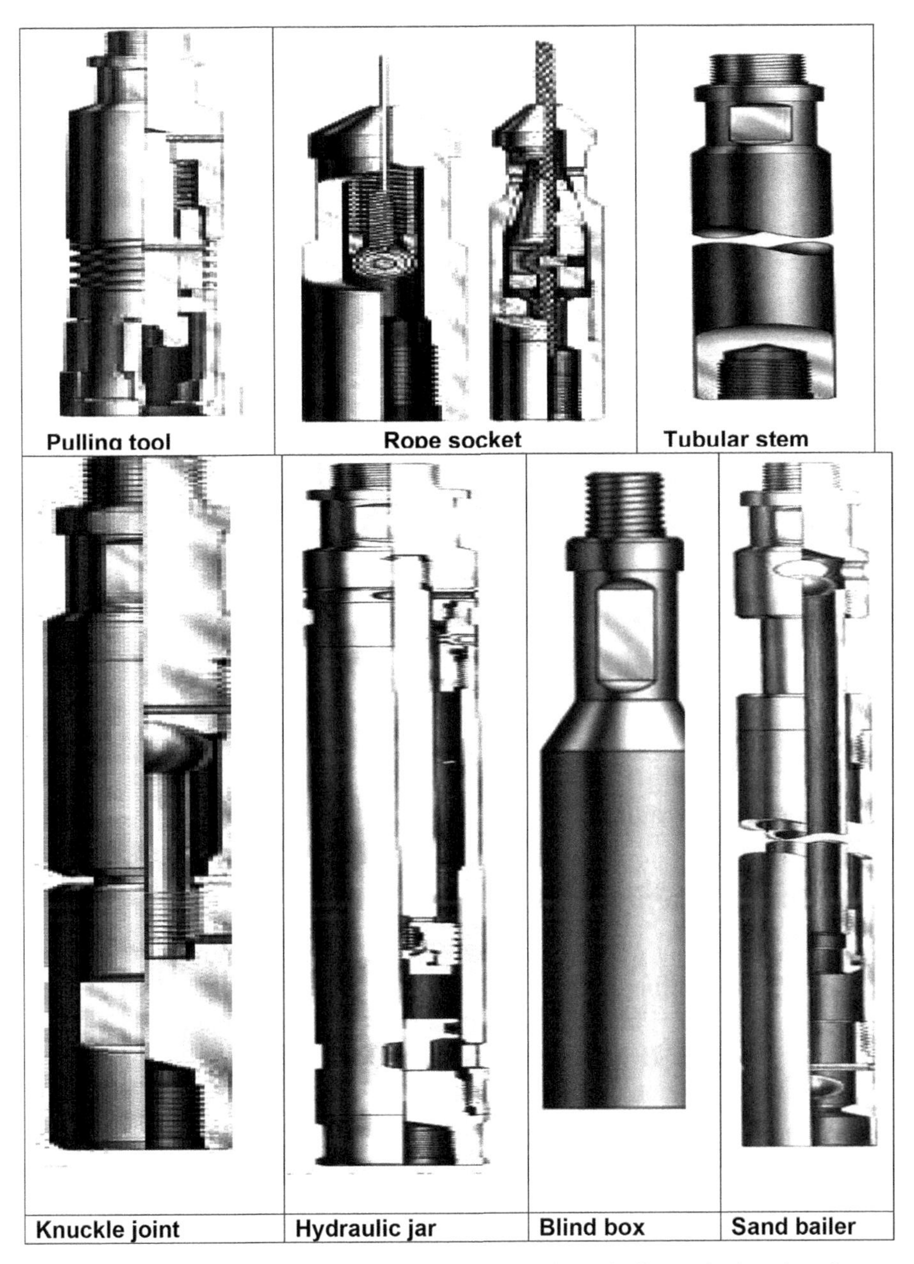

Figure 4.9 Pictorial representation of all wireline tools. (*continued*)

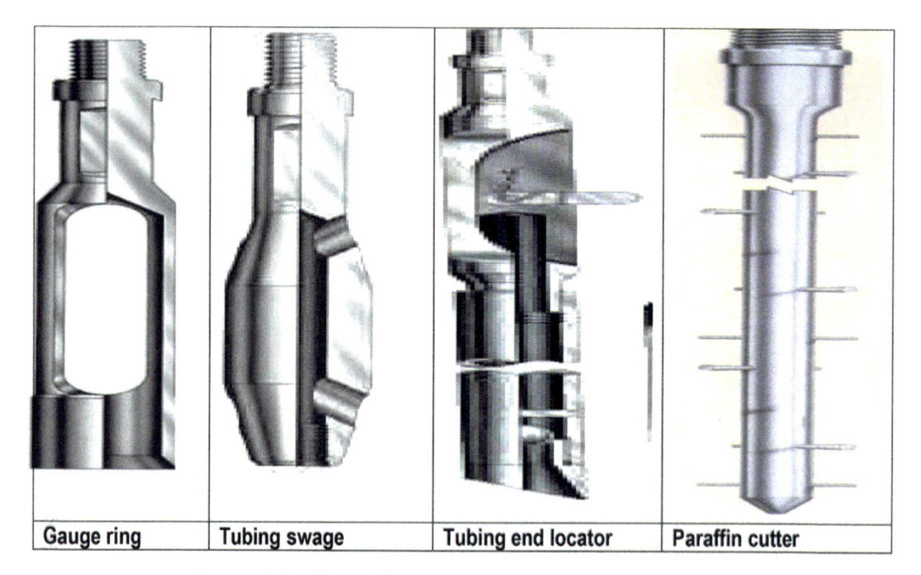

Figure 4.9 Pictorial representation of all wireline tools.

- Impression block
- Blind box
- Paraffin cutter/scraper/ cutter
- Tubing end locator

4.10 Coiled Tubing Units

Well servicing using coiled tubing (C/T) has grown significantly with the development of tooling and tubing technology. In recent years, the size of tubing available has increased from the original 1 ins. through 1 1/4, 1 1/2, 1 ¾, and now 2 ins. Even larger sizes are now being used as siphon strings, etc., but these are not yet generally used as workstrings. Along with this increase in the size of tubing has come material improvements to give higher performance. C/T units have largely replaced snubbing units for operations on completed wells and their versatility and, due to new tooling developed, have extended their range of capabilities in recent years.

The range of services now provided includes:

- Drilling and milling using hydraulic motors.
- Casing cutting.

- Circulating.
- Tubing clean-outs (sand or fill).
- Cementing.
- Through-tubing operations.
- Tubing descaling.
- Running, setting, and pulling wireline pressure-operated type tools.
- Fishing wireline tools.
- Logging (stiff wireline).
- Nitrogen lifting.
- Selective zonal acidizing.
- Perforating.

Much of the recent increase in capability is due to the increased performance of downhole motors, which provided the ability to rotate enabling drilling and milling operations, etc. The limitation of C/T is usually the pressure rating of circa 5000 psi and the depth to which it can be run, constrained by its relative low strength.

C/T units are constructed similarly and consist of (Figure 4.10):

- Operator's control cabin.
- Tubing reel.
- Powerpack.
- Goose neck.
- Injector head.
- Stripper.
- BOP system.

4.10.1 Operators control cabin

The cabin houses all of the controls for the reel and the injector head, and also all electronic logging systems and instrumentation. The controls operate the hydraulic valves and pressure supplied from the power pack. It is placed directly behind the reel to provide the operator with a full view of all activities especially the spooling of the tubing off and on the reel.

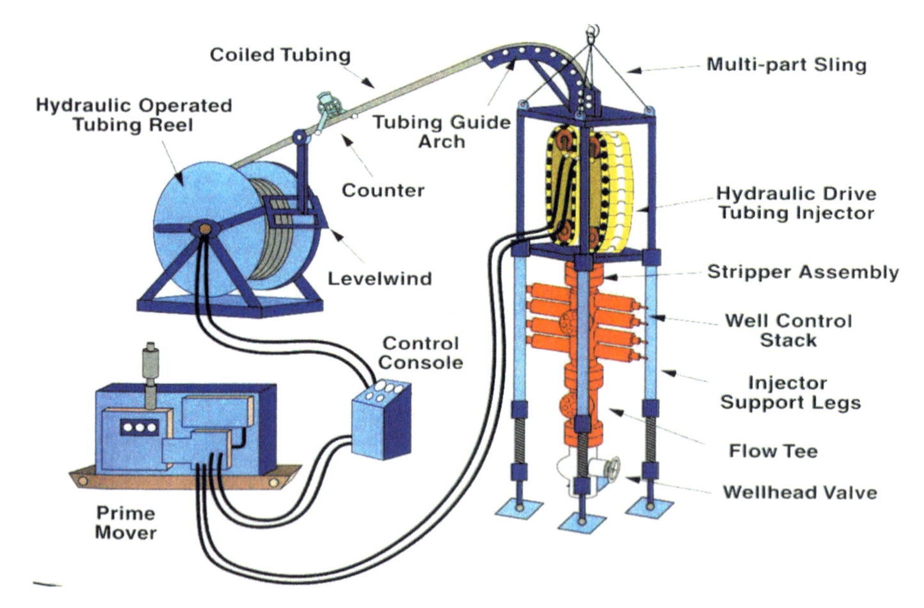

Figure 4.10 Schematic diagram of the CTU system.

4.10.2 Tubing reel

The reel stores the tubing that is coiled around the core of the reel. Ideally the core should be as large a diameter as possible to prevent severe bending of the tubing but must be of a manageable size for transporting to and from well sites. The radius of the core of the reel is sharper than that of the goose neck, e.g., 24 ins. (4 ft. diameter) versus 72 ins. for 1 1/4 ins. tubing; hence, most tubing fatigue is caused at the reel. The reel is driven by chain from a hydraulic motor controlled from the control cabin. The tubing is pulled off the reel up over the goose neck by the injector. The reel holds constant back tension to prevent the spool unravelling and to keep the tubing steady.

4.10.3 Power pack

The power pack is the provider of all hydraulic power. It consists of a skid-mounted diesel engine and hydraulic pumps and supplies regulated pressure for all the systems in the reel, injector head, BOPs, and the control cabin.

4.10.4 Goose neck

The goose neck is simply a guide that accepts the tubing coming from the reel and leads it into the injector chains in the vertical plane. The goose

Figure 4.11 Gooseneck.

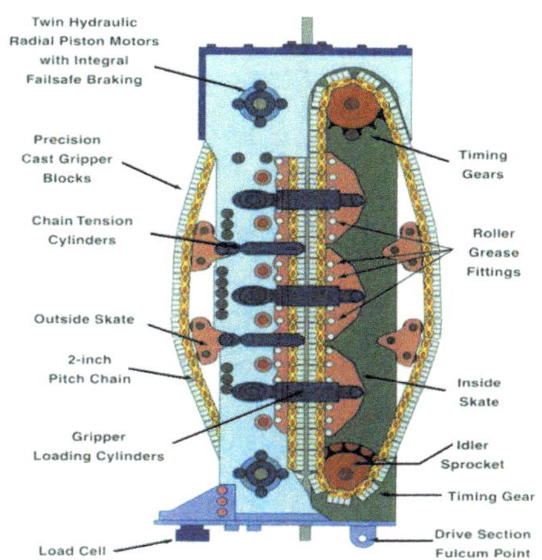

Figure 4.12 Injector.

neck guides the pipe using sets of rollers in a frame spaced on the recommended radius for the tubing being run, i.e., 72 ins. with 1 1/4 ins. Tubing, etc (Figure 4.11).

4.10.5 Injector

The injector is the motive device that imparts upward or downward movement to the tubing and is mounted above the BOPs on the wellhead. It must be supported as the connection to the BOPs is not designed to absorb the weight and lateral forces caused by the tension in the tubing from the reel. This support can be a crane for land wells (provided the lifting gear and pad eyes are rated for the weight of equipment and forces encountered) or to a mast or derrick offshore. Freestanding frames with hydraulic jacking legs are also available where no other means of rigging up is available. Movement is imparted to the tubing by sets of traveling chains equipped with gripper blocks that are hydraulically driven.

The gripper blocks grip by friction, which is adjustable through a hydraulic piston applying pressure across the chains. This pressure must be sufficiently high enough to grip the tubing eliminating slippage but not excessively high enough to crimp the tubing (Figure 4.12).

4.10.6 Stripper

The stripper is situated below the injector head in the injector head frame. It is designed to be as close as possible to the gripper chains to prevent buckling

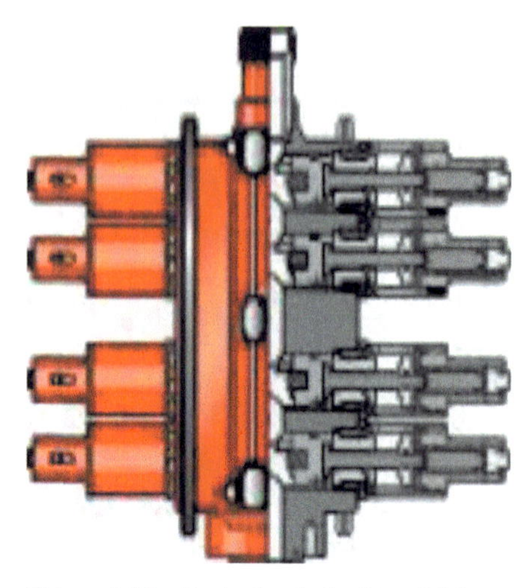

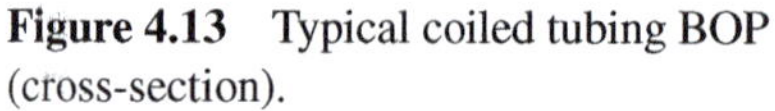

Figure 4.13 Typical coiled tubing BOP (cross-section).

Figure 4.14 Typical coiled tubing BOP.

due to snubbing forces. The stripper is hydraulically controlled to press the rubber element against the tubing to create a seal (Figure 4.13).

The stripper rubber is exposed to wear from the roughness of the pipe OD and will need to be changed from time to time, which can be done on the wellhead by closing the BOPs and removing the well pressure (Figure 4.14).

4.11 BOP system

The BOPs are very similar in function to wireline BOPs and are mounted above a wellhead adapter. They usually have four sets of rams dressed as follows, top to bottom:

- Blind.
- Shear.
- Slip.
- Pipe.

The shear rams usually can cut stiff wirelines, i.e., C/T with electric line cable inside it, used on C/T logging operations. In some areas of the world, an additional shear/seal valve is installed between the BOPs and the wellhead adapter as a tertiary barrier. The shear seal valve can cut the tubing and affect the seal. It is generally tied into a higher volume hydraulic pressure supply than available from the C/T unit such as a rig Koomey or independent system, etc.

Figure 4.11 Gooseneck.

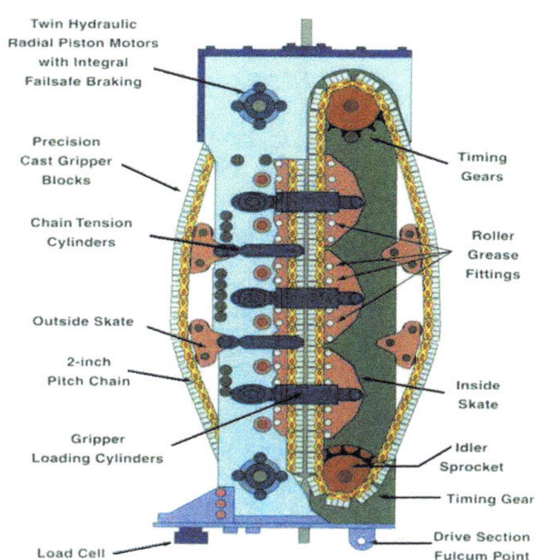

Figure 4.12 Injector.

neck guides the pipe using sets of rollers in a frame spaced on the recommended radius for the tubing being run, i.e., 72 ins. with 1 1/4 ins. Tubing, etc (Figure 4.11).

4.10.5 Injector

The injector is the motive device that imparts upward or downward movement to the tubing and is mounted above the BOPs on the wellhead. It must be supported as the connection to the BOPs is not designed to absorb the weight and lateral forces caused by the tension in the tubing from the reel. This support can be a crane for land wells (provided the lifting gear and pad eyes are rated for the weight of equipment and forces encountered) or to a mast or derrick offshore. Freestanding frames with hydraulic jacking legs are also available where no other means of rigging up is available. Movement is imparted to the tubing by sets of traveling chains equipped with gripper blocks that are hydraulically driven.

The gripper blocks grip by friction, which is adjustable through a hydraulic piston applying pressure across the chains. This pressure must be sufficiently high enough to grip the tubing eliminating slippage but not excessively high enough to crimp the tubing (Figure 4.12).

4.10.6 Stripper

The stripper is situated below the injector head in the injector head frame. It is designed to be as close as possible to the gripper chains to prevent buckling

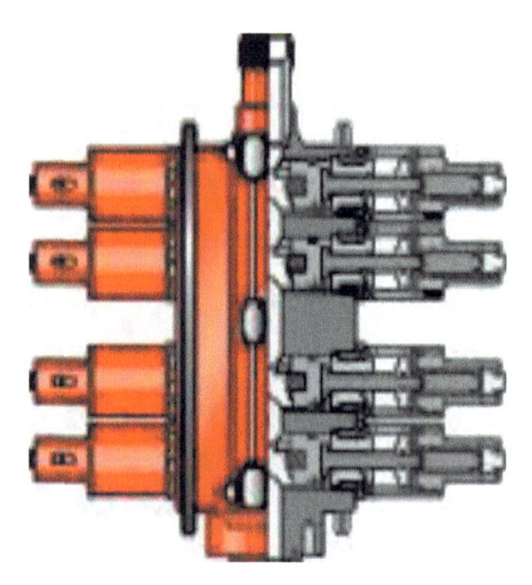

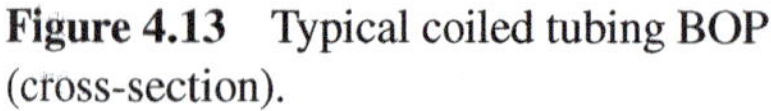

Figure 4.13 Typical coiled tubing BOP (cross-section).

Figure 4.14 Typical coiled tubing BOP.

due to snubbing forces. The stripper is hydraulically controlled to press the rubber element against the tubing to create a seal (Figure 4.13).

The stripper rubber is exposed to wear from the roughness of the pipe OD and will need to be changed from time to time, which can be done on the wellhead by closing the BOPs and removing the well pressure (Figure 4.14).

4.11 BOP system

The BOPs are very similar in function to wireline BOPs and are mounted above a wellhead adapter. They usually have four sets of rams dressed as follows, top to bottom:

- Blind.
- Shear.
- Slip.
- Pipe.

The shear rams usually can cut stiff wirelines, i.e., C/T with electric line cable inside it, used on C/T logging operations. In some areas of the world, an additional shear/seal valve is installed between the BOPs and the wellhead adapter as a tertiary barrier. The shear seal valve can cut the tubing and affect the seal. It is generally tied into a higher volume hydraulic pressure supply than available from the C/T unit such as a rig Koomey or independent system, etc.

4.11.1 Tubing

There are several coiled tubing manufacturers, but they are mainly US or Japanese companies. Some of the US companies use Japanese-supplied steel for tubing manufacture. The normal method of tubing manufacture is to produce rolled plate steel, which is cut into long flat strips. Each strip is then progressively folded round with rollers and formed into a long spiral. When it is completely formed into a round tube, the edges, now abutting, are welded. These individual lengths are then welded together to produce the length required to be contained on a shipping reel. Continuously milled tubing has now been introduced but is much more costly.

4.11.2 C/T unit accessories

In conjunction with the C/T unit, many of the services require additional auxiliary equipment such as pumping or nitrogen services. These may require cryogenic converter pumps, tankage, hoppers, filtration units, and interconnecting piping. These are connected up to the tubing reel inlet swivel, which allows the reel to rotate while still pumping.

4.12 Advantages and Limitations

4.12.1 Advantages of the CT system

1. CT can be RIH or POOH under pressure, which eliminates the cost of kill fluid and the risk of formation damage.
2. Fluids can be circulated continuously while RIH or POOH.
3. Units are highly mobile, self-contained, compact, and requires less footprint and less site preparation.
4. RIH or POOH speed can be as fast as 250 ft/min, and eliminates the need for making and breaking joints.
5. Manned by only 2–3 operators, whereas 4 or more operators are required for conventional workover rigs.
6. CT usually outperforms other methods of performing workover completions and remedial services in terms of control, accuracy, and efficiency.
7. Faster speed reduces formation damage like produced sand and clay swelling.

8. Increased size and strength of CT along with associated equipment allows it to service deeper and high-pressure wells.
9. The use of nitrogen with CT to “activate” has eliminated the need for highly risky operation like swabbing.
10. CT-conveyed specialized downhole tools allow services like selective acidizing, cementing, slim-hole drilling, wireline work in deviated wells, and cleaning of flow-lines.
11. With the development of completion tools for CT, it can be installed for a variety of purposes like producing, well treating, and gas lifting.
12. Highly deviated and horizontal wells can be logged and perforated only with the help of CT.

4.12.2 Limitations

1. Size and strength: Initially, the sizes available were 1. through 1 1/2 ins. with minimum yield strength of 50,000–60,000 psi. CT industry has come a long way and presently 1 through 3 1/4 ins. CT with 75,000 100,000 psi. yield strength is used routinely.
2. Maximum allowable pumping pressure: With the increase in strength, CT has overcome its initial pressure limitation of 5000–10,000 psi or more. Of course, the maximum working pressure is a function of tube condition and is determined by the user.
3. Wellhead pressure: Previously, WHP more than 2500 psi used to cause a problem for the CT as well as for the injector. With present CT strength and injectors having 60,000–100,000 lb. pulling capacity, servicing of high-pressure wells (WHP +/– 10,000 psi) are performed routinely, sometimes replacing snubbing.
4. Rotational ability: This limitation has been partially overcome with the development of downhole PD motor for light-duty drilling.

5

Formation Damage – Prevention and Remedy

5.1 Introduction

Formation damage is considered as any process that impairs the permeability of reservoir formations such that production or injectivity is decreased. Formation damage can be recognized by lower-than-expected productivity and accelerated production decline on affected wells. This is due to a reduced permeability in the near-wellbore vicinity, which has been affected by the damage mechanism. This area of reduced permeability results in an additional pressure drop imposed on the producing system, which is proportional to the rate of production.

Formation damage can occur at any time during a well's history from the initial drilling and completion of a wellbore through the depletion of a reservoir by production. Operations such as drilling & completion (Figure 5.2), workovers (Figure 5.5), Production (Figure 5.3) and stimulations (Figure 5.1 & 5.4), which expose the formation to a foreign fluid, may result in formation damage due to adverse wellbore fluid/formation fluid or wellbore fluid/formation reactions. While withdrawing fluids from a formation during normal production operations, formation damage may occur due to mechanisms such as fine migration or scaling, which result from imposing a pressure drop on the formation and the formation fluids.

For many years, the drilling industry focused on practices that gave high rates of penetration and minimum wellbore problems. The cementing industry focused most of the time on designing slurries, which will not bridge up or prematurely set within the casing. As a consequence, drilling and cementing fluids were often formulated to drill and cement the well cheaply and quickly with little thought of the impact on well productivity. However, other considerations such as the increase in small and marginal field development should stress the importance of understanding and controlling formation damage.

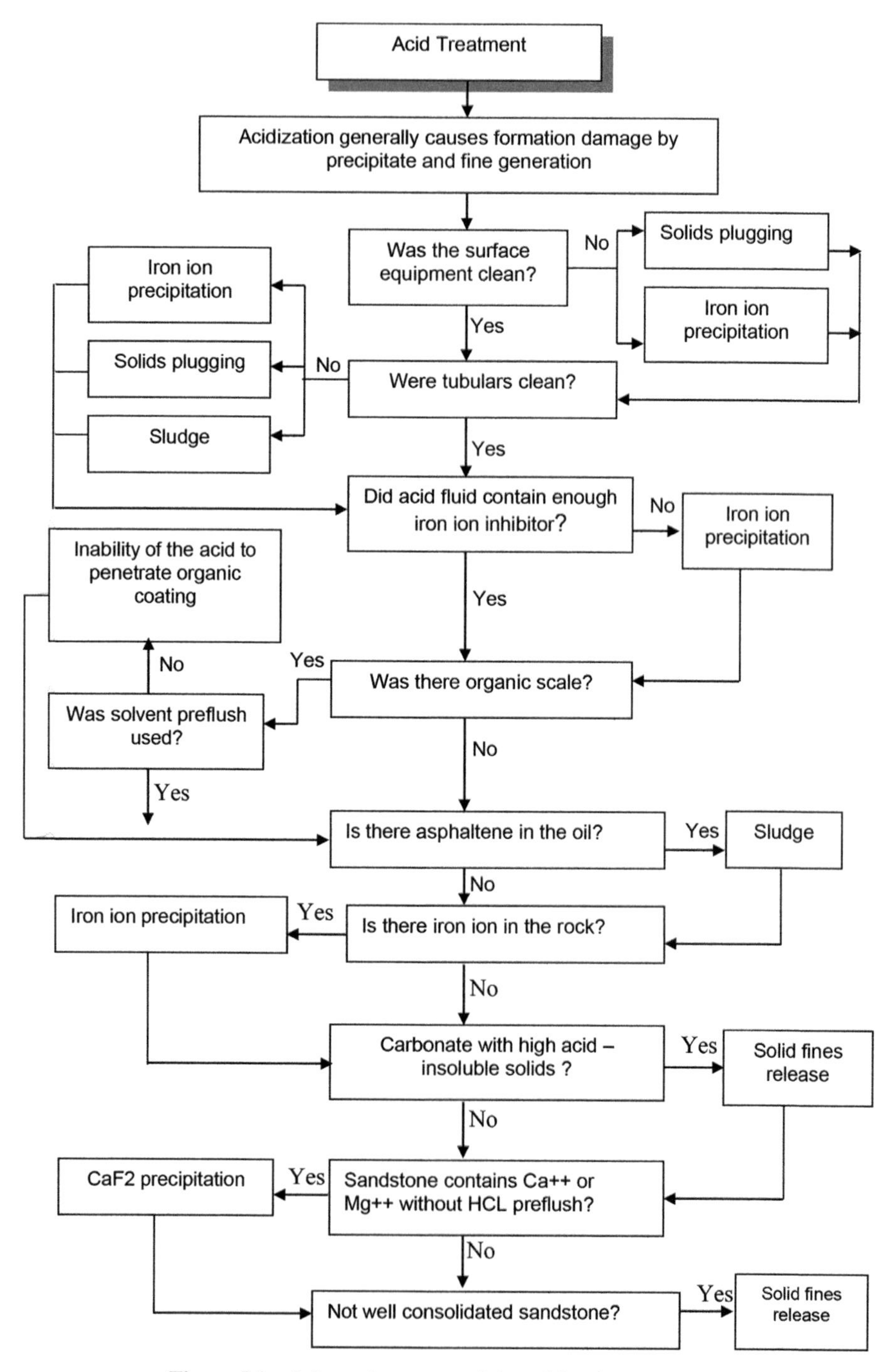

Figure 5.1 Schematic process of the acidization treatment.

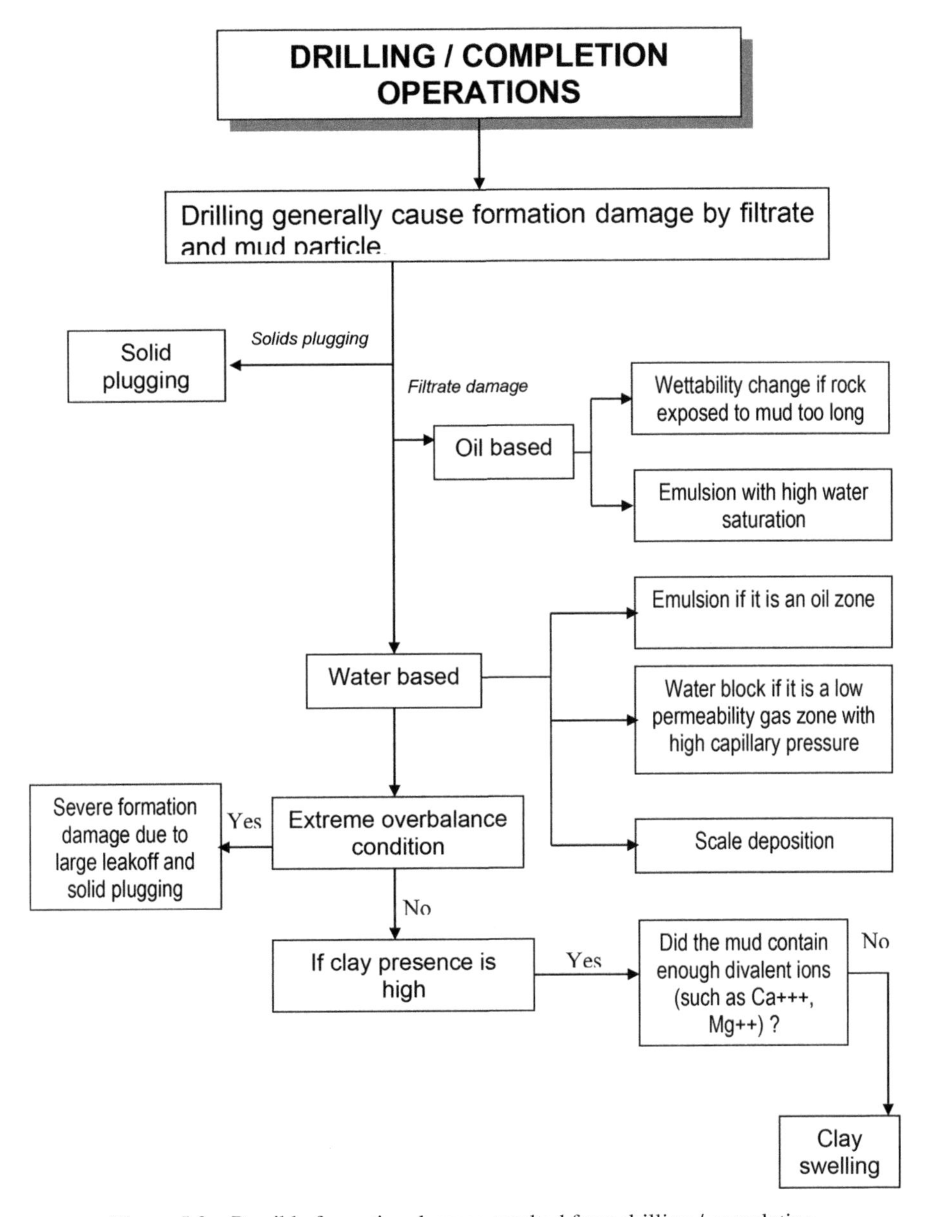

Figure 5.2 Possible formation damage resulted from drilling / completion.

Drilling departments focus the design of drilling fluid on volume and cost minimization. For the cementing operation, special care is given to reducing additive and spacer preflush usage. The production department has to deal with maximizing production. Those objectives are very often not complementary and even sometimes opposite. The drilling department will

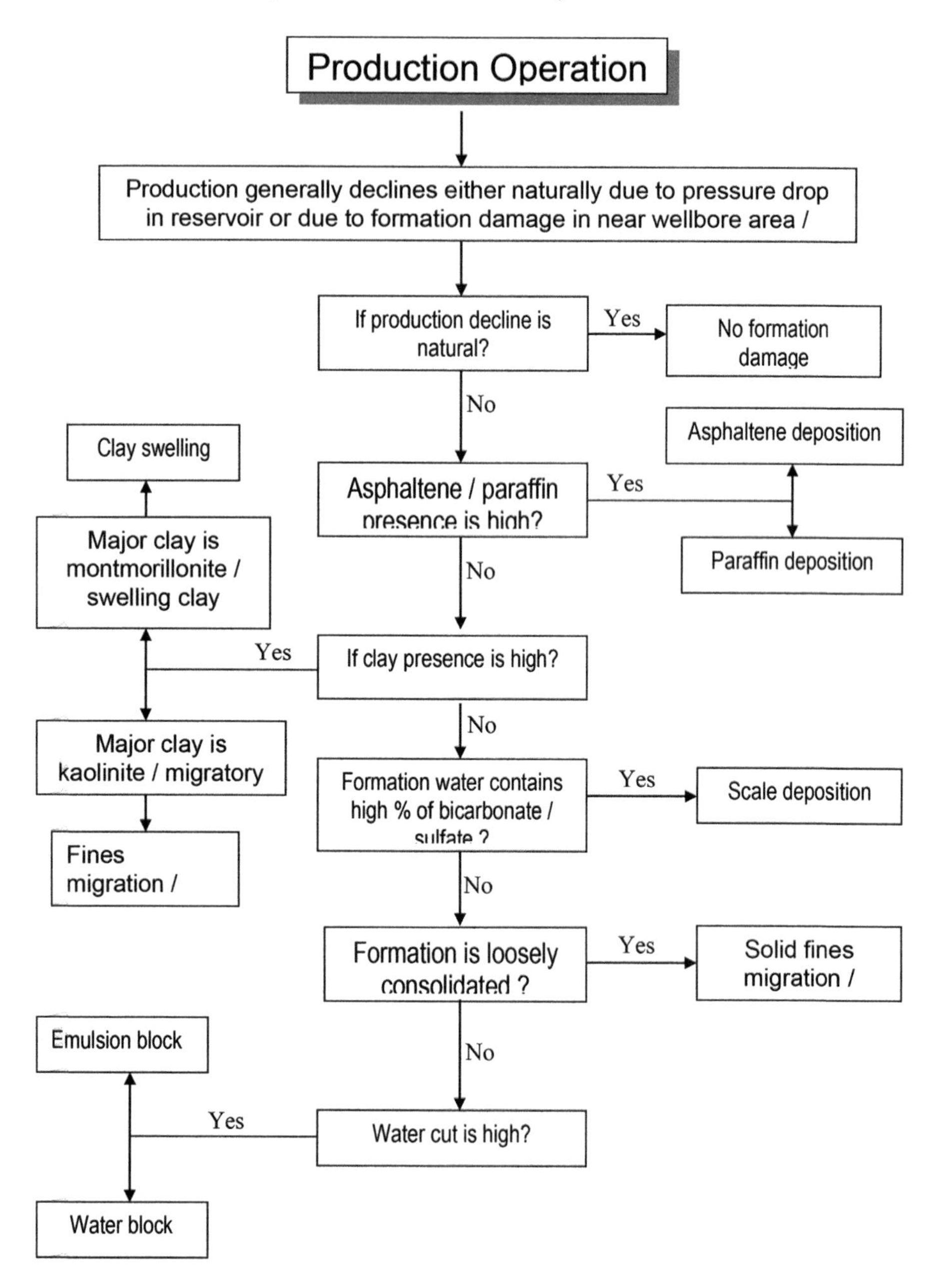

Figure 5.3 Possible formation damage resulted during production operations.

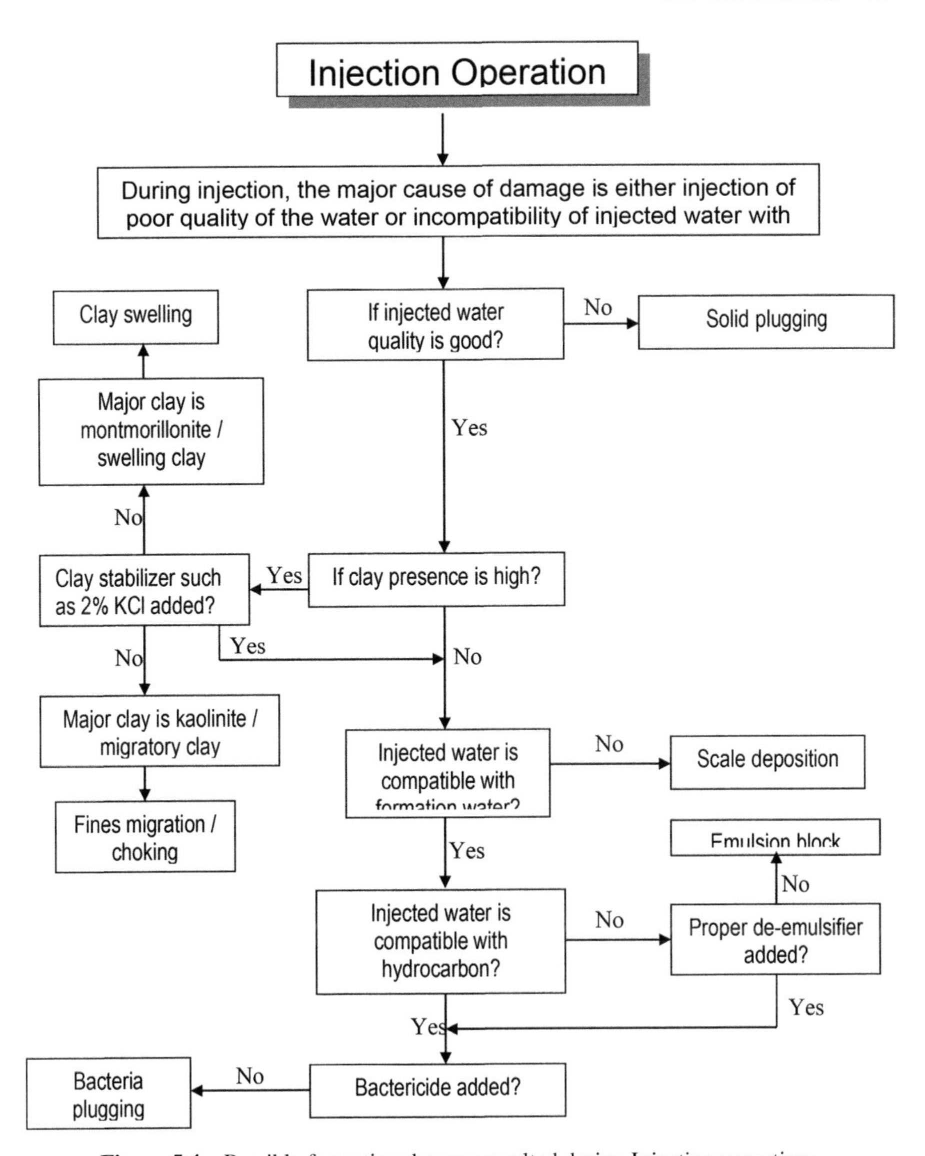

Figure 5.4 Possible formation damage resulted during Injection operations.

therefore be successful if the well is drilled and cemented at minimal costs without any major problems. However, when the well is handed over to the production department, any fluids or solids invasion will affect production dramatically and the same well could be seen as a failure.

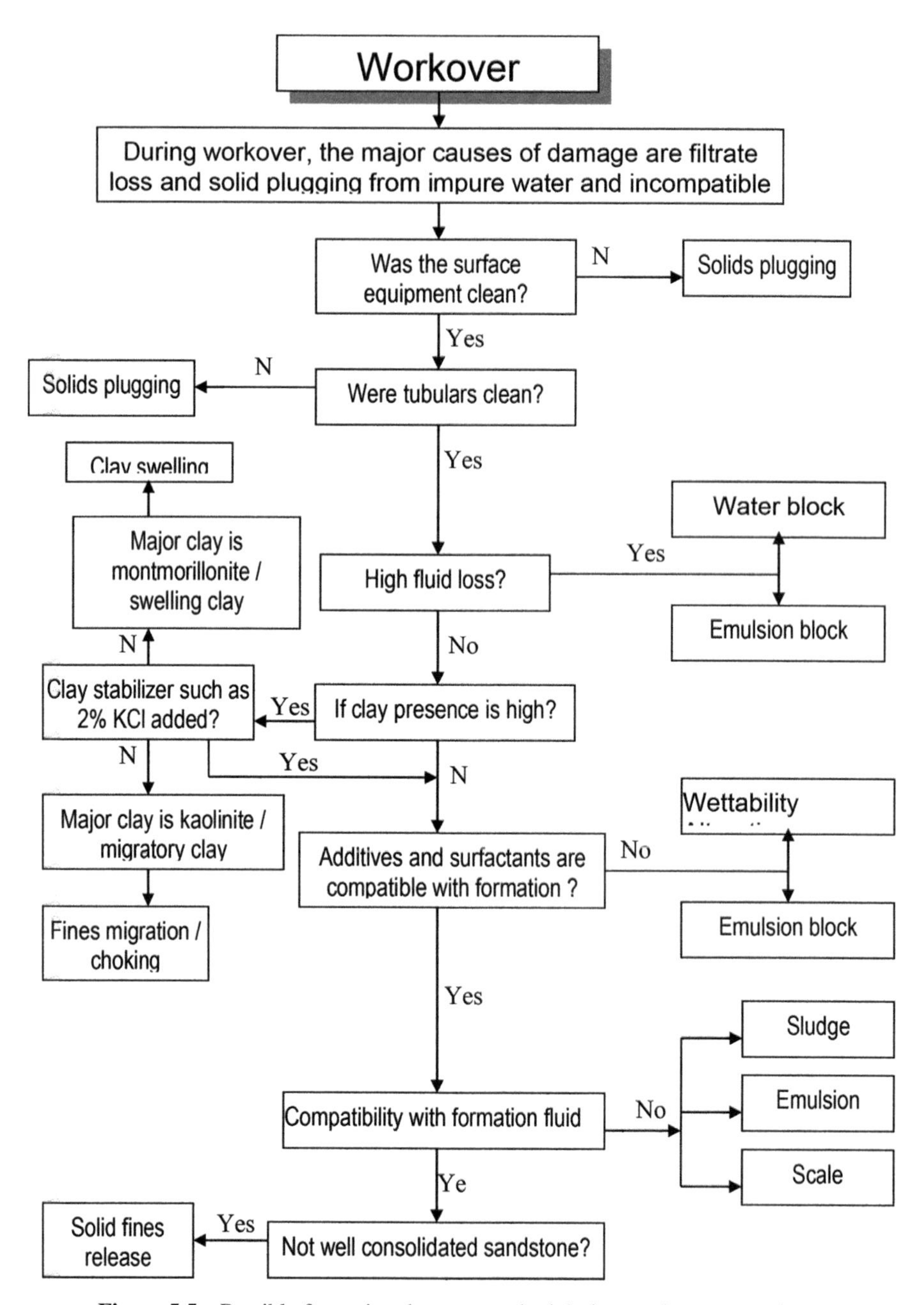

Figure 5.5 Possible formation damage resulted during workover operations.

Other reasons, which should increase the industry awareness of formation damage, could be:

- The increase in small and marginal field developments. This often means drilling only a small number of wells, which must produce to their full potential.
- Stimulation treatment could cure formation damage. However, prevention of damage is better than cure and could make the well cheaper.

5.2 Course Objectives

When a well is producing below its optimum productivity, the source of the problem must be identified before corrective measures can be taken. In some instances, this may require a systematic study of the entire producing system.

In this regard, some questions arise automatically, which need to be addressed if the situation demands. A systematic and comprehensive approach then can be adopted once the answers of the following questions are correctly evaluated.

i) How much formation damage is caused by drilling mud, preflushes, and cement slurries?

ii) Which fluid phase is the most damaging?

iii) Is the degree of damage a reason for concern?

iv) If so, what can be done to minimize or eliminate formation damage?

v) Should fluid designs or procedures be changed?

If formation damage is suspected to be the cause of a well's low productivity, there are many techniques available to evaluate a well in order to identify this problem. Once the presence of formation damage is confirmed, additional measures can be taken to assure that a stimulation attempt has a good probability of success.

In order to assure that the greatest return on investment dollars is realized, a well's productivity must be maximized. It is for this reason that effective formation damage identification, evaluation, and treatment techniques are considered essential to the working knowledge of the practicing engineer. This course addresses all the above points, which concern the production as well as non-production engineers about formation damage.

5.3 Wellbore Damage

5.3.1 Mechanical damage from drilling

The drilling process itself modifies the local stresses around the wellbore, generating a zone of reduced permeability in the near-wellbore area (Dusseault and Gray, 1992). It has been shown that such damage affects primarily soft formations where the difference between the minimum and maximum stresses orthogonal to the wellbore is large. In the worst cases, the extent of the permeability decrease can be as large as 21.2 wellbore diameters (Morales *et al.*, 1995), and perforations do not bypass the damaged zone. Because permeability impairment in this case is the result of rock compaction, acidizing is ineffective. Short proppant fracturing treatments are apparently the only cure, though extreme overbalance perforating may give positive results in some cases (Petitjean *et al.*, 1995).

5.3.2 Pipe problems

Whenever well production is reduced, the first determination should be to establish that the tubing is open and the lift system is working. Numerous pipe problems from leaks to collapsed pipe can occur, and fill in the tubing is also a possibility. Well conditions change over time, and an effective completion at the start of the well's life may not be effective after several years of production as the reservoir pressure declines.

5.3.3 Poor perforations

The usual treatment for poor perforations is to add additional perforations. In zones that are extremely laminated, such as the shaly sands of the U.S. Gulf Coast and other areas, 8–12 spf is considered adequate, but perforation breakdowns (i.e., small fractures) may be required for complete linking. Lower perforation density is possible if the well will be fractured. Fracturing will cross the barriers of laminations and in many field cases has provided extensive productivity increases. Adding perforations is easy, but the typical 0° phased, small through-tubing guns deliver only small holes and short penetrations. The newer downhole-deployable guns that provide minimum clearance and phasing are preferred, especially when hydraulic fracturing will be performed.

5.3.4 Hydrates

Hydrates are mixtures of ice and other elements, principally natural gas, that may resemble a dirty ice deposit. Unlike ice, they can form at temperatures greater than 32 °F. The formation of hydrates is usually associated with a drop in temperature or a reduction in pressure that may accompany the production of fluids. Hydrates may also form in gas-cut drilling mud, particularly when the mud is circulated near the seafloor in cold locations. Hydrate plugging of chokes and valves can be a serious problem. Hydrate particle abrasion of equipment is also possible. The most common occurrence of hydrates is in gas wells with a small amount of water production. The quantity of water relative to the quantity of gas production is critical. As the water cut increases, many hydrate problems disappear.

Adding a freezing-point depressant such as alcohol or glycol below the hydrate formation point prevents hydrates. They may also be controlled by temperature preservation in the produced fluid or the elimination of severe pressure drops that allow expanding gas to chill the liquids to their freezing points.

5.3.5 Fill

Debris from formation spalling into the perforation or wellbore can be one of the most serious detriments to production. Fill in the wellbore is easily identified with a sinker bar on the wireline and is usually easily removed using tubing or coiled tubing and N_2 or foam unloading practices. Fill in the perforations is more difficult to identify and much more difficult to remove. When fill in the perforations is suspected, reperforating the well is generally the most direct method of testing the theory and restoring the well to productivity. Where the fill is acid soluble, acid injection may be useful; however, injecting acid into a perforation that is filled with small debris is usually difficult.

5.3.6 Water problems

Water production is not only a major economic problem in surface separation, but it also causes a major reduction in the relative permeability of oil and gas. Water production from the well can lead to significant problems such as corrosion, backpressure, emulsions, and movement of the formation or fines. Water may flow from the bottom (coning), rise through fractures, or flow from the edge in fractures through the matrix or in high permeability streaks. Because of its low viscosity, water flows much easier than oil, and once in the pores of the rock, it is difficult to displace with low-viscosity fluids such as gas. Shutting off water (water control) is a special technique and is discussed elsewhere in the literature.

5.3.7 Microporosity

Microporosity is created by a number of clays and a few minerals. It is simply a condition where a large surface area exists for water coating or trapping. Microporosity rarely presents a problem except when it occupies the pore throat area of the formation. In these cases, it may trap either debris or water and obstruct flow. The removal of microporosity can generally be accomplished with HF, or deep problems can be bypassed by fracturing.

5.4 Induced Particle Plugging

5.4.1 Mud solids

To remove shallow mud damage in natural fractures, a solvent or cleaner that will disperse the mud should be selected on the basis of tests of a field sample of the mud. Energizing the fluid with N_2 can assist in removing large masses of drilling mud from a fracture system. Experience with drilling mud cleanup from natural fracture systems shows that slugs of drilling mud may flow back on initial treatment, and damage can often reassert itself as mud moves from the outer reaches of the fracture system into the wellbore. This condition can require repeated treatments of the same high efficiency cleaner, plus N_2, to get good cleanup of the well. Acid may help, but tests of the acid's effect on the field mud sample are required. When extremely large volumes of heavyweight mud are lost, it may be beneficial to sidetrack the well and redrill the pay zone. Whenever possible, the drilling mud overbalance should be minimized, and the mud should be conditioned to reduce solids before the pay zone is drilled.

Experience with drilling highly fractured formations has led to experimentation with underbalance drilling in some zones. Underbalance drilling can result in only minimal damage in producing wells in comparison with the damage created by traditional drilling methods. There are dangers, however, in underbalance drilling, and the risk versus benefit must be evaluated carefully. Mudcakes are usually damaging only in openhole completions without significant fractures (Burton, 1995). In vertical wells, they are usually easily mechanically removed to a great extent by pressure drawdown. In long horizontal wellbores, the necessary drawdown is almost impossible to impose on any section other than the heel, particularly when a compressible fluid is in the hole. Circulations for mudcake removal should be conducted with minimum clearance between the wash pipe and the borehole to promote turbulence. Residual mudcake in prepacked screens or slotted liners completions is particularly problematic because it can plug the screen (Browne *et al.*, 1995; Ryan *et al.*, 1995).

5.4.2 Dirty fluids

When particle damage is known to have occurred because of the use of unfiltered or poor-quality fluids, cleanup depends on finding a solvent or acid that can either remove the particles or break the structure of the bridges formed in the formation or fracture system. Surfactants, acids, and mutual solvents are usually the most beneficial materials. The addition of N_2 to provide a high-energy boost may also be beneficial. The decision of which surfactant or mutual solvent to use should be based on core tests or field response. Including a gas such as N_2 or CO_2 is based on fluid and solids recovery requirements and wellbore unloading ability. For designing cleanup operations for particulate damage, flowing the well back quickly after the treatment helps in the removal of the particles. Lower pressure formations may require a gas boost. In higher pressure formations, natural flow is usually adequate to unload these solids, especially when a properly designed fluid has been used and the solids are no deeper than the surface of the wellbore face. Mechanical scraping and cleaning can exert influence only as far as the wellbore wall.

5.4.3 Acidizing

The leading edge of an otherwise effective mutual solvent and acid system can be loaded with debris cleaned off the walls of the tanks and tubing. For this reason, the leading edge of the acid job is usually circulated out of the well using a process called pickling the tubing. In this treatment, acid and solvents are injected down the tubing to disperse and dissolve iron, pipe dope, mud, and other debris from the tubing and are then circulated or reversed out of the well without being injected into the formation. These jobs are extremely effective when the tubing has not been cleaned or its condition is unknown. Volumes of both acids and additive treatments range from 1 to 21.2 tubing volumes depending on the condition of the tubulars. Minimum acid and solvent volumes typically range from 250 to 500 gal. Coated tubing can reduce iron scale significantly, but other contaminants, such as scale and pipe dope, may still be present. If load-fluid recovery influences well production, surfactants or mutual solvents that reduce surface and interfacial tension are usually beneficial. The treatment volumes depend on the fluid, formation, and amount of load fluid lost.

5.4.4 Water floods

The removal of particles injected during water-flood operations depends on the identity of the material and use of a cleaner and an acid to disperse the material. One of the best techniques for cleaning up injection wells or disposal wells is to backflow the well as hard as possible prior to the treatment. This usually removes enough mass from the wellbore to eliminate the need for stimulation. However, if backflowing does not adequately clean the wellbore, acid and a mutual solvent in volumes ranging from 50 to 100 gal/ft are usually necessary. When large amounts of solids are expected, the well should be backflowed after acidizing. If oil carryover and emulsions are the problem, acid and a mutual solvent can be injected and displaced permanently with injection water behind the acidizing job.

5.5 Oil-based Drilling Fluids

The prevention of OBM emulsions is relatively easy. Either a surfactant-base cleaner that is mixed after specific OBM testing or a more general xylene wash of the zone must be done before contact with either high-salinity brine or acid. After the cuttings and mud fines have been cleaned and totally water wetted, the remaining damage problems of wettability can be reversed with a formation cleaner or mutual solvent. Acid is usually used as a following stage after cleaning to remove mud particles and clean up formation debris. Removal of known OBM emulsions resulting from mixing with high-salinity brine or acid usually requires an aromatic solvent wash or a specialized surfactant treatment that targets the silt stabilized emulsion. Evaluation of any cleanup mechanism or treatment using laboratory samples of OBM should be avoided. Only field samples of the mud are appropriate for designing the removal treatment.

Treatment fluid volumes range from 15 to 50 gal/ft of aromatic solvent or surfactant mixture, and the agitation and soak times are critical to the success of the treatment. Application difficulties include trapping the treating fluids across the pay in a column of heavyweight fluids where density segregation may be rapid. Packers and gelled plugs are the first line of isolation.

5.6 Water Blocks

Removal of a water block can be accomplished using a surfactant or alcohol applied as a preflush to reduce surface tension, followed by a postflush of N_2 or CO_2 to remove the water from the near-wellbore area and reestablish gas saturation. Once the water has been mixed with the surface-tension-lowering materials, removal is easier. The difficulties in this type of operation are placement of the fluid and getting an even distribution of the fluid around the wellbore. Repeated treatments are usually necessary, and selective injection devices are beneficial.

5.7 Wettability Alteration

Wettability alteration damage is removed by injecting (mutual) solvents to remove the oil-wetting hydrocarbon phase and then injecting strongly water wetting surfactants. Again, a surfactant by itself will not work. The oil phase, which is usually precipitated asphaltenes or paraffins, must first be removed with a solvent. (The same applies to an adsorbed oleophilic surfactant.) Then, a strongly water-wetting surfactant can be injected and adsorbed onto the rock minerals. This reduces the tendency for new hydrocarbon precipitates to stick to the mineral surfaces and oil-wet them again. For retrograde condensation problems, the most appropriate treatment technique is the injection of neat natural gas in a periodic "huff and puff" operation. Condensate is picked up by the gas and transported into the reservoir. Reprecipitation requires the retrograde of the process after several months of production.

5.8 Fines and Clays

5.8.1 Migrating fines

The treatment of moveable fines can be accomplished by either prevention (using a clay-control process) or removal. Removal of migrating fines in sandstone formations is best accomplished by treatment with a fluid containing HF and HCl mixtures – these are the commonly used mud acids.

Deeply penetrating acid systems, containing fluoboric acid, show a good possibility for particle destruction and extend some potential for clay stabilization.

Fracturing the formation is also a treatment possibility because the effect from linear flow in the walls of the fracture has a less detrimental effect on production than inward radial flow in an unfractured well. The success of both clay control and fine removal depends on the depth extent of the fines movement problem. In many cases, tip-screenout (TSO) fracture design using a short fracture for damage bypass is a better alternative.

HCl systems are typically used to remove fine damage in a carbonate formation. Because the fines are not dissolved but are dispersed in natural fractures or the wormholes that are created, N2 is usually recommended to aid fine removal when the well has a low bottomhole pressure.

5.8.2 Swelling clays

The removal of smectite is usually accomplished with HF or fluoboric acid, depending on the depth of penetration. In the event of deep clay-swelling problems (more than 2 ft), the best treatment is usually a fracture to bypass the damage.

5.8.3 Unconsolidated formations

Two basic problems determine the method of treatment for unconsolidated formations. If the formation moves as discrete large particles (i.e., the building blocks of the formation are moving), then the problem is a lack of cementation between the grains for the applied production forces, and the formation is classified as a low-strength formation.

Treating low-strength formations can be difficult if the cementing materials are reactive with the fluid that is injected to remove formation damage or to improve permeability. Fortunately, the cementing materials in most formations have a small surface area and are less reactive with acids than with fines or clay particles in the pores of the rock. When formations expel large grains into the wellbore, it may be beneficial to add additional perforations to reduce the velocity across the sandface or to design a fracture to reduce the drawdown. (It is common to fracture formations with permeabilities higher than 100 md.) These fractures are usually TSO designs that provide short, highly conductive fractures that can reduce the drawdown and control sand movement by both pressure reduction and use of the proppant at the interface contacts of a gravel pack as an "information" gravel pack.

The treatment of spalling problems is extremely difficult. Propped fractures may help contain the formation and spread out the drawdown to reduce the spalling force, although totally halting spalling may be impossible. One of the keys to treatment selection is whether the spalling is caused by high initial pressures that will quickly deplete or by cyclic mechanical loads that will recur. If high initial pressure is the problem, a cleanout may suffice. If cycling is the problem, a permanent control method is the best solution. Control methods include gravel packing, fracture packing, selective perforating (along the fracture axis), and some plastic-bonding methods.

5.8.4 Scales

Various solvents dissolve scales, depending on their mineralogy. The most common treatments for the scales in a well are as follows:

- **Carbonate scale** ($CaCO_3$ and $FeCO_3$) – HCl will readily dissolve all carbonate scales if the acid can penetrate to the scale location (Tyler *et al.*, 1985).
- **Gypsum** ($CaSO_4$ and $2H_2O$) or anhydrite ($CaSO_4$) – These calcium sulfate scales are removed with compounds that convert the sulfate to a hydroxide or other ion form followed by acid or by direct dissolvers such as ethylene diamene tetra acetic acid (EDTA) or other types of agents. Following a calcium sulfate dissolver with acid may double the amount of scale dissolved because most scales are mixtures of materials and HCl has some ability to dissolve the finest particles of calcium sulfate. The tetrasodium salt of EDTA is preferred because its dissolution rate is greater at a slightly alkaline pH value; the more acidic disodium salt has also been used, as well as other strong sequestrants of the same family, although they do not show a marked difference from the EDTA performance. Care must be used not to over-run the spent scale dissolver or converter solutions with acid because massive reprecipitation of the scale will occur.
- **Barite** ($BaSO_4$) or celestite ($SrSO_4$) – These sulfate scales are much more difficult to remove, but their occurrence is more predictable. Barium and strontium sulfates can also be dissolved with EDTA if the temperature is high enough and contact times are sufficient (typically a 24-hour minimum soaking time for a 12,000-ft well with a bottom-hole temperature of about 212 °F [100 °C]; Clemmit *et al.*, 1985). Barium and strontium sulfate removal methods are usually mechanical. Most chemical removers are only slightly reactive, especially in thick deposits, but mixtures of barium sulfate and other scales can usually be removed by properly formulated dissolvers with sufficient soak times. Thick deposits should be removed by mechanical or abrasive methods. Care must be exercised when analyzing well debris to avoid mislabeling barite from drilling mud residue as barium sulfate scale.
- **Sodium chloride** (NaCl) – Sodium chloride scale is readily dissolved with fresh water or weak acidic (HCl, acetic) solutions. Redesigning the mechanical system to avoid heat loss and water drop-out are also treatment possibilities.

- **Iron scales**, such as iron sulfide (FeS) or iron oxide (Fe_2O_3) – HCl with reducing and sequestering (EDTA) agents dissolves these scales and prevents the reprecipitation of by-products, such as iron hydroxides and elemental sulfur (Crowe, 1985). Soak times of 30 minutes to 4 hours are usually beneficial in removing these scales when using acid. Where iron sulfide is a thick deposit, mechanical action such as milling is suggested. Water jetting typically will not cut an iron sulfide scale except where it is dispersed with other scales or exists as a thin coating.
- **Silica scales** – Silica scales generally occur as finely crystallized deposits of chalcedony or as amorphous opal and are readily dissolved by HF.
- **Hydroxide scales:** Magnesium ($Mg(OH)_2$) or calcium ($Ca(OH)_2$) hydroxides – HCl or any acid that can sufficiently lower the pH value and not precipitate calcium or magnesium salts can be used to remove these deposits. Contact time is an important factor in the design of a scale removal treatment. The major concern in treating scale deposits is allowing sufficient time for the treating fluid to reach and effectively dissolve the bulk of the scale material. The treating fluid must dissolve most of the scale for the treatment to be successful.

5.9 Organic Deposits

Organic deposits are usually resolubilized by organic solvents. Blends of solvents can be tailored to a particular problem, but an aromatic solvent is an efficient, general-purpose fluid. Environmental concerns have led to the development of alternative solvents (Samuelson, 1992).

- *Paraffin* removal can be accomplished using heat, mechanical scraping, or solvents. Heating the tubing with a hot oiler may be the most common type of treatment. It may also be the most damaging and least effective in some cases. Injection of hot oil from the surface will melt the paraffin from the walls of the pipe, but the depth to which the injected fluid stays hot is a function of the well configuration. If the well is allowed to circulate up the annulus while the hot oil is injected down the tubing, the heat will not penetrate more than a few joints of tubing from the surface. The heat is quickly transferred through the steel tubing to the fluids rising in the annulus and little, if any, heat reaches deep in the well. As the hot oil cools, the paraffin picked up in the upper part of the well can precipitate. If hot oiling is required at depths greater than 150 ft, an alternate method of placement must be used. Deeper application of heat is available with other processes that feature heat generation as part of an exothermic chemical reaction. The processes require close control and are generally expensive.

Mechanical scraping can be useful in cases where extensive deposits of paraffin must be removed routinely. Scraping is usually accomplished with slickline and a cutter. In wells that utilize a rod string, placing scrapers on the string may automatically scrape the tubing walls.

Solvent treating to remove paraffin may be based around a straight- or aromatic-chain solvent. The most appropriate solvent depends on the specific paraffin and the location of the deposit. Heat (at least to 130 °F [55 °C]) and agitation significantly increase the rate of removal.

- **Asphaltenes:** Removal treatments for *asphaltenes* use aromatic solvents such as xylene and toluene or solvents containing high percentages of aromatics. Naphtha is usually not effective as a solvent. Some materials being tested provide dispersant benefits without stabilizing the entire mass of the asphaltene. Solvent soak time, heat, and agitation are important considerations for treatment.

5.10 Mixed Deposits

Combined deposits require a dual-solvent system, such as dispersion of a hydrocarbon solvent (usually aromatic) into an acid.

- **Emulsions**
  Emulsions are stabilized by surface-active materials (surfactants) and by foreign or native fines. Generally, mutual solvents, with or without de-emulsifiers, are used for treating emulsion problems. De-emulsifiers, which may work well in a laboratory or in a separator or tank because of the large number of droplets in contact per unit volume, may not work by themselves in a porous medium because of mass transport phenomena in getting the product to where it should work. Another reason they may not work alone is the mechanism involved in breaking emulsions, which should provoke the coalescence of droplets and then phase separation.

  Asphaltic iron-catalyzed sludges are the most difficult emulsions to break. These emulsions are catalyzed by dissolved iron in the acid or water and resemble a cross-linked oil polymer in some instances. Prevention is the best treatment. An effective antisludge treatment for the area and an iron reducing agent in the acid are the best methods. Removal of an existing asphaltene sludge is usually accomplished by dispersing it in a solvent and attacking the components of the sludge with additives designed for cleanup and removal.

5.11 Bacteria

Prevention of polymer destruction by bacteria is usually handled with biocides and tank monitoring. Control of bacteria downhole is more difficult and involves scraping or treatments with sodium hypochlorite or other oxidizers followed by acidizing and then treatment with an effective biocide at a level at least 1.2 times the minimum kill level. Frequent rotation of the type of biocide is also necessary to prevent the development of biocide-resistant strains of bacteria.

6

Sand Control

6.1 Introduction

The incursion of the formation of sand into a well is one of the oldest problems plaguing the oil and gas industry because of its adverse effect on well productivity and equipment. It is normally associated with shallow, geologically young formations that have little or no natural binding material to hold the individual sand gains together. As a result, when the wellbore pressure is lower than the reservoir pressure, drag forces are applied to the formation sands because of fluid production. If the formation restraining forces are exceeded, sand will be drawn into the wellbore. The sand either plugs the well or is produced. Several operational problems can arise if a well produces sand. All are troublesome and costly, but the degree of severity varies widely. Probably the less severe problems are solved simply by periodic removal of sand from sub-surface facilities, flow lines, manifolds, and separators. The more serious consequences relate to erosion of surface equipment, causing damage and productivity loss.

Controlling formation sand is costly and usually involves either decreasing the production rate or using gravel packing or sand consolidation techniques or other sand control techniques.

6.1.1 Types of formation sands

1. Consolidated sand

    a. Cementing agent between sand grains prevents sand movement
    b. Pore spaces between grains permits oil or gas flow

2. Unconsolidated sand

    a. Quick sand
        - No cementing agent, sand flows readily

b. Packed sand or partially consolidated
   - Little cementing agent
   - Cavities form around wellbore
   - Formation collapse around casing
c. Friable sand
   - Sand will crumble
   - Sand gets eroded by fluid or gas production forming large cavities
   - Formation collapse around casing
d. Clayey sands
   - Extremely small in size
   - Can swell
   - Reduce permeability or completely plug

6.1.2 Reasons for sand production

- Totally unconsolidated formations
- Onset of water production
- High production rates
- Depletion in reservoir pressure
- Improper well completion practices

6.2 Sand Control Methods

There may be different ways to combat the problem of sand production, which may differ from field to field and operator to operator. Some of the methods, that have been used, are:

6.2.1 Regular cleanout and bailing

Sand control by frequent wellbore cleanout may be the oldest sand control method and is still found to be in use today. As long as the well can produce and can be periodically cleared out successfully, the operator may feel that a sand problem does not crest. Unfortunately, when the operator realizes that a problem does exist and is beyond control, it is probably too late to successfully control the sand and maintain. As the well depletes and production rates are lower, it becomes more difficult to carry out sand to the surface and sand accumulates more rapidly in wellbore. Lower production rate and more frequent cleanouts accelerate the reaching of the economic limit of the well.

6.2.2 Limit production to maximum sand-free rate

Sand control by limitation of production to max sand-free rate can be initially successful, however; the sand-free rate may not be economically acceptable. The sand-free rate tends to decline with the depletion of the well. In particular, the beginning of water production may affect the max sand-free rate. When the sand-free rate is not economical, other sand control methods will have to be considered.

6.2.3 Chemical consolidation

Sand control by "in-situ" consolidation of the formation sand is not new and has been in use for more than 50 years. The consolidating material is typically a plastic resin, and various resin systems have been developed including phenolics, epoxies, and furanes. Plastic consolidation can be successful; however, for various reasons, it is not popular at present. Problems include permeability reduction near wellbore, inability to consistently treat all the perforations, and cost involved thereof. All the methods, which require injection, suffer from the same deficiency of inability to consistently treat the entire production internal.

6.3 Perforation Optimization and Selective Perforation

By increasing the perforation density and/or perforation hole size, sometimes the sand production can be avoided. This will basically decrease the drawdown across the perforation and hence the drag force on the reservoir rock. Also the rock strength from the logs may be used to interpret the sanding tendency of that formation sand, if *G*/*Cb* ratio was more than 0.8×10^{12} psi, no sanding is expected (*G* & *Cb* being shear modulus and bulk compressibility as computed from sonic and density logs). It has been suggested that the sonic log is the basic indicator and thus in itself may be a predictive tool, and where ▲T (sonic transit time) exceeds 90 microseconds per foot, sand control is frequently required. Thus, selective perforation may be done to avoid probable sand production as far as possible.

6.3.1 Screen or slotted liner without gravel

For many years, we used liners and screens alone to control formation sand, with some success. We best determined the openings of these screens by using the 10-percentile formation sand size as the criterion for wire spacing.

The liner openings should be equal to the 10-percentile sand grain size. The openings of the screen will trap the larger 10% of the sand grains, thereby preventing the remaining 90% of the formation sand from passing through. The screen diameter should be as large as possible when using this technique to control the formation of sand.

We do not recommend using a liner without gravel as a good sand control technique because it will almost always reduce the production capacity of wells. The intermixing of sands, clay, and shale results in significantly lower permeability compared to the native formation sand.

Another common issue with screens without gravel is that the sand often bridges around the screen before the screen openings erode. As a result, the eroded screen is unable to control the formation of sand.

6.4 Gravel Packing

Gravel packing is the most widely used and well-accepted method of sand control in the world. Water wells used this mechanical sand control technique long before oil or gas wells did. It involves running a mechanical device, such as a screen or slotted liner, in the well and placing accurately sized gravel around the screen or slotted liner. This placement allows the entry of fluids through the gravel but filters the formation sand from the flow stream so that sand-free production is possible. However, most gravel packs produce a finite amount of solids, which consist of very fine particles that can move through the gravel pack. When placed properly, gravel packs yield a long life and high productivity. The fundamental challenge is to manage the formation sand without unduly lowering the well's productivity.

Regardless of the completion configuration, the three primary objectives of any gravel pack are:

- Sand-free production
- High productivity
- Completion longevity

The implications of achieving these objectives are that operators must have the knowledge and capabilities of properly performing gravel packs under a wide variety of field conditions. Hence, gravel packing must be approached from the total-system standpoint that involves many interrelated technologies. Effective control reduces a good gravel pack design and execution, including obtaining representative samples of the formation sand, analyzing the formation grain size distribution, selecting an optimum gravel size in relation to formation sand size to control formation sand movement, and using the optimum screen slot width to retain the gravel. Others items that affect the gravel pack are proper well-preparation procedures, choice of an effective placement techniques, and implementation of procedures that will not impair productivity.

The importance of packing the perforations has been critical to gravel pack productivity and completion longevity. As a can sequence, quality management in gravel packing field procedures has been developed to ensure that the perforations are effectively packed with gravel.

Experience has shown that sand control should be initiated before the formation is seriously disturbed by sand production. As the volume of produced sand is allowed to increase, the greater the probability of formation sand disturbance, the more difficult it is for sand control to be successful. It has been reported that when the core sample was subjected to deliberate disturbance, a large decrease in permeability occurred but there was little change in porosity. Such a loss in formation permeability could occur downhole and become increasingly more severe as sand is allowed to produce. Another bad effect of allowing sand production prior to sand control is the removal of support for shale streaks. The shale could fall across some of the perforations after the removal of support. This may increase the water cut also. Impairment of production and prevention of satisfactory placement of remedial sand control materials could then result. Even with proper placement of sand control chemicals, production may still be reduced by shale across the producing interval.

It is not surprising that the initial sand control installation has proved to be more successful than remedial treatments. It is also primarily common for remedial installations for several reasons, some of which are not fully understood to impair productivity.

6.4.1 Saucier's design criteria

The gravel should be selected based on 50-percentile grain size of the smallest productive formation sand of the core within the interval that is to be gravel packed. The selected 50-percentile size of the gravel should be less than six times the 50-percentile size of the smallest grained productive formation sand size.

Saucier's plot:

- Plot grain size vs. cumulative weight % on semi-log paper
- Determine the median grain size, i.e., 50% point
- Recommended median grain size of gravel –

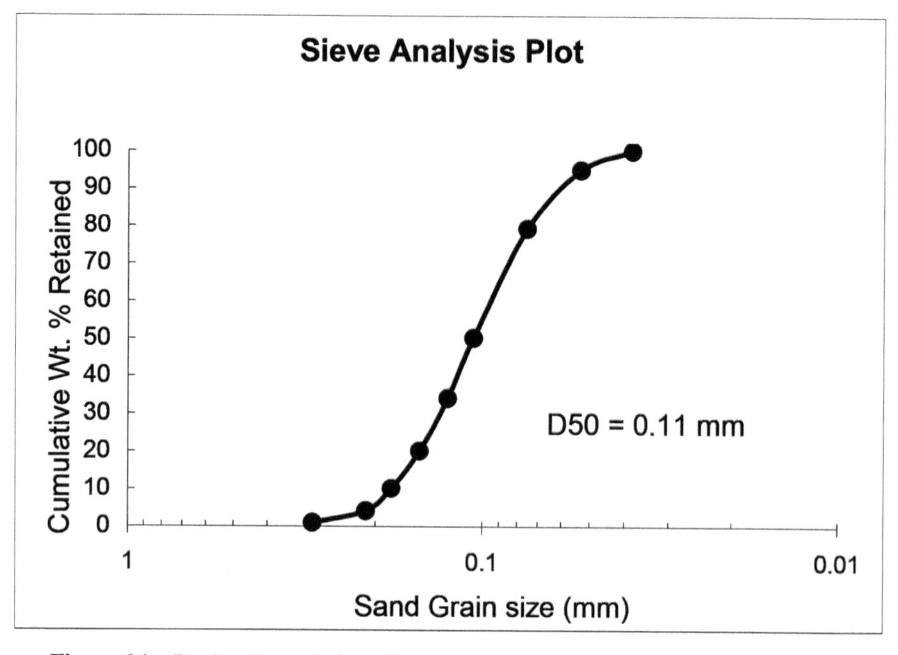

Figure 6.1 Depicts the variation of grain percentage retained concerning grain size.

Six (6) times larger than or equal to the median grain size of the formation sand (Figure 6.1)

i.e., $\frac{\text{D50 gravel}}{\text{D50 formation}}$ < or = 6

6.4.2 Selection of gravel size

- D50 point on sand grain : 0.11 mm
- Uniformity coefficient (D40/D90) : 1.71
- Largest gravel size (8 × D50) : 0.88 mm
- Smallest gravel size (4 × D50) : 0.44 mm
- Corresponding gravel size : – 20 + 40 US mesh
- Selected commercial gravel : – 20 + 40 US mesh

Median size:

- 20/40 US mesh gravel, D50 = 0.025”
- Formation sand, D50 = 0.00433” (0.11 mm)
  (from Sieve analysis using Saucier’s plot)

Ratio:

$$\frac{\text{Median gravel size}}{\text{Median sand size}} = \frac{0.025}{0.00433} = 5.77$$

Ratio is less than 6; hence, 20/40-mesh gravel is suitable and will effectively stop formation sand.

All the gravel that is used for gravel packing should meet or preferably exceed all API RP58 specifications for gravel packing sand, including the specification that no more than 2% of the gravel grains should be larger than the designated gravel size range and not more than 2% smaller than the designated gravel size grain. A random sample of gravel on site during the G. P. job may be taken up and tested for particle size distribution to ensure the correct gravel size.

6.4.3 Sand control screens

In gravel packing, accurately sized gravels are used as a filtering medium. The gravel is retained in an annular region within the wellbore by a mechanical device called gravel pack screen. Single wire wrapped screens with keystone-shaped wire have been used to control sand production in oil and gas wells since the 1930s. The keystone tapers toward the center, which avoids the liability of plugging once the particulate passes through the screen slots. The wire is stainless 304. They have the advantage over pre-packed screens in that they do not become plugged as easily by drilling mud. Furthermore, they function as a surface filter, where the plugging material is easily removed, whereas pre-packed screens are depth tilters where the plugging material tends to get trapped inside the pre-pack.

The design criterion of a single wire wrapped screen is a function of the relationship between the gravel particle size and screen slot width. Ideally, slots should be as wide as possible while retaining sand grains without restricting the flow of fluids and minute fines. Because all the gravel must be tightly packed and retained, the screen slot width for a gravel pack should be about one-half the smallest gravel diameter. The slot should not be wider than 70% of the smallest gravel size diameter to avoid the production of gravel. Theoretically, a slot width greater than 70% of the smallest gravel size will retain it. Because the flow capacities of screens are high, sizing the slot width on the lower side will not restrict productivity and will tend to ensure that the gravel remains outside the screen. The screen slot opening should be like this (Table 6.1):

Table 6.1 Scree slot opening requirements for different gravel size.

Gravel	Mesh (inch)	Gauge	Inch	mm
12-20	0.0661*0.0331	20	0.020	0.50
20-40	0.0331*0.0165	12	0.012	0.30
40-60	0.0165*0.0098	8	0.008	0.20

6.5 Frac–pack

Frac–pack is an innovative well completion technique that combines the stimulation advantages of hydraulic fracturing with sand control techniques of gravel packing. Historically, the installation of sand control hardware and treatment in wells do not bring in the full production potential. Flow efficiencies of 25% or less are common and accepted as an unavoidable consequence of sand control. The primary purpose of frac–pack completion is to eliminate high skins often associated with cased hole gravel packs by providing highly conductive flow path through the near-wellbore formation damaged area. It is therefore an alternate method to bypass near-wellbore damages mechanically when matrix acidizing is ineffective, controls sand when gravel packing or other sand exclusion techniques result in low productivity, provides complete zonal coverage either in highly lan1inated reservoir or in non-perforated zones, reduces drawdown thereby eliminating associated problems, and also eliminates non-Darcy flow effects.

A frac and pack can be achieved by two methods. The most common method is to do the frac–pack with screen and liner in place against the perforated internal. Another technique is to first fracture the well, wash out the well, run in the screen and liner, and then pump the annular gravel pack. The second approach is used only when the first method cannot be implemented for operational reasons. The impact of hydraulic fracturing on well productivity depends on conductivity (propped width) and length. In medium to high permeability reservoirs, the key to successful frac and pack treatment is to maximize fracture conductivity. This can be achieved with a treatment aggressive enough to induce an early tip screen out (TSO). The execution of such fracture treatments consists of two distinct stages:

- Creation of hydraulic fracture and TSO
- Fracture inflation and packing

The first stage of frac and pack treatment can be designed with the aid of fracture design simulator. The tip screen out consists of establishing a pumping design that will propagate the fracture to a desired length and cause the slurry to dehydrate at the perimeter of the fracture. The slurry dehydration will allow the proppant to bridge, stopping the fracture propagation at the tip.

Additional slurry is pumped into the fracture for fracture inflation and packing, causing the net pressure in the fracture to increase and form a wider fracture. The wider fracture allows placement of a much higher proppant concentration. The design of the second stage is based on material balance calculations using the fracture parameters at the end of the first stage as input variables. The above two-stage design processes critically depend on predicting the moment of TSO accurately. It improves very aggressive proppant pumping schedules. This can be achieved only with high-quality input data, which can be derived from a minifrac test wherein the actual fracturing fluid to be used is pumped inside the well.

Frac pack should not be attempted where the purpose is too close to water/oil, water/gas or gas/oil contact, wells having small ill of casing & in wells with long perforation intervals.

6.6 High Rate Water Pack

High-pressure/rate packing of gravel using water as carrier fluid has recently been field-tried in a few wells over the world and is now almost established as a technique of gravel packing. Industry concerns about the ability of water carrier fluids to "Turn the comer" and pack perforations tunnels have been alleviated by field results. In high rate water pack jobs, gravel/water slurry is pumped above fracture pressure, which creates short narrow fractures. These fractures fill up and the net pressure in the wellbore/fracture increases until another portion of the interval breaks down. If the gravel concentration in the wellbore is low enough and the wellbore does not get completely filled, then this process is repeated until the complete interval that can be fractured at the available pressure and rate is treated. This process will fill open non-damaged perforations and creates conductive paths through damaged perforations by breaking them down. Numerous pressure increases and break-backs are observed during high-pressure water packing. Pressure increase of 200–500 psi above the fracture extension pressure is normally observed.

Fluid flow is the primary mechanism responsible for the deposition of sand in gravel packing. If there is no flow into a perforated tunnel, the only way for the perforation to be filled is through the force of gravity. The flow into the perforation tunnel is controlled by the leakage of fluid through the perforation and into the formation. During the process of gel packing, the slurry is introduced into the perforated tunnels and achieves complete packing through the leakage of the viscous HEC polymer into the formation.

In the context of a water pack, the sand is conveyed into the perforated tunnel by leak-off, bed building, and turbulence. The perforated tunnel is filled by the infiltration of water with significantly lower viscosity and the accumulation of sand in the bed formations. As the sand bed moves through the perforations tunnel, fluid enters the formation and fills the perforations from the sand bed and slurry.

6.7 Gravel Pack Well Preparation

Effective prepping of a well for gravel packing is crucial for achieving successful completion. Precise strategizing, thorough organization, and meticulous implementation are necessary to enhance productivity and ensure long-term success. Proper preparation of the well increases the likelihood of it being placed on undamaged and sand-free production.

The impact of drilling operations on a gravel pack is like their impact on traditionally perforated wells. To ensure the stability of the borehole, it is necessary to drill the well and use drilling fluids that do not cause damage to the formations. This is important to prevent the formation damage depth from exceeding the perforation penetration. The filtrates of drilling fluid must be compatible with completion fluids and must not disrupt completion operations. It is preferable to use a drilling fluid that is sufficiently dense to create a somewhat higher pressure in the well, has minimal fluid loss, and is compatible with the clays in the productive formation. Before perforating a well, it is necessary to scrape the casing and circulate a series of clean-up solvents to eliminate mill scale, mud cake, cement residue, rust, and other solid substances from the casing. This is done to prevent these materials from entering the perforation tunnels and becoming trapped by the gravel.

6.8 Casing Tubing Cleanout

It is crucial to utilize work strings and casing that have been meticulously cleaned. To do this, it is necessary to utilize the casing tubing cleanout (CTC) approach. This procedure involves the use of slugs consisting of a sand slurry with a particle size of 20-40, a clean gel, xylene (400 liters in 3 cubic meters), a 5% caustic wash, a mild acid solution, an acetic acid solution, and a mixed surfactant. This solution is injected immediately before the gravel is pushed into the well. Additionally, take the lead in performing the gravel slurry treatment using a solution consisting of 7.5% hydrochloric acid (HCL) and 0.5% hydrofluoric acid (HF). It is important to completely clean any surface lines, pumps, gravel packing equipment, and tanks to prevent any contamination of the fluid, particularly after it has been adequately filtered.

6.9 Completion Fluid

The oil industry has long recognized the need for clean work-over and completion fluids. Fluids that are free of solids and compatible with formation material are essential to successful quality sand control. Formation damage caused by incompatible or solid-laden wellbore fluids can mean excessive pressure drawdowns and stress levels to produce desired fluid rates.

Water-based fluids are more commonly used as gravel pack completion fluids as they are more flexible than oil-based fluids. Their densities, viscosities, and formation compatibilities are more easily controlled. The source of completion fluid can vary. Regardless of its origin, the fluid should contain minimum particulate materials and its chemistry must be compatible with the rock formation and connate water. Water that is too fresh may cause clays to swell or disperse and in the presence of some ions may cause precipitation when in contact with the formation water.

The density of completion/work-over fluids should be regulated using soluble salts such as sodium chloride, calcium chloride, calcium bromide, and other similar substances. The most often utilized substances for gravel packing activities are sodium and calcium chlorides. In specific scenarios with low bottom hole pressure, foams have been utilized on the side with low density. Viscosity is typically regulated using both natural and synthetic polymers. Hydroxyethyl cellulose (HEC) is widely used as a viscosity builder due to its exceptional ability to produce high viscosity while leaving minimal residue. Fluid loss management is frequently required during gravel packing of wells to prevent excessive loss of costly or hard-to-remove fluid, as well as to avoid damage to the formation caused by particle matter, hydrocolloidal viscosity builders, or ionic species present in the work-over fluid. Efforts should be made to regulate fluid loss by modifying the density of the completion fluid, if feasible. When there is a need for fluid loss materials, it is important to employ them in a manner that does not hurt the gravel pack.

6.10 Filtration

A gravel pack completion fluid must be sufficiently clean so that suspended particles do not plug or reduce the permeability of the gravel pack or the producing formation. Filtration to 2 micron is recommended. Filtration is carried out by cartridge filters. The clean lines of the fluid are the key issue in any gravel pack job. Cartridge filter in series probably gives the cleanest fluid possible in terms of cost and time. When it is suspected that the brine is too dirty, then before going through the 2-micron filter, it should be filtered through a coarser filter. This will avoid the choking of the main filter and low pressures during filtration. Filtration should be conducted before viscosities are added to a work-over fluid with polymers because at this point, filtration tends to strip the polymers from the fluid and to plug the filters. To achieve the desired level of polymer mixing, fluids should be sheared with intermixing devices that disperse the polymer uniformly in the completion fluid to eliminate un-hydrated cluster, commonly called “Fisheyes.”

6.11 Perforation Density

Sand control wells can be either cased and perforated or left entirely open holes. Open hole completions tend to be more productive than cased and perforated wells when the same gravel pack is used in both scenarios. However, the utilization of cemented casing and perforation is often necessary in wells that have several and/or productive sections. This is done to prevent the intrusion of interbedded water, gas, or unwanted shale streaks. It is also done in cases where there is a potential for deeper completion in the future.

Increasing the hole density in certain instances resulted in a reduction of sand generation. The primary reason for this was the reduction in pressure drop across the sand face of the formation and the wellbore. As the hole area rises, the velocity of the fluid passing through it drops. This could be considered as a method for sand control. However, the productivity of a gravel packed well is significantly influenced by the density and size of perforations. Enhancing the perforation density and diameter of the well completion, as well as the size of the perforation, considerably improves the performance of the gravel-packed well. By increasing the perforation density/size in gravel packed wells, it is anticipated that there will be a productivity gain of 2-3 times.

6.11.1 Perforating/perforation cleaning

Most cased hole gravel packed wells are jet perforated, and all the cased holes that are to be gravel packed in adequately pressured zones should be perforated with an underbalanced perforating technique preferably with a tubing conveyed perforating system. The underbalanced pressure during perforating should be > 500 psi obtained by displacing fluid in the work string with nitrogen. The well should be open to flow when the perforation gun fires and the maximum flow rate continued for an equivalent of 4 lts per perforation. If underbalanced perforation cannot be done, a surge tool should be used to clean perforations with a surge chamber volume. Underbalanced perforation can be applied where:

- The presence of a stable formation allows for the application of enough pressure difference and reverse flow volume to effectively remove debris and mud from all perforations without causing sand to enter the wellbore.
- Reservoir pressure is near original hydrostatic so that underbalance can be obtained with a diesel column.

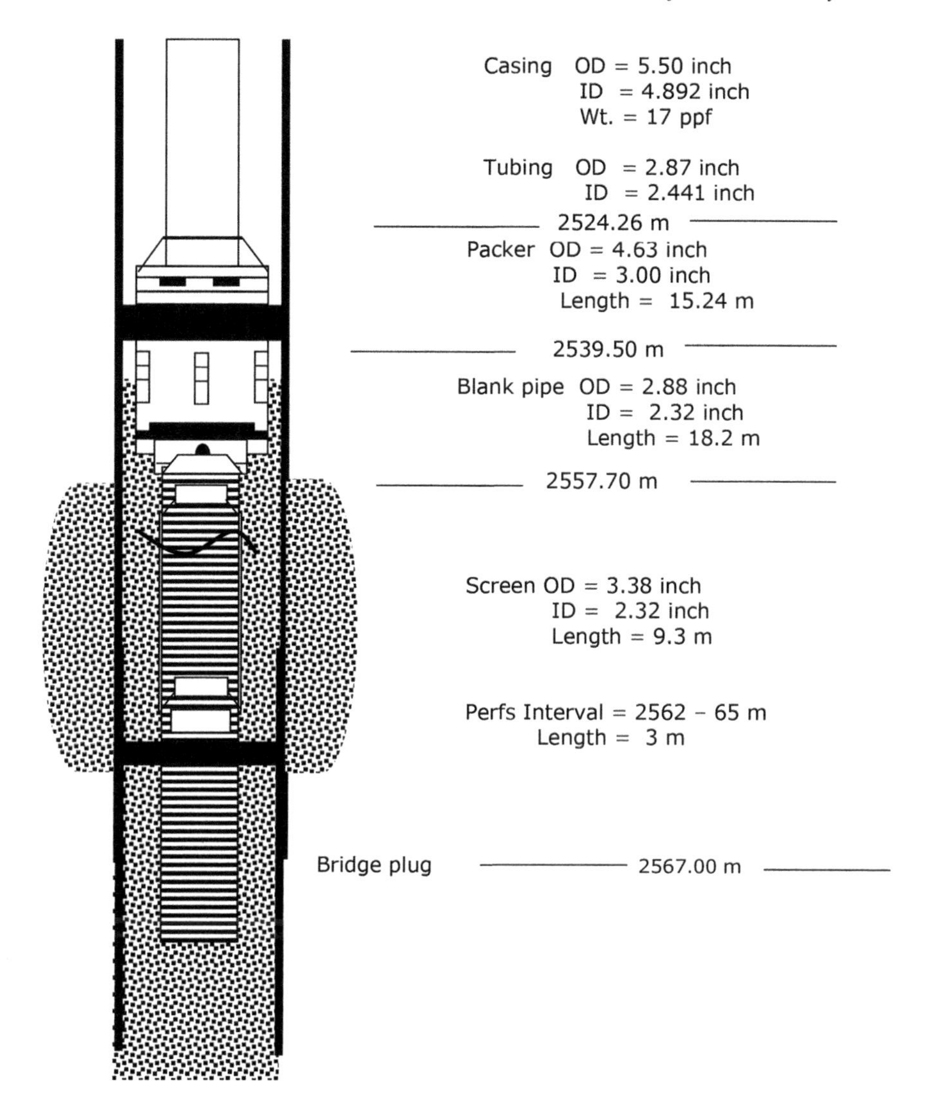

Figure 6.2 Gravel pack well completion schematic.

- Perforation is performed via -tubing and is restricted to a single run of guns, which is then followed by consolidation therapy.

Perforation washing is perhaps the most positive method of removing perforation debris, mud, and formation sand from the perforation tunnel and from behind the casing. In the positive method, washing fluid is forced out of perforations isolated by opposing cups and the fluid is circulated to the other perforations above the packer cups. During circulation outside the casing, the fluid carries out perforation debris and some formation sand (Figure 6.2).

Back surging is another effective way of perforation cleaning method. It exposes perforations instantaneously to essentially atmospheric pressure to draw out a limited volume of perforation debris, mud, formation sand, and reservoir fluid. The back surge tool is used for the purpose, where the back surge volume of at least 1 gallon per perforation may improve productivity greatly. It is an excellent perforation cleanout technique for relatively short intervals. Back surging is generally preferred to perforation washing when reservoirs are low pressured and excessive fluid loss would occur.

7

Matrix Stimulation

7.1 Sandstone Acidizing

Stimulation of petroleum wells by acid injection in sandstone reservoirs (almost all of our onshore fields) by mud acid injection is accomplished by selective dissolution of part of formation rock to reduce the resistance to fluid flow in the vicinity of the wellbore. Because of the radial flow geometry, the productivity of the well may be increased greatly.

7.2 Brief History of Sandstone Acidizing

Sandstone acidizing with hydrofluoric acid (HF) was practiced in Texas in 1933 following the issuance of a patent to the Standard Oil Company; however, the field tests were not successful because of plugging of the formation. Commercial application of HF acidizing of sandstones occurred in the Gulf Coast of Mexico in 1940 when Dowell introduced mud acid, a mixture of HCl and HF. Dowell's research indicated that the HCl helped maintain a low pH and decreased the precipitation of damaging precipitates. Following this event, the application of sandstone acidizing grew rapidly.

As the application of acidizing expanded, several chemical and mechanical problems were addressed. Numerous acid additives and systems were developed to solve the problems of acid sludging, acid-induced emulsions, spent acid cleanup, live acid penetration, and fines migration. Parallel to this development was the development of methods to improve zone coverage during acidizing.

7.3 Technology and Application

Matrix acidizing in sandstone reservoirs can be very beneficial to many damaged oil, gas, and water injection wells, but not all matrix treatments are successful even when the well is severely damaged. A complete and accurate well and formation analysis, treatment design, well preparation, job supervision, and follow-up evaluation all are required to achieve maximum benefit from matrix acidizing. The words of great pioneers in the field, Mr. P. E. Fitgerald of Dowell Inc, 1934, 'The value of repeated acid treatments' – it is recognized now that every well is a problem in itself and must be analyzed individually in order to obtain the best results.

Injection of acid in the matrix of producing formation for dissolution of damaging materials or generation better near wellbore permeability through dissolution of the clay minerals can give rise to other interactions, such as:

7.3.1 Precipitation of reaction products

Secondary reactions occur in acidizing, particularly in sandstones that can result in the precipitation of reaction products from the bulk liquid phase. Obviously, precipitated solids may block pore spaces and work against the goal of matrix acidizing.

7.3.2 Acid fluid–reservoir fluid interactions

The acid solution injected in matrix acidizing may interact physically and/or chemically with the reservoir fluids as well as with the minerals. These interactions can result in changes in wettability, phase saturation distribution, precipitation of solids, or emulsification.

7.3.3 Variations in reservoir permeability or the distribution of damage

A successful acidizing treatment requires contacting all damaged regions around the well with acid. This is usually complicated by variations in the injectivity to acid along the wellbore, which leads to the use of techniques to affect good acid coverage (acid diversion).

In view of this, the primary design considerations for matrix acid treatment are:

- **Fluid selection:** Acid type, concentration, and volume
- **Injection schedule:** Planned rate schedule and sequence of injected fluids.
- **Acid coverage and diversion:** Special steps taken to improve acid contact with the formation.
- **Real-time monitoring:** Methods to evaluate the acidizing process as it occurs.

- **Additives:** Other chemicals included in the acid solution to enhance the process or to protect tubular goods.

7.4 Fluid Selection

Hydrofluoric acid has been widely used in stimulation treatments since 1935, when mud acid was introduced to the petroleum industry. Originally, this hydrochloric–hydrofluoric acid mixture was intended to remove mud filter cake, but it has since been successfully applied to many other oilfield problems. Hydrofluoric acid's reactivity with silica makes it unique in application. Other mineral acids such as hydrochloric, sulfuric, or nitric are unreactive with most silicious materials, which comprise sandstone formations. A typical sandstone reservoir may contain 50%–85% sand or quartz. Chemical reactions of different sandstone minerals with basic mineral acids used in sandstone acidizing are given below.

7.4.1 Hydrochloric acid (HCl)

Most acid treatments of carbonaceous formations overwhelmingly employ hydrochloric acid. Usually, it is used as a 15% (by weight) solution of hydrogen chloride gas in water. The acid reacts with calcite ($CaCO_3$) or dolomite ($CaMg(CO_3)_2$) to form carbon dioxide, water, and a calcium or magnesium salt. Typical reactions are:

$$2HCl + CaCO_3 \longleftrightarrow CaCl_2 + H_2O + CO_2$$
$$4HCl + CaMg(CO_3)_2 \longleftrightarrow CaCl_2 + MgCl_2 + 2H_2O + 2CO_2$$

The effects of HCl on matrix acidization of sandstone formations are threefold:

- It shifts the equilibrium concentration of reaction products to prevent precipitation of such compounds as calcium and magnesium fluoride in the sandstone pores.
- As the HCl concentration increases, the equilibrium shifts such that fewer fluorine atoms are bound up in the aluminum and silicon-fluoride reaction product complexes. Therefore, the increase in HCl concentration reduces the stoichiometric coefficient so that a lower consumption of hydrofluoric acid occurs during acidization.
- HCl catalyzes the reaction between HF and the dissolvable minerals (e.g., feldspars) in the porous matrix by being adsorbed on the reaction surface.

Mud acid: hydrochloric – hydrofluoric acid (HCl + HF):

The mud acid formulations commonly used are:
3% HF – 12% HCl, regular mud acid
6% HF – 12% HCl, super mud acid
1% HF + 61/2 % HCl half-strength mud acid.

This acid mixture is used because hydrofluoric acid is reactive with clay minerals, sand, drilling mud, or cement that may be restricting near-wellbore permeability and can dissolve silica and increases the permeability. The HCl does not react with these materials but is needed to keep the pH low, thereby reducing the precipitation of HF reaction products.

Reaction with silica:

$$SiO_2 + 4HF \longleftrightarrow SiF_4 + 2H_2O$$
$$SiF_4 + 2HF \longleftrightarrow H_2SiF_6$$

Reaction with aluminio silicates:

$$36\ HF + Al_2Si_4O_{10}\ (OH)_2 \rightarrow 4H_2SiF_6 + 12\ H_2O + 2H_3AlF_6$$

Reaction with silicates:

$$Na_4SiO_4 + 8HF \longleftrightarrow SiF_4 + 4NaF + 4H_2O$$
$$2NaF + SiF_4 \longleftrightarrow Na_2SiF_6$$
$$2\ HF + SiF_4 \longleftrightarrow H_2SiF_6$$

Reaction with calcite:

$$CaCO_3 + 2HF \longleftrightarrow CaF_2 \downarrow + H_2O + CO_2$$

Reaction with quartz:

$$4HF + SiO_2 \leftrightharpoons SiF_4 \text{ (silicon tetrafluoride)} + 2H_2O$$
$$SiF_4 + 2HF \leftrightharpoons H_2SiF_6 \text{ (fluosilicic acid)}$$

Reaction with albite (sodium feldspar):

$$NaAlSi_3O_8 + 14HF + 2H^+ \leftrightharpoons Na^+ + AlF_2^{\ +} + 3SiF_4 + 8H_2O$$

Reaction with orthoclase (potassium feldspar):

$$KAlSi_3O_8 + 14HF + 2H^+ \leftrightharpoons K^+ + AlF_2^{\ +} + 3SiF_4 + 8H_2O$$

Reaction with kaolinite:

$$Al_4Si_4O_{10}(OH)_8 + 24HF + 4H^+ \leftrightharpoons 4AlF_2^{\ +} + 4SiF_4 + 18H_2O$$

Reaction with montmorillonite:

$$Al_4Si_8O_{20}(OH)_4 + 40HF + 4H^+ \leftrightharpoons 4AlF_2{}^+ + 8SiF_4 + 24H_2O$$

The type and strength (i.e., concentration) of acid used in sandstones are selected primarily on the basis of field experience with particular formations. For years, the standard sandstone acidizing formulation consisted of a 12% HCl–3% HF mixture, preceded by a 15% HCl preflush. In fact, the 12% HCl–3% HF mixture has been so common that it is referred to generically as mud acid. In recent years, however, the trend has been toward the use of lower strength HF solutions. The benefits of lower concentration HF solutions are a reduction in damaging precipitates from the spent acid and lessened risk of unconsolidation of the formation around the wellbore. The selection of acidizing fluids should always begin with an assessment of the formation damage present; in general, the damaging material must be soluble in the treating fluids. Geochemical models can be used to guide acid selection.

Although HF can dissolve many types of minerals, there are conditions that can cause the dissolved ions to subsequently precipitate from the solution. Because many of these precipitates are gelatinous, a net decrease in formation permeability can result where reaction product precipitation is allowed to occur.

Contacting HF with calcium ions will produce precipitation of insoluble calcium fluoride.

$$2HF + Ca^{++} \rightarrow CaF_2(ppt) + 2H^+$$

In order to lessen the presence of calcium ions in the formation, the HCl is injected into the formation as preflush to dissolve the calcium ions before the HF is injected. The main source of calcium ions is from carbonate minerals (calcite) present in the formation rock. By initially injecting a volume of plain HCl to dissolve this calcite prior to the arrival of the HF, the problem of calcium chloride precipitation can be avoided. Some carbonates may remain after preflushing, either because of the initial amount of carbonate cementing material in the sandstone or as a result of the carbonates' initial protective siliceous coating. Also, slightly soluble, fine crystalline CaF_2 readily forms when calcite contacts HF. This can lead to substantial damage.

The sodium (Na^+) and potassium (K^+) salts of fluosilicic acid are very insoluble. The introduction of Na^+ or K^+ ions into a spent hydrofluoric acid solution containing fluosilicic acid will cause immediate precipitation of the salt.

$$H_2SiF_6 + 2Na^+ \rightarrow Na_2SiF_6(ppt) + 2H^+$$
$$H2SiF_6 + 2K^+ \rightarrow K_2SiF_6(ppt) + 2H^+$$

There are generally two sources of Na^+ and K^+ ions in sandstone formations. First, feldspars and clay minerals contain Na^+ and K^+ ions in their crystal structures. As these minerals dissolve, some Na^+ and K^+ ions are put into solution and precipitation occurs. Second, both Na^+ and K^+ are common constituents of formation brine. Mixing of formation brine and spent HF acid will always result in precipitation. When designing a sandstone acidizing treatment, it is important to avoid contacting the HF with formation brine.

As the HF is consumed in the formation, the HF concentration becomes very small. Under these conditions, fluosilicic acid can decompose to yield HF and silicon tetra-fluoride. The silicon tetra fluoride then undergoes hydrolysis to yield silicic acid. This precipitation reaction is almost the reverse of silica dissolution, except that silicic acid is precipitated instead of quartz. Silicic acid is a hydrated form of silica and is deposited as a gelatinous precipitate in the pore space.

$$H_2SiF_6 \rightarrow SiF_4 + 2HF$$
$$SiF_4 + 4\ H_2O \rightarrow Si(OH)_4(ppt) + 4HF$$

The limitations related to the use of mud acid to remove damage in sandstone formations include the following:

- Rapid spending provides only a short penetration, especially at high temperatures (maximum depth about 12 inches).
- Fines, composed of either mostly quartz or mostly clay minerals, can be generated during the acid reaction and can migrate with the fluid flow. The destabilization of fines can lead to a quick production decline after treatment. Gravel-packed gas wells can exhibit a 50% productivity reduction.
- The high dissolving power of mud acid destroys rock integrity at the formation face.

In view of this, HCl:HF ratio and concentration are selected to prevent or reduce the formation of damaging precipitates. McLeod (1984) provides guidelines for deciding the HCl:HF ratio for different formations, which are summarized in Table 7.1.

To overcome the above-listed drawbacks of mud acid, special acid systems are also developed. Although the reacting acid, i.e., HF remains the same, however, its availability for reaction at the formation face is reduced by different chemical mechanisms to achieve greater penetration of the live acids. Some of these systems are listed below.

Table 7.1 Acid use guidelines for sandstone acidization.

Condition or Mineralogy	Acid Strength (blend)
HCl solubility > 20%	HCl Only
High permeability (> 50 md)	
High quartz (> 80%), Low clay (< 5%)	12%HCl – 3% HF
High feldspar (> 20%)	13.5% HCl – 1.5% HF
High clay (> 10%)	10% HCl – 1% HF
High iron clay (> 15%)	10% acetic acid – 1% HF
Low permeability (<= 10 md)	
Clay (< 10%)	6% HCl - !% HF
Clay (> 10%)	6% HCl – 0.5% HF

[1]Preflush with 15% HCl
[2]Preflush with 10% HCl
[3]Preflush with 10% acetic acid

7.4.2 Fluoroboric acid

Fluoroboric acid does not contain large amounts of HF at any given time and thus has a lower reactivity. However, it generates more HF, as HF is consumed by its hydrolysis. Therefore, its total dissolving power is comparable to a 2% mud acid solution. Fluoboric acid solutions are used as a preflush before treating formations sensitive to mud acid; this avoids fine destabilization and subsequent pore clogging. They are also used as a sole treatment to remove damage in a sandstone matrix with carbonate cement or in fissures that contain many clay particles. Another use is as an overflush after a mud acid treatment that has removed near-wellbore damage (up to 0.5 ft) to allow easier penetration of the fluoboric acid solution (a few feet). Fluoboric acid is recommended when the sandstone contains potassic minerals to avoid damaging precipitates and in the case of fine migration owing to its fine stabilization properties.

In the field, fluoboric acid is easily prepared by mixing boric acid (H_3BO_3), ammonium bifluoride, and HCl. Ammonium bifluoride, an acidic salt of HF, reacts first with HCl to generate HF:

$$NH_4HF_2 + HCl \rightarrow 2HF + NH_4Cl$$

Tetrafluoboric acid is formed as a reaction product of boric acid with HF, according to

$$H_3BO_3 + 3HF \rightarrow HBF_3OH + 2H_2O \text{ (quick reaction)}$$
$$HBF_3OH + HF\ HBF_4 + H_2O \text{ (slow reaction)}$$

The unique advantage of fluoboric acid is that it provides efficient stabilization of clays and fines through reactions related to borate and fluoborate ions.

7.4.3 Sequential mud acid

The sequential mud acid system involves the *in-situ* generation of HF, occurring from the alternate injection of HF and ammonium fluoride. The reactions of HF are thought by some to take place at the rock surface by adsorption followed by ion exchange, but the yield of this heterogeneous process seems highly doubtful for several reasons:

- If HF were generated through such a process, it would be a small quantity, hardly enough to etch the surface of the clay material.
- Because this process is based on the CEC of the clays, migrating kaolinite would hardly be touched.

7.4.4 Alcoholic mud acid

Alcoholic mud acid formulations are a mixture of mud acid and isopropanol or methanol (up to 50%). The main application is in low-permeability dry gas zones. Dilution with alcohol lowers the acid–mineral reaction rate and provides a retarding effect. Cleanup is facilitated; acid surface tension is decreased by the alcohols while the vapor pressure of the mixture is increased, which improves gas permeability by reducing water saturation.

7.4.5 Mud acid plus aluminum chloride for retardation

An acidizing system to retard HF–mineral reactions has been proposed in which aluminum chloride ($AlCl_3$) is added to mud acid formulations to complex some of the fluoride ions in the injected mixture, according to the reactions.

$$AlCl_3 + 4HF + H_2O \rightarrow AlF_4^- + 3HCl + H_3O^+$$
$$AlF_4^- + 3H_3O^+ \rightarrow AlF_2^+ + 3HF + 3H_2O$$

7.4.5.1 Organic mud acid

Because total acidity speeds mineral dissolution with mud acid, organic mud acid involves replacement of the 12% HCl component with organic acids (9% formic acid, a weak acid that only partially dissociates), mixed with 3% HF, to retard HF spending. This system is particularly suited for high-temperature

Table 7.2 Chemical composition of typical sandstone minerals.

Classification	Mineral	Chemical composition
Quartz		SiO_2
Feldspar	Microcline	$KAlSi_3O_8$
	Orthoclase	$KAlSi_3O_8$
	Albite	$NaAlSi_3O_8$
	Plagioclase	$(Na,Ca)Al(Si,Al)Si_2O_8$
Mica	Biotite	$K(Mg,Fe^{2+})_3(Al,Fe^{3+})Si_3O_{10}(OH)_2$
	Muscovite	$KAl_2(AlSi_3)O_{10}(OH)_2$
	Chlorite	$(Mg,Fe^{2+},Fe^{3+})AlSi_3O_{10}(OH)_8$
Clay	Kaolinite	$Al_2Si_2O_5(OH)_4$
	Illite	$(H_3OK)_y (Al_4.Fe_4 .Mg_4 .Mg_6)(Si_{8-y}.Al_y)O_{20}(OH)_4$
	Smectite	$(Ca_{0.5}Na)_{0.7}(Al,Mg,Fe)_4(Si,Al)_8O_{20}(OH)_{4\,n}H_2O$
	Chlorite	$(Mg,Fe^{2+},Fe^{3+})AlSi_3O_{10}(OH)_8$
Carbonate	Calcite	$CaCO_3$
	Dolomite	$CaMg(CO_3)_2$
	Ankerite	$Ca(Fe,Mg,Mn)(CO_3)_2$
	Siderite	$FeCO_3$
Sulfate	Gypsum	$CaSO_4 .2H_2O$
	Anhydrite	$CaSO_4$
Chloride	Halite	NaCl
Metallic oxide	Iron oxides	FeO, Fe_2O_3, Fe_3O_4

wells (200° to 300 °F [90° to 150 °C]), for which pipe corrosion rates are diminished accordingly. This system also reduces the tendency to form sludge (Table 7.2).

The system is especially recommended in a zeolite environment. Zeolite minerals are sensitive to HCl and strong mineral acids. Several core studies have shown that the use of HCl alone causes significant damage, whereas weak organic acid reduces the damage. The problem is that the weak organic acid does not necessarily remove the damaging mineralogy to restore permeability. The solution to the problems associated with zeolites is to recognize the presence of these minerals before a treatment is performed.

Utilizing an organic acid as part of a two-step preflush process, followed by a low-concentration mixture of hydrofluoric acid (HF) that matches the residual minerals in the formation, has demonstrated remarkable efficacy in restoring permeability and eliminating damage. It is necessary to include an organic acid in all injected fluids to maintain a low-pH environment. Operators have discovered that utilizing an all-organic-acid system, followed by an organic acid-HF formulation, is an effective approach in high-temperature environments including zeolites. It is crucial to identify the presence of these minerals before doing the treatment. Utilizing an organic acid as part of the initial two preflush steps, and subsequently employing a low-concentration mixture of hydrofluoric acid (HF) that matches the remaining minerals in the formation, has demonstrated remarkable efficacy in the restoration of permeability and the elimination of damage. It is necessary to include an organic acid in all injected fluids to maintain a low-pH environment. Operators have discovered that utilizing an all-organic-acid system, followed by an organic acid-HF formulation, is successful in high-temperature conditions.

7.4.5.2 SGMA (self-generating mud acid)

The first retarded sandstone-acidizing system to be used extensively was developed by Shell Oil Company. It involves pumping ammonium fluoride and an organic ester, methyl formate, into the formation. (Methyl formate has a very low flash point and should be pumped with caution.) In time, ester hydrolysis produces formic acid. This acid reacts with ammonium fluoride to form HF, which then dissolves clays or any siliceous minerals it contacts.

7.4.6 Damage characterization and the type of acid

Selection of a chemical for any particular application will depend on which contaminants are plugging formation permeability. HCl will not dissolve pipe dope, paraffin, or asphaltenes. Treatment of these solids or plugging agents requires an effective organic solvent (usually aromatic solvents like toluene, xylene, or orthonitrotoluene). Acetic acid effectively dissolves calcium carbonate scale; however, it will not dissolve ferric oxide (iron oxide) scale. HCl dissolves calcium carbonate scale quite easily but has little effect on calcium sulfate scales. Calcium sulfate can be converted to calcium carbonate or calcium hydroxide by treatment with potassium hydroxide or sodium carbonate. HCl then can be used to dissolve the converted scale. Calcium sulfate also can be dissolved in one step with sodium salt of ethylene diamine tetra acetic acid (EDTA). HCl will not dissolve formation clay minerals or drilling mud. Hydrofluoric acid (HF) must be used to dissolve these aluminosilicates in rock pores around the wellbore.

Because different plugging solids require different solvents for their removal, there is no universal solvent for wellbore damage. It is important to know the specific material that is damaging the formation around the wellbore.

- Production increases are most significant for matrix acidizing of damaged formations. Production increases resulting from HF treatment of undamaged formations would not, in most cases, be significant enough to justify the cost of the stimulation treatment.
- In the formations with drilling mud damage resulting from clay particle invasion, the volume of acid sufficient to remove only the clay contained within a 1-inch radius of the wellbore should yield the most economical production increases. This applies only if no natural damage has occurred because of mud filtrates contacting water-sensitive clays.
- When treating formations with natural clay damage, the production increase realized is directly dependent upon the distance at which live HF can be pumped into the formation. The penetration of this live HF is dependent upon the following factors: (a) clay concentration, (b) formation temperature, (c) Initial HF concentration, (d) rate of HF reaction, and (e) pump rate.
- As the clay concentration is increased, the penetration radius of unspent acid is decreased.
- As the formation temperature is increased, the penetration radius of unspent acid is decreased.
- Greater depths of penetration will be obtained by increasing the initial HF concentration.
- Retarding the rate of reaction of HF on silicates will facilitate greater penetration of live acid into the formation. With greater depths of penetration, a higher production increase for formations with natural clay damage will be achieved. Undamaged formations or formations with damage resulting from mud invasion will not benefit from acidizing with retarded HF as much as will a naturally damaged formation.
- As the depth of damage increases, the need for a retarded acid system becomes more desirable.
- Increasing the pump rate will slightly increase the penetration radius of the live acid.

7.5 Typical Sandstone Acid Job Stages

A preflush stage should be used ahead of the HCl especially when high sulfate ion or high bicarbonate ion concentrations exist in the formation connate water or seawater or when $CaCl_2$, KCl, or CaBr completion fluids have been used and calcium carbonate is a formation mineral. HCl dissolution of the calcite generates high calcium ion concentrations that mix with the incompatible formation water and generate scale (calcium sulfate or calcium carbonate).

7.5.1 Tubing pickle

One of the first items to be addressed when matrix treatments are considered should always be a tubing pickle (cleaning). Tubulars, regardless of how new, have scale, rust, and other debris that result from handling, installation, and production and that can be loosened by the solvents and acid injected into the well. The pickling process may be multiple-staged and may involve expensive solvent packages. Typically, a small treatment containing solvent and acid stages will greatly improve, if not eliminate, the problems associated with tubular debris. The pickling process should be included in the procedure and time allotted for job execution.

The purpose of the pickling process is to:

- remove rust, iron oxides, and scale;
- dissolve oily films and pipe dope that could plug the downhole equipment and perforations;
- limit the amount of iron that gets into the formation and contacts the crude oil.

7.5.2 Preflush

The sequence of fluids in sandstone treatments is dependent largely on the damage type or types. The use of multiple stage preflush should functionally address the different types of damage and thereby prepare the surfaces for the main treatment of fluids. Hydrocarbon solvents are used to remove oil films and paraffin deposits so that the aqueous acid systems can contact the mineral surfaces. Acid-compatible brines (e.g., brine containing ammonium chloride) can be used as an excellent preparatory flush to help remove and dilute acid-incompatible species (e.g., potassium or calcium). An example of a preflush sequence is preceding the HCl portion of the preflush with a large quantity of brine containing ammonium chloride followed by a hydrocarbon-based surfactant mixture. The purpose of the brine preflush is to dilute the incompatible species to soluble levels.

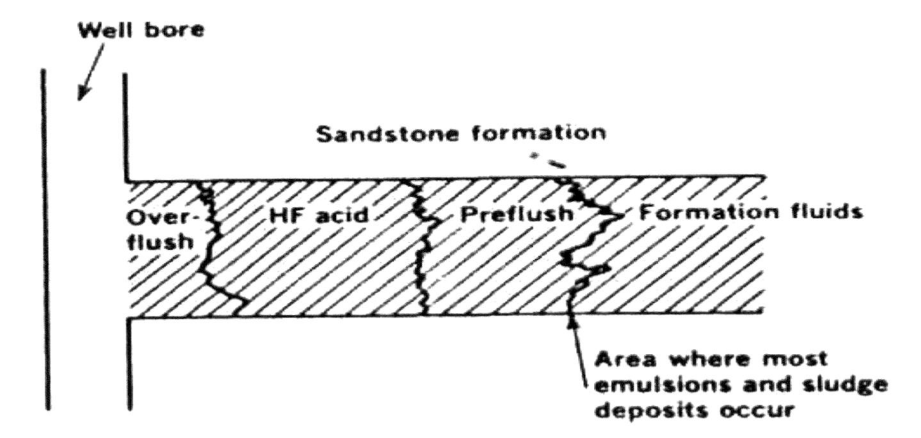

Figure 7.1 Fluid entry in formation.

The next consideration for preflushes is compatibility with formation fluids. Certain crude oils have a high sensitivity to acidic mixtures. These situations may require dilution with hydrocarbons or other isolating or buffering fluid systems (e.g., foams). Further compatibility consideration should be given to the iron content of the initial injection fluids that contact the crude or condensate because even low iron concentrations can cause sludge formation. Displacement of the fluids away from the near-wellbore region reduces the potential for problems that can reduce production success and limit or halt the injection process. HCl pre-flushes in sandstone acidizing are extremely important. Their function is to remove as much of the calcareous material as possible before injection of the mud acid. The HCl preflush step should never be neglected when using mud acid mixtures. A few systems containing HF can be injected without an HCl preflush, but these are systems with extremely low HF concentrations, such as fluoboric acid. These systems can be used without an HCl preflush because the HF concentration in fluoboric acid is low enough not to present a precipitation potential (Figure 7.1).

7.5.3 Main fluid stage

The HCl–HF mixture used in each treatment should be cautiously decided. Low HF concentrations should be used to avoid the precipitation of AlF_3 or CaF_2 if the remaining calcite cannot be quantified. The mud acid (12% HCl + 3% HF) can be used even in low-calcite environments without a precipitation problem. Some significant problems that may occur in high-clay-content formations include compromised formation integrity and excessive fines generation. These conditions can be the result of too high HF concentrations. The volumes should be determined using a field-validated simulator to sensitize the severity of the damage.

The percentage of acidizing successes increases as the volume of mud acid increases for gas wells, whereas a maximum of 100–125 gal/ft of perforations is required to maximize the success of oil wells. If diversion is maximized and the damage is known or perceived to be shallow, then smaller quantities per foot can be used. Acid strength is important, because precipitation potential and formation matrix collapse are problems that can be irreversible.

7.5.4 Overflush stage

The purpose of the overflush is twofold. First, it should displace the main fluid stage more than 3–4 ft away from the wellbore, which is the critical matrix area for radial flow. Second, the portion of the main stage that is not displaced should be diluted. Both of these factors help to eliminate damage in the near-wellbore area caused by the precipitation potential of the spent main fluid stage. Overflush fluids must be chosen carefully to avoid creating damage during the treatment flowback. Overflush systems should meet the following criteria. The portion of the overflush immediately following the main fluid stage should be aqueous based, have a low pH value, and have dilution potential for the spent mud acid. The remainder of the overflush should be miscible and compatible with the previous stages. The total minimum overflush volume must completely displace the main fluid stage at least 4 ft away from the wellbore. Any anisotropy of the formation permeability can warrant doubling or tripling the overflush volume if the energy in the reservoir is sufficient to unload the injected fluid.

7.6 Matrix Acidizing Design Guidelines

Matrix acidizing is the process of injecting acid into the formation in radial flow below fracturing pressure to remove damage and restore the permeability to the original reservoir permeability or higher. The below steps for treatment design should be followed:

1. Estimate safe injection pressures:

   a. determine present fracturing gradient;

   b. determine present bottomhole fracturing pressure;

   c. determine allowable safe injection pressure both at the wellbore and at the surface.

2. Estimate safe injection rate into the damage-free formation.
3. Estimate safe injection rate into the damaged formation.
4. Select stages required for fluid compatibility.
5. Calculate volume of each stage required:
    a. crude oil displacement;
    b. formation brine displacement;
    c. HCl stage or acetic acid stage;
    d. mud acid stage;
    e. overflush stage.
6. Select acid concentrations according to formation mineralogy.

7.7 Flowback and Cleanup Techniques

Choosing the appropriate flowback process is crucial. The flowback that occurs during multiphase transition periods can result in permanent harm. The fines that were released during the acid job are consistently returned to the area near the wellbore. The fines can be eliminated by diluting their concentrations and allowing them to pass through the completion. This process can be facilitated by creating modest, gradual pressure reductions. The following are the key factors to consider for flowback in sandstone formations:

- The fluids flowing back are more viscous than those injected. They are capable of carrying natural formation fines and other partially dissolved solids at lower velocities, which can cause plugging before the well cleans up completely.
- The spent acid usually has a higher density than the formation water. The tubing pressure should be lower than when connate water is produced, owing to the higher hydrostatic pressure of the spent acid.
- Spent acid has an equilibrium established of potential precipitants, held in place by dissolved gases and dissolved salts. Should these gases (e.g., CO_2) be removed from the spent fluid as a result of creating an excessive pressure drop, precipitation will occur.
- A minimum velocity is necessary for liquid to be voided from the tubing without slippage occurring. The minimum velocity to the unload tubing can be calculated. The flow rate and tubing pressure in this

calculation should include the heavier liquid density. The flow rate should be achieved gradually but sufficiently soon to avoid precipitation in the formation. The rate should then be maintained until all injected fluids are returned and both the tubing pressure and production rate are steady.

- HF systems should be flowed back immediately after injection of the overflush. The potential damaging precipitates that are generated form when the pH increases as the HCl is spent. If the acid is returned quickly, then the pH change may not reach the range for precipitation. Many iron precipitates also drop out when the pH increases. The exception is fluoboric acid treatments. The shut-in time required for complete HF generation and fine stabilization varies on the basis of temperature.
- The majority of the additives that are injected are produced back. Because the acidizing additives are by design water soluble, they are partitioned into the water phase. This can cause separation and floatation equipment problems. The return fluids are also acidic, which creates problems for chemical-electric detection devices in the separation equipment.

7.8 Additives in Acidizing Fluid Composition

Proper fluid selection is critical to the success of a matrix treatment. The treatment may be a failure if the proper additives are not used. The treating fluid is designed to effectively remove or bypass the damage, whereas additives are used to prevent excessive corrosion, prevent sludging and emulsions, prevent iron precipitation, improve cleanup, improve coverage of the zone, and prevent precipitation of reaction products. Additives are also used in preflushes and overflushes to stabilize clays and disperse paraffins and asphaltenes. The functions of some frequently used are as follows:

7.8.1 Organic acid

7.8.1.1 Acetic acid

It is commonly used as a 10% weight solution in water. At this concentration, the products of the reaction (calcium and magnesium acetates) are generally soluble in spent acid. Sometimes, it is used as a mixture with hydrochloric acid in hybrid acids. Based on cost per unit of dissolving power, acetic acid is more expensive than either hydrochloric or formic acids.

7.8.1.2 Formic acid

It is substantially stronger than acetic acid, though appreciably weaker than hydrochloric acid. It is less costly than acetic acid but is more corrosive.

7.8.1.3 Corrosion inhibitors

The most important acid additives are corrosion inhibitors. A corrosion inhibitor is a chemical that slows the attack of acid corrosion on drillpipe, tubing, or any other metal that the acid contacts during treatment. The major kinds of corrosion are corrosion of metals, pitting types of acid corrosion, and hydrogen embrittlement. The degree of dissociation of hydrogen ions from the acid molecule, along with the acid concentration, determines the hydrogen ion activity, which is directly proportional to its corrosivity on steel. The relative degree of dissociation for some common acids is hydrochloric > formic > acetic. Therefore, hydrochloric acid (HCl) is more corrosive on steel than formic acid, which is more corrosive than acetic acid. Inhibitors function by interfering with the chemical reactions that occur at the anode or cathode of the corrosion cell. The two basic types of corrosion inhibitors are inorganic and organic.

Inorganic corrosion inhibitors include the salts of zinc, nickel, copper, arsenic, antimony, and various other metals. Of these, the most widely used are the arsenic compounds. When these arsenic compounds are added to an acid solution, they "plate out" at cathodic sites of exposed steel surfaces. The plating decreases the rate of hydrogen ion exchange, because the iron sulfide that forms between the steel and the acid acts as a barrier. It is a dynamic process in which the acid reacts with iron sulfide, rather than the metal.

Organic corrosion inhibitors are composed of polar organic compounds capable of adsorbing onto the metal surface, thereby establishing a protective film that acts as a barrier between the metal and the acid solution. They usually serve as a cathodic polarizer by limiting hydrogen ion mobility at cathodic sites. Organic inhibitors are composed of rather complex compounds, with one or more polar groups made of sulfur, oxygen, or nitrogen.

7.8.2 Surfactants

Surfactants, or surface-active agents, are used in acidizing to break undesirable emulsions, reduce surface and/or interfacial tension, alter wettability, speed cleanup, disperse additives, and prevent sludge formation.

The use of surfactants requires careful selection of an appropriate molecule. Remarkably, in the design of most well treatments, surfactants are selected with little or no laboratory data to support the choice and sometimes

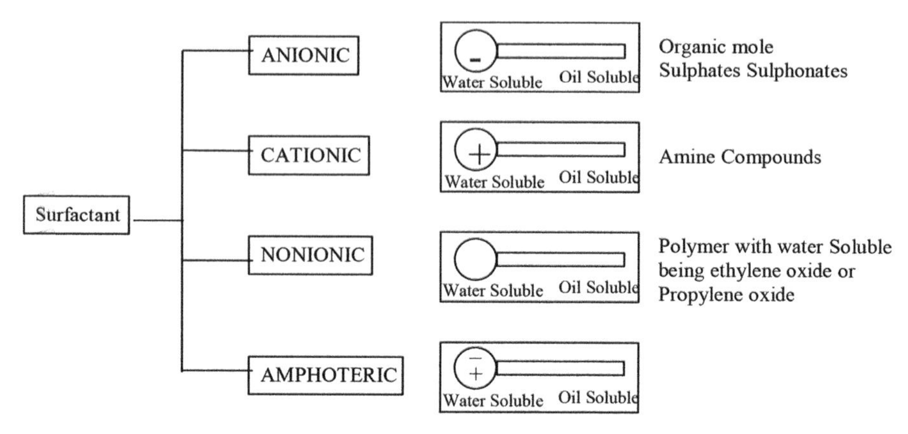

Figure 7.2 Different types of surfactants and their solubility.

without full knowledge of their properties at the conditions in which they will be applied. Improper surfactant selection can lead to results contrary to those intended and may be detrimental to the success of the treatment. Surfactants owe their properties to their "dipolar" composition. The surfactant molecule consists of a water-soluble (hydrophilic) group and an oil-soluble (lipophilic) group, which are separated from each other although linked by a strong covalent chemical bond (Figure 7.2).

The molecules are classified into five groups according to the ionic charge carried by the water-soluble group:

- **Anionic**: Anionic surfactants are commonly added to well treatment fluids. These surfactants carry a negative charge when they ionize in an aqueous solution. Anionic surfactants are used primarily as nonemulsifying agents, retarding agents, and cleaning agents. Examples of anionic surfactants are sulfates, sulfonates, phosphates, and phosphonates.

- **Cationic**: Cationic surfactants carry a positive charge when they ionize in aqueous solutions. Cationic and anionic surfactants are generally incompatible. When mixed, they tend to precipitate in aqueous solutions.

- **Nonionic**: Nonionic surfactants have no charge at all in the hydrophilic group and a long-chain organic (R) for the lipophilic group. Nonionic surfactants obtain their water solubility by attaching the long hydrocarbon chain to a highly soluble molecule such as polyhydric alcohol or by reacting it with ethylene oxide. These surfactants are used as nonemulsifiers and foaming agents.

- **Amphoteric**: Amphoteric surfactants have a hydrophilic group that changes from cationic to nonionic to anionic with increasing pH. In other words, if the solution is acidic, the amphoteric surfactant acts like a cationic surfactant; if the solution is neutral, it acts like a nonionic surfactant; and if the solution is basic, it acts like an anionic surfactant.
- **Fluorocarbons**: Fluorocarbons form surfaces of lower free energy than hydrocarbon surfaces. Consequently, fluorocarbon surfactants lower the surface tension of solutions to a greater extent than hydrocarbon surfactants.

The main properties of a fluid or a mineral affected by surfactants are the surface and interfacial tensions, emulsification tendency, wettability, micelle formation, and dispersibility. In recent years, the uses of surfactants have grown quickly. This unique class of chemicals has found application in almost all phases of acidizing. They are applied as emulsifiers, demulsifiers, nonemulsifiers, silt suspending agents, antisludge agents, surface tension reducers, corrosion inhibitors, bactericides, clay treaters, foaming agents, etc.

7.8.3 Clay stabilizer

Chemicals used to stabilize clays and fines function by being adsorbed, usually by electrostatic attraction or ion exchange, on the minerals to be stabilized. Because silicates above their pzc values have a negative charge, the most effective stabilizer has a positive charge (cationic). Common clay stabilizers are highly charged cations, quaternary surfactants, polyamines, polyquaternary amines, and organosilane.

7.8.4 Mutual solvents

Mutual solvents, as the name implies, are chemicals that are mutually soluble in both hydrocarbons and water. The most efficient mutual solvents are glycol ethers, a reaction product of alcohols and ethylene oxide. These chemicals are relatively safe and easy to use in the field. The preferred glycol ethers contain at least a butyl or higher molecular weight group. The use of mutual solvents in the acid stimulation of a sandstone reservoir is a common practice. Mutual solvents are used in acid solutions and overflushes to:

- aid in reducing water saturation around the wellbore by lowering the surface tension of the water to prevent water blocks;

- solubilize a portion of the water into a hydrocarbon phase to reduce the water saturation;
- aid in providing a water-wet formation to maintain the best relative permeability to oil;
- help to prevent insoluble fines from becoming oilwet and stabilizing emulsions.

7.8.5 Iron control additives

When appreciable quantities of iron in the form of Fe3+ (ferric ions), rather than the usual Fe2+ (ferrous ions), are dissolved by the acid, iron precipitation and permeability reductions can occur after acidizing. The three methods currently used to help keep iron in solution are pH control (low pH with weak acid), sequestering agents (citric acid, ethylene diamine tetra acetic acid (EDTA), and nitrilotriacetic acid (NTA)) and reducing agents (also effective as oxygen scavengers). These may be used individually or in combination, depending on the source and amount of iron dissolution expected.

7.8.6 Alcohols

Alcohols are used in acidizing fluids to remove water blocks, enhance fluid recovery, retard acid reactivity, and decrease water content. The most common alcohols used in acidizing are isopropanol and methanol. Isopropanol is normally used at a maximum of 20% by volume. Methanol is used at various concentrations, but a typical concentration may be 25% by volume.

8

Hydraulic Fracturing

8.1 Introduction

Hydraulic fracturing is one of the primary engineering tools for improving well productivity, which is achieved by:

- placing a conductive channel through near-wellbore damage and thus bypassing this crucial zone;
- extending the channel to a significant depth into the reservoir to further increase productivity;
- placing the channel such that fluid flow in the reservoir is altered.

The question that may be asked is why or when we should hydraulic fracture a well. All the plausible reasons can be categorized as follows:

- to bypass near-wellbore damage and return a well to its "natural" productivity;
- to extend a conductive path deep into a formation and thus increase productivity beyond the natural level;
- to alter fluid flow in the formation, a powerful tool for reservoir management.

These conditions are explained briefly below:

8.1.1 Damage bypass

Near-wellbore damage reduces productivity. This damage can occur from several sources, including drilling-induced damage resulting from fine invasion into the formation while drilling and chemical incompatibility between drilling fluids and the formation. Various damage mechanisms have been explained in the chapter on formation damage. Whatever the cause, the result is undesirable. Matrix treatments are usually used to remove the damage chemically, restoring a well to its natural productivity. In some instances, chemical procedures may not be effective or appropriate, and hydraulic fracture operations are used to bypass the damage.

8.1.2 Improved productivity

Unlike matrix stimulation procedures, hydraulic fracturing operations can extend a conductive channel deep into the reservoir and stimulate productivity beyond the natural level of the well.

8.1.3 Reservoir management

Long fractures in tight reservoir ($K < 0.1$) can help in developing a field with lesser number of wells. Fracturing (more specifically Frac Pack) is used to control sand production without comprising on well productivity. Figure 8.1 shows the schematic of a hydraulic fracture in a well.

8.2 Fracturing Process

Figure 8.2 below portrays a typical fracturing operation. It consists of large tanks to store fracturing fluid (a viscous fluid), blending equipment to mix sand in the fluid, and then pumps for pumping the blended fluid into the pay zone at very high rates to create and wedge the fracture. Initially, a neat fluid called "pad" is pumped to initiate the fracture and to establish propagation. This is followed by a slurry of proppant and frac fluid. This slurry further extends the fracture and also keeps it propped. After the planned schedule of pumping, the fracturing fluid chemically breaks back to lower viscosity and flows back out of the well. The well now has a deep, highly conductive path from the wellbore into the reservoir.

The well normally has two wings extending in the opposite direction from the well and is oriented in the vertical direction. Fractures with "horizontal" orientation are also possible, but at a very shallow depth (<600 m).

8.3 Fracturing Mechanism

If fluid is pumped into a well faster than the fluid can escape into the formation, inevitably pressure rises, and at some point the rock breaks, resulting in the wellbore splitting along its axis as a result of tensile hoop stresses generated by the internal pressure. The wellbore breaks – i.e., the rock fractures – owing

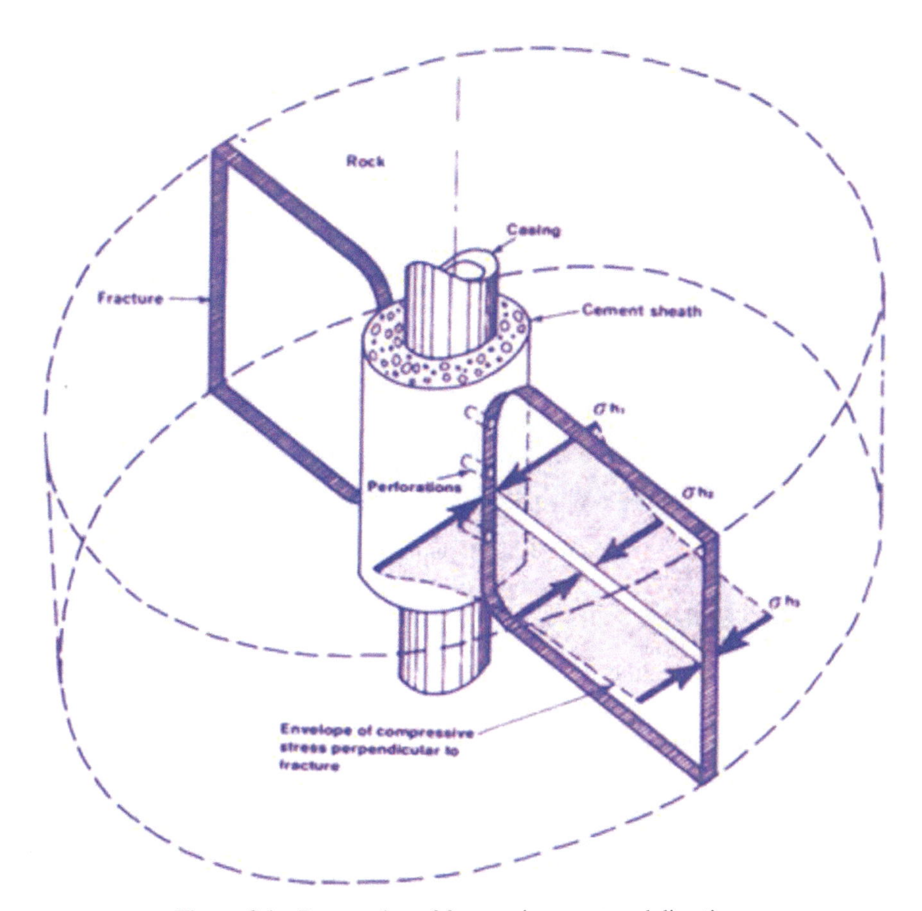

Figure 8.1 Propagation of fracture in an outward direction.

to the action of the hydraulic fluid pressure, and a "hydraulic" fracture is created. Because most wells are vertical and the smallest stress is the minimum horizontal stress, the initial splitting (or breakdown) results in a vertical, planar parting in the earth (Figure 8(3a)).

If pumping is continued at a rate more than the leak off, then the fracture can be propagated further (Figure 8(3b)). At this point, if the pumping is stopped, then the fracture will "heal." To prevent this proppant, laden slurry is pumped (Figure 8(3c)) in the fracture. Thus, when the pumping is stopped and fluid leaks off, the proppant in the fracture keep the fracture open. The result is improved flow into the well through the newly created path in the reservoir (Figure 8(3d)).

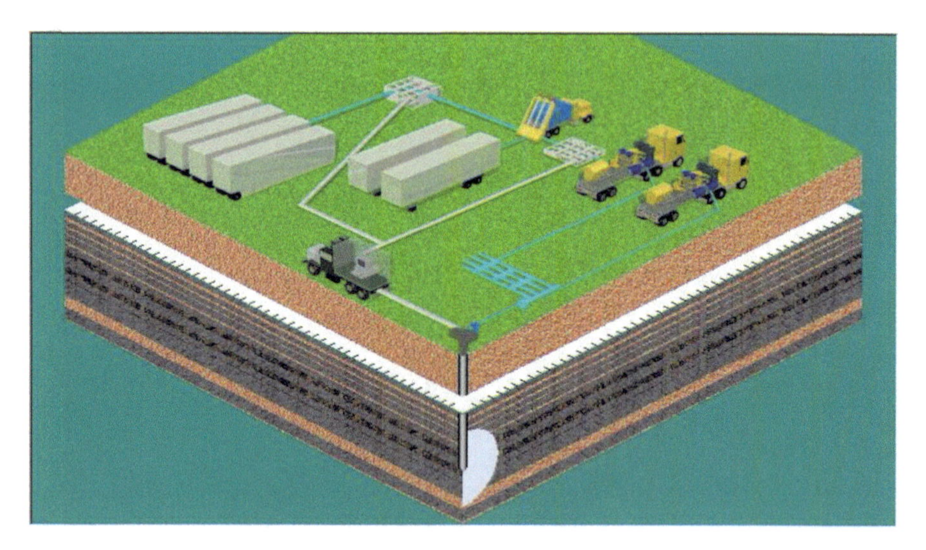

Figure 8.2 Fracturing process.

8.4 *In Situ* Stress

In situ stress, in particular the minimum *in situ* stress (also termed as fracture closure pressure), is the dominant parameter controlling fracture geometry. For relaxed geologic environments, the minimum *in situ* stress is generally in horizontal direction; thus, a vertical fracture that formed when a vertical wellbore broke remains vertical and is perpendicular to this minimum stress. Hydraulic fractures are always perpendicular to the minimum stress, except in some complex cases, and even for those cases any significant departure is only at the well. This occurs simply because that is the least resistant path. Opening a fracture in any other direction requires higher pressure and more energy.

The minimum stress controls many aspects of fracturing:

- At very shallow depths or under unusual conditions of tectonic stress and/or high reservoir pressure, the weight of the overburden may be the minimum stress and the orientation of the hydraulic fractures will be horizontal; for more normal cases, the minimum stress is generally horizontal and the maximum horizontal stress direction determines whether the vertical fracture will run north–south, east–west, etc.
- Stress differences between different geologic layers are the primary control over the important parameter of height growth.
- Through its magnitude, the stress has a large bearing on material requirements, pumping equipment, etc., required for a treatment. Because the bottomhole pressure must exceed the *in situ* stress for

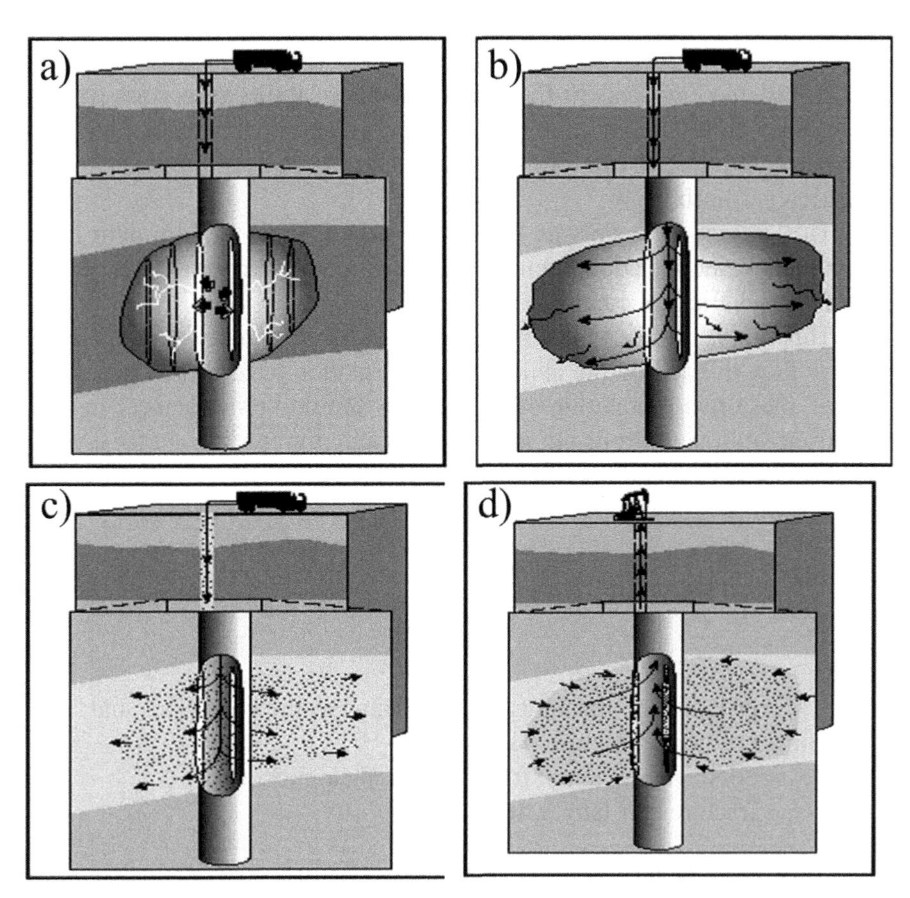

Figure 8.3 (a) Injection pressure breaking the wellbore. (b) Propagation of fractures. (c) Proppant in the fracture. (d) Improved flow into wellbore.

fracture propagation, stress controls the required pumping pressure that the well tubulars must withstand and also controls the hydraulic horsepower (hhp) required for the treatment. After fracturing, high stresses tend to crush the proppant and reduce *kf*; thus, the stress magnitude dominates the selection of proppant type and largely controls post-fracture conductivity.

Estimation of minimum *in situ* stress is possible by:

- MiniFrac test in which a small fracturing job is done without the proppant.
- Calculate the horizontal stress using Poisson's ratio, depth, and reservoir pressure (Poisson's ratio can be obtained through Lab or Well logging).
- Other wells of the same reservoir

8.5 Design Goals

Historically, the emphasis in fracturing low-permeability reservoirs was on the productive fracture length x_f. For higher permeability reservoirs, the conductivity k_{fw} is equally or more important, and the two are balanced by the formation permeability k.

This critical balance was first discussed by Prats (1961), more than 10 years after the introduction of fracturing, with the important concept of dimensionless fracture conductivity.

This dimensionless conductivity ratio (C_{fD}) is the ability of the fracture to carry flow divided by the ability of the formation to feed the fracture. In general, these two production characteristics should be in balance. In fact, for a fixed volume of proppant, maximum production is achieved for a value of C_{fD}.

The dimensionless fracture conductivity C_{fD} is shown below:

$$C_{fD} = \frac{k_p w}{k_{EH} x_f}$$

where

k_p = proppant permeability at producing closure conditions, md;
w = producing fracture width, ft;
k_{EH} = effective horizontal formation permeability, md;
x_f = fracture half length, ft.

As can be noted in Figure 8.4, a dimensionless fracture conductivity (C_{fD}) of 10 or greater provides almost the same production performance when dimensionless times are greater than 0.1. It is seen that dimensionless fracture conductivity (C_{fD}) of 10–500 give almost the same production performance after around 250 days. Therefore, the fracture should be optimized for a dimensionless fracture conductivity (C_{fD}) of 10. As longer fractures are created to increase production, the fracture conductivity must be increased to maintain the dimensionless fracture conductivity (C_{fD}) value equal to or larger than 10.

Prats also introduced another critical concept, the idea of the effective wellbore radius $r_{w'}$. As shown in Figure 8.5, a simple balancing of flow areas between a wellbore and a fracture gives the equivalent value of $r_{w'}$ for a propped fracture (qualitative relation only).

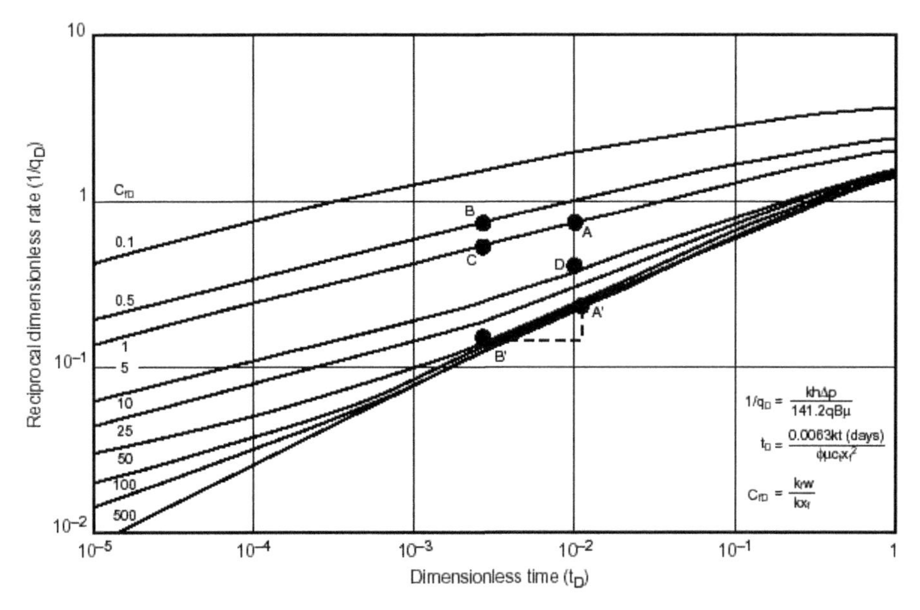

Figure 8.4 Dimensionless fracture conductivity.

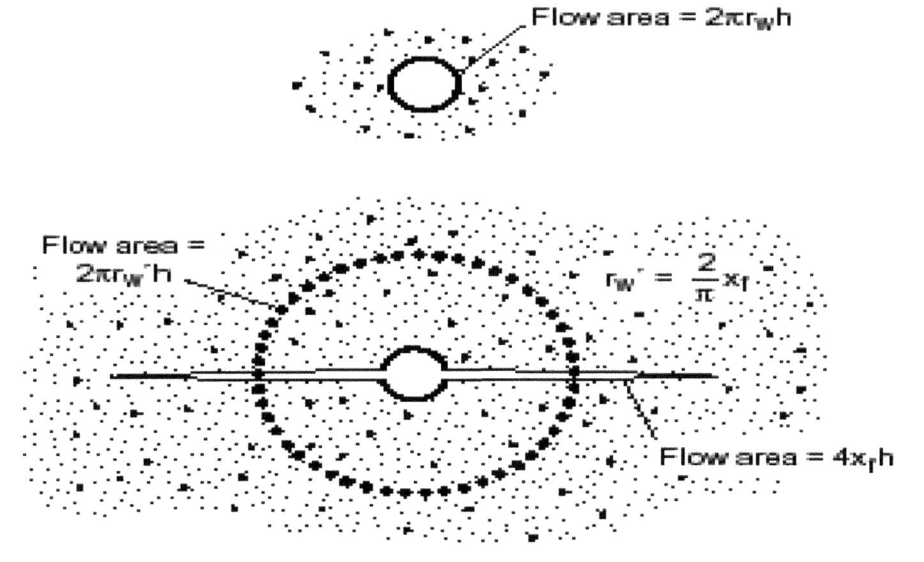

Figure 8.5 Equivalent wellbore radius.

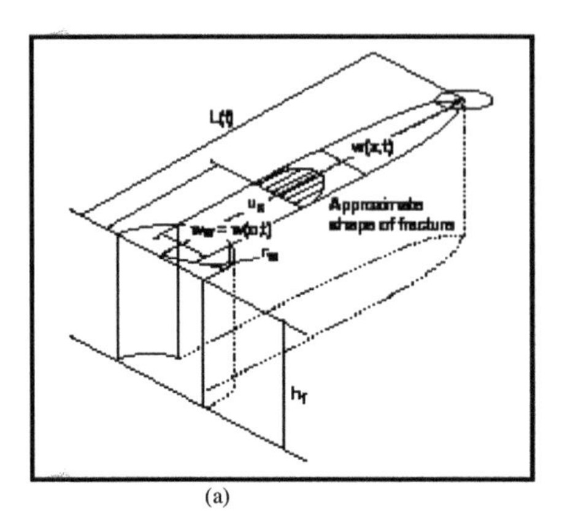

(a)

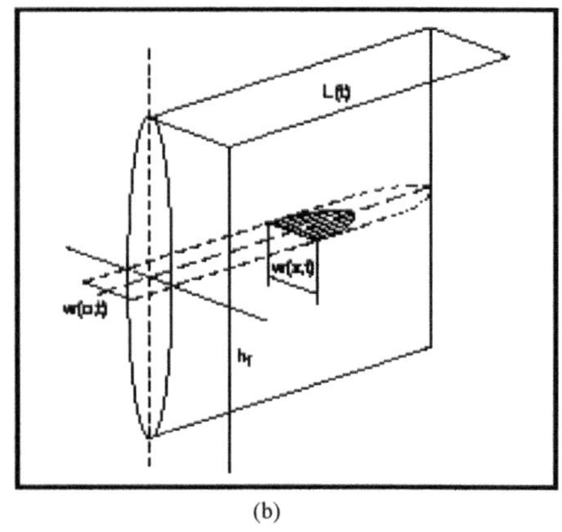

(b)

Figure 8.6 (a) KGD model. (b) PKN model.

8.6 Hydraulic Fracture Modeling

Fracture modeling is necessary to design and analyze fracture treatments and also monitor fracturing operations.

8.6.1 2-D models

The first work on hydraulic fracture modeling was performed by several Russian investigators. The other major contribution was the work of Perkins and Kern (1961). These models were developed to calculate the fracture geometry, particularly the width, for a specified length and flow rate, but did not attempt to satisfy the volume balance. Carter (1957) introduced a model that satisfies volume balance but assumes a constant, uniform fracture width.

This approach was made obsolete by extensions to the Khristianovich and Zheltov and Perkins and Kern models developed by Geertsma and de Klerk (1969) and Nordgren (1972), respectively. These two basic models, generally known as the KGD and PKN models after their respective developers, were the first to include both volume balance and solid mechanics. The PKN and KGD models, both of which are applicable only to fully confined fractures, differ in one major assumption: the way in which they convert a three-dimensional (3D) solid and fracture mechanics problem into a two-dimensional (2D) (i.e., plane strain) problem (Figure 8.6 (a & b)).

8.6.2 3-D and pseudo 3-D models

The simple models discussed in the previous sections are limited because they require the engineer to specify the fracture height or to assume that a

Figure 8.7 Pumping unit.

radial fracture will develop. This limitation can be remedied by the use of planar 3D and pseudo-3D (P3D) models.

The three major types of hydraulic fracture models that include height growth are categorized according to their major assumptions.

- General 3D models make no assumptions about the orientation of the fracture.
- Planar 3D models are based on the assumption that the fracture is planar and oriented perpendicular to the far-field minimum *in situ* stress.
- P3D models attempt to capture the significant behavior of planar models without the computational complexity.

8.6.3 Advantages of 3-D models

1. It is closer to reality.
2. Net pressure matching can serve as diagnostic.
3. Predicts fracture height and thereby removes the guessing part.

8.7 Execution

Unlike matrix stimulation, fracturing is a much more complex procedure performed on a well. This is due in part to the high rates and pressures, a large volume of materials injected, continuous blending of materials, and large amount of unknown variables for sound engineering design. The fracturing pressure is generated by single-action reciprocating pumping units that have between 700 and 2000 hydraulic horsepower (Figure 8.7).

These units are powered by diesel, turbine, or electric engines. The pumps are purpose-built and have not only horsepower limits but also job specification limits. These limits are normally known (e.g., smaller plungers provide a higher working pressure and lower rates). Because of the erosive nature of the materials (i.e., proppant), high pump efficiency must be maintained or pump failure may occur. The limits are typically met when using high fluid velocities and high proppant concentrations (+18 ppg). There may be numerous pumps on a job, depending on the design.

Mixing equipment blends the fracturing fluid system, adds the proppant, and supplies this mixture to the high-pressure pumps. The slurry can be continuously mixed by the equipment or batch mixed in the fluid storage tanks. The batch-mixed fluid is then blended with the proppant in a continuous stream and fed to the pumps.

A stimulation engineer controls the whole process from an FRAC van, which provides all the fracturing parameters. Pumps and the blenders can also be remotely controlled from this van.

8.8 Real-time Monitoring

In all of these stimulation operations, two parameters viz. flowrate and pressure, continuously vary. The variation in these parameters reveals a lot about the progress and success of the treatments. Earlier, these parameters were grossly monitored, but nowadays these parameters are fed into computers, which uses complex mathematical modeling (mostly 3D models) for analyzing the treatment. This process is known as real-time monitoring. It basically involves sensing devices (flow, pressure, and density), data acquisition systems, and computers hosting interpretation software.

In stimulation, for real-data analysis of fracturing treatment, the real data being used is generally the measured surface treating pressure, measured flowrate, and the proppant concentration (slurry density). The use of real-time monitoring can be categorized into two categories. The first is estimation of fracture geometry (length, height, and width) using a fracture model. This is done using a fracturing simulator, which takes as input the measured pumping rate and proppant concentration, and predicts its own net pressure, which is compared with the actual measured net pressure.

The second category of real-time data is for diagnostic purpose (also called MiniFrac test) to estimate parameters before a fracturing job for successful completion of the main fracturing job. Diagnostic injection tests (shown in Figure 8.8) include: breakdown, MiniFrac, proppant slug, and step-down.

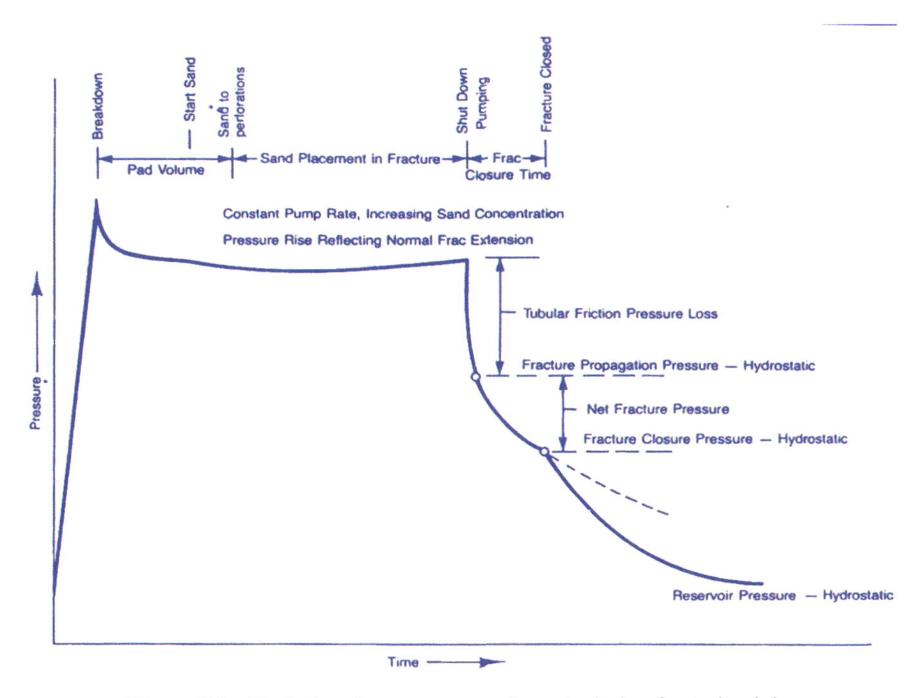

Figure 8.8 Typical surface pressure and events during fracturing jobs.

Since all stimulation operations are targeted toward formation, which are very deep, any claim about success or failure of the job can only be ascertained by indirect means, i.e., interplay of flowrate and pressure.

8.9 Fracturing Fluid Systems

The basic functions of fracturing fluid as explained in the fracturing mechanism are (Figure 8.8):

- Transmit pressure to the formation.
- Transport proppant into the fracture.
- Additionally, it must break completely (reduce to water-like consistency) after the job is completed.

The basic types of fracturing fluid are:

- Newtonian – water, liner gels, low cps oils.
- Non-Newtonian – cross-linked gels (most commonly used).

The main fracturing fluid additives are as follows:

- Primary components:
  - Gelling agents (Gellants)
  - Cross-linkers
  - Breakers
- Secondary components :
  - Buffers
  - Surfactants
  - Clay stabilizers
- Other components:
  - Foamers
  - Friction reducers
  - Fluid loss control

8.9.1 Gellants

Gellants are the base materials that provide viscosity. In early days, only linear gels were used, which have gellant only. The different types of gellants are as follows:

- Water gellants
  - Guar Gum (normally used by ONGC)
  - HPG, CMG, CMHPG, HEC, CMHEC (imported)
- Methanol gellants
  - HPG
- Oil gellants
  - phosphate ester
- Acid gellants
  - nonylphenol, alkylphenol

8.9.2 Cross-linkers

The cross-linkers significantly increase the viscosity of linear gels by increasing molecular weight of the base polymer by cross-linking multiple molecules together. Thus, it requires lower polymer loading for creating high viscosity fluid and thereby less residue is formed. The result is less expensive viscosity.

Borates form non-permanent x-link gallant. Metal cross-linkers form a more permanent site that requires more breaker. Most of the cross-linkers are pH- and shear-dependent.

Examples of various cross-linkers:

- Zirconium
- Titanium
- Organo-borate
- Borax (used by ONGC)
- Boric acid

8.9.3 Breakers

After completion of the job, it is also necessary that the frac fluid breaks down so as to allow hydrocarbon to move through the proppant pack. Breakers controllably degrades viscous gelled fluids back to thin base fluids. Most of them are pH-dependent. Lab studies are required to ascertain the concentration (loading) of breakers. The loading depends on:

- desired break time;
- bottom hole temperature;
- gel loading;
- laboratory break test;
- specific base fluid.

Types of breakers:

- catalyzed oxidizers (low temp)
- conventional oxidizers, ammonium per-sulfate

- delayed activated oxidizers
- polymer-specific enzymes
- encapsulated breaker (generally for high temp.)

Oxidizers are free radicals and will react with tubulars, tanks, and formation (very corrosive). The conventional enzymes are limited to 180 °F. Their important properties are as follows:

- React immediately
- Limited pH range
- "De-natured" by high temperature and high pH
- React with several polymers
- Do not react with tubular or formation
- Environmentally friendly

8.9.4 Buffers

It adjusts and maintains the pH to allow the gellant to hydrate and maximize viscosity and are used in water-based fluids and can be used to control hydration/cross-link.

8.9.5 Surfactants

These are used to modify wettability of the formation. It can be used to create, break, prevent, or stabilize emulsions. It disperses additives in oil and water and lowers surface and interfacial tension from 73 dyne/cm^3 down to 18–28 dyne/cm^3. It helps to suspend fines.

8.9.6 Clay stabilizers

These can minimize permeability impairment from clay swelling and some can control migrating clays. These are helpful in water-sensitive formations. Normally 2% KCl is used in the base fluid as clay stabilizer.

8.9.7 Fluid loss additives

Typically, these are mechanical bridging agents such as silica flour. They plug the formation fracture face to minimize fluid loss into the formation.

They generally produce high residue (undesirable). Normally cross-linked fluids have very good leakoff prevention characteristics and thus would not require any fluid loss additives.

8.9.8 Friction reducers

These reduce friction pressures of base fluid often reduced by 30%–70% and used in the flush to reduce pumping pressures. It reduces friction by suppressing turbulence. But these should not be used in emulsions (breaks the emulsion). Gels in single phase have low friction and thus do not require these additives.

8.9.9 Various frac fluid systems

The frac fluid systems normally used in industry are as follows:

- Water-based
- Hydrocarbon-based
- CO_2 fluids
- Emulsion-based
- N_2 foams
- Methanol
- Acid systems

Among these frac fluid systems, the most commonly used is water-based.

The water-based fracturing fluids are cheaper as the base fluid water is cheaper. It uses both linear and cross-linked gellants.

- Guar, HPG, CMHPG, HEC, CMHEC polymers
- VES – viscous elastic surfactants
- Various cross-linkers

8.9.10 Polymer-free frac fluid

Residue from conventional polymer-based fluid may clog pore spaces in the proppant pack, reducing fracture permeability. Polymer-free frac fluid is the latest development in the industry and offers solution by not using any polymer. This is achieved by using viscoelastic surfactant fluid (also known as VES). This fluid requires no chemical breaker but cleans up when it mixes with influx from reservoir. The fluid is easy to mix and pump in the field. The only disadvantage is its higher cost as compared to guar based frac fluid.

8.10 Proppants

The objective of the proppant is to prop the newly created fracture open to provide a highly conductive path for fluids to flow and create a "pipeline" for reservoir fluid flow. Fractures that are not propped would heal as pressure dissipates, thus losing its conductivity. It is important to use the right kind of proppant since it is only proppant that remains in the reservoir during the wells' life.

Following parameters are considered during proppant selection:

- Closure stress
- Proppant crushing
- Proppant pack conductivity
- Tortuosity/near-wellbore restrictions
- Economics
- Proppant flow-back control
- Sphericity/angularity

8.10.1 Crush resistance

The rounded particles more crush-resistant than angular particles and small particles more crush-resistant than large.

- Angularity creates stress concentrations.
- Reduces angularity and stress concentrations.
- 30/50 is more crush-resistant than 20/40.

8.10.2 Proppant types

- Sand (quartz)
- Ceramic proppant

- Resin-coated proppant (RCP)
  - Either sand or ceramic
- Sintered bauxite

8.10.2.1 Properties of sand

The true density of sands are around 22.1 lb/gall or 2650 kg/m^3. These are of low cost and low crush resistance. Jordan, Ottawa, or Northern White, Brady, Hickory, or Texas Brown, Colorado, Indigenous (Saurashtra) are the different types of proppants that are named after their location.

8.10.2.2 Properties of ceramic

The true density is around 22.1–26.7 lb/gall or 26.50–3200 kg/m^3. It has intermediate strength and is crush-resistant. But it is of high cost and it has better sphericity and angularity than sand, and thus has better conductivity. Econoprop, Carbolite, Carboprop, Valuprop, and Interprop are the different types of proppants available.

8.10.2.3 Properties of resin-coated proppants

It has true density around 22.1–26.7 lb/gall or 26.50–3200 kg/m^3. It is pre-cured or curable types. It improves sphericity and angularity of the base proppant, thus improving conductivity. Super LC, DC, OptiProp, Tempered LC, DC, Ac Frac PR. - 6000, PR - 4000, SB - 6000, etc. are the different types of resin-coated proppants.

8.10.2.4 Properties of sintered bauxite

It has true density around 29.6 lb/gall or 3550 kg/m^3. It has got very high strength and is most expensive. It is seldom required – only for very deep fracs. When crushed, bauxite splits instead of shatters and split bauxite still provides decent conductivity.

8.11 Frac Equipment

The various types of frac equipment in use are:

- Blender
- Pumpers
- Tanks (storage of frac fluid)
- Dumpers (for proppant)

- Connection and valves
- Chemical injection pump
- Monitoring unit

8.11.1 Blender

The blender is the heart of the treatment; if it goes down, the job is over. The blender's functions include:

- Mixing base fluids
- Adding chemicals at proper ratios during the treatment
- Adding proppant at proper ratios during the treatment
- Feeding the pumpers with slurry

The blender takes fluid from the tanks (containing premixed frac fluid). Dumpers add proppant into blenders tub at set rate. Chemicals like breaker are also added at this stage. The blender finally pumps this slurry to the pumpers (normally more than one).

8.11.2 Pumpers

The pumper's only function is to be able to pump fluids/slurries at various rates. If one pumper goes down, then the other pumpers can raise the gears to compensate for the rate drop. Pumpers come in various horsepower ratings (ONGC has 800, 1200, and 2000 HP) and can be fitted with various size fluid ends to match pressure and rate designs.

8.11.3 Monitoring units

The monitoring unit includes facilities to measure the following parameters:

- rates (clean, dirty)
- pressures (STP, BHTP)
- density (sand concentrations)
- liquid additive rates (buffers, complexes)
- actual bottom hole pressure (static or gauge)

9

Well Analysis

9.1 Introduction

Wells are our ultimate source of clues about the reservoir that lies buried thousands of feet below the surface. The idea of well analysis is to look into crucial well data and conceptualize the reservoir based on these. In other words, it tries to tie up seismic, production, geological, petrophysical, logging, and other data to fit the observed facts. Concern of any oilfield manager is to maximize returns from its assets, and this calls for effective management. Well analysis is a tool in the hands of such managers to assess and evaluate a field's performance and adopt measures toward the enhancement of production. Through well analysis, it is possible to look at a reservoir in a more comprehensive manner. Studying individual well performance histories speaks volumes about the character of a reservoir. The objectives of well analysis can be classified as follows.

- Analyze each well in terms of production, injection, pressure, and problems faced (wellbore, AL, de-bottlenecking, etc.).
- Study reservoir behavior.
- Integrate well analysis with reservoir behavior to identify areas of improvement.
- To indicate hitherto unknown problems, if any.
- Provide well-wise/field action plan.

A comprehensive sick well analysis is always considered an ideal approach before the decision is taken to bring a workover rig at the well location for carrying out the sequence of jobs required. A periodic well analysis of all the wells in a field not only helps in increasing oil production by identifying the production problem and taking the requisite corrective measure but also helps in a better resource planning (e.g., annual planning for services of workover rig, MSV, etc.), which could be an important aspect in the management of oilfields.

Well analysis is one of the most practical aspects of petroleum engineering. By properly analyzing the well, it can be inferred whether the production behavior of the well is normal or subnormal. In the latter case, the causes for the subnormal behavior can be identified so that remedial measures can be planned and executed. Some basic background information about the reservoir goes a long way in understanding well behavior; hence, it is necessary to outline these.

9.2 Geological Aspects

The following maps must be understood well to conceptualize the reservoir. The engineers are not expected to construct the maps. However, when the maps prepared by geologists are available, the engineers must be able to interpret the important features and draw their inferences.

9.2.1 Structure contour maps

These maps help in the visualization of the structure of the reservoir. The dip, faults, oil–water contacts, and the gas–oil contacts help in understanding individual well behaviors. Other important maps are iso-porosity maps, iso-permeability maps, iso-saturation maps, iso-pay maps, and iso-bar maps.

9.2.2 Drilling aspects

Drilling records of the wells have to be screened to pick up zones at which abnormal drilling conditions were faced such as:

i) Mud loss

ii) Stuck drill pipe

iii) High- and/or low-pressure formations

Zones in which the above-mentioned problems were encountered while drilling may later on prove to be harmful to casing, giving rise to corrosion, micro deformation, etc.

9.2.3 Drilling fluids

The following properties have to be tabulated from drilling records.

Weight: Mud weight used while drilling productive formations influences the productivity of the well. For example, high weight muds may damage the formation resulting in low productivity.

Composition: As the composition of the mud such as fresh water base or saline or oil base affects the log characteristics, the engineer has to be careful while interpreting the logs for well analysis.

The physico-chemical reactions between the mud, mud filtrate, and the formations have to be studied to evaluate the cause of formation damage, if any.

9.2.4 Cementing

Composition of the cement and the bonding properties play a great role to prevent extraneous fluids entering the well.

9.2.5 Openhole log interpretation

The following types of logs are used for well studies.

i) Resistivity logs
ii) Spontaneous potential logs
iii) Gamma ray logs
iv) Gamma ray absorption logs
v) Neutron logs
vi) Sonic porosity logs

The following information is obtained from log interpretation:

i) Identification of markers
ii) Estimation of thickness
iii) Depositional trends
iv) Determination of porosity and water saturation

Since log analysis is a highly technical skill demanding considerable experience, it is always convenient and desired to obtain interpreted data from the log analysts.

9.3 Reservoir Fluid Analysis

Some of the properties of crude, such as bubble point pressure, viscosity, composition, etc., help in comparing the oil produced from different wells and also for production log interpretation.

9.3.1 Productivity index

The productivity of a well is a measure of its oil producing potential for a specific pressure distribution between the well and the reservoir. The productivity of a well is largely controlled by reservoir properties and by the properties and distribution of fluids in the reservoir. The oil productivity of a well is reduced when gas and oil are flowing simultaneously in the reservoir. If the flowing bottomhole pressure is below the bubble point, the well productivity is reduced by gas evolution resulting from pressure drawdown near the wellbore. The reservoir geometry and the characteristics of the production string and well equipment also influence the productivity of a well. Furthermore, productivity will change with time if the pressure distribution or fluid distribution in the reservoir changes.

The productivity index of a well is a commonly used measure of a well's productivity. The productivity index, *J*, for an oil well is the production rate in STB/D per psi of drawdown. As long as the bottomhole flowing pressure is above the bubble point, the following commonly used equation is applicable.

Productivity index:

$$J = q_o/(p_e - p_{wf}), \tag{9.1}$$

where

q_o = well producing rate, STB/D;
p_e = pressure at the external drainage radius, psig
p_{wf} = bottomhole wellbore pressure while producing, psig

The productivity index as expressed by eqn (9.1) suggests that the producing rate of a well will equal the productivity index times the pressure drawdown.

$$q_o = J(p_e - p_{wf}) \tag{9.1a}$$

Eqn (9.1a) is strictly applicable only for the steady-state flow of a single, incompressible fluid. From a practical standpoint, we can use eqn (9.1a) for an oil reservoir only if P_{wf} is above the bubble point pressure.

If the reservoir pressure is below the bubble point, the following equation is used for inflow performance:

$$q_o/q_{max} = 1 - (0.2 * a) - (0.8 * a^2), \text{ where } a = p_{wf}/p_r.$$

9.4 Well Test Analysis

Methods of analyzing transient pressure tests are:

i) Classical graphical methods

ii) Type curve matching

iii) Computer-aided methods

Two parameters, very useful for well analysis, which can be obtained from well test analysis, are:

i) Permeability

ii) Skin

The total skin is evaluated from well tests. This can be resolved into various components like skin due to non-Darcy flow, partial penetration, perforation, and formation damage.

9.4.1 Cased hole logs – types and uses

Cased hole logging may be used for following:

1. Formation evaluation (gamma ray, pulsed neutron capture, neutron log, sonic/acoustic logs)
    a. Shale content
    b. Clay type
    c. Type of rock/hydrocarbon and their saturations
    d. Mechanical properties of rock/mineralogy
2. Determining wellbore integrity (acoustic cement evaluation, USIT, mechanical calipers, acoustic casing inspection tools, borehole videos)
    a. Evaluation of cement sheath, fraction of annular fill
    b. Casing condition in terms of the extent of damage
3. Studying fluid movement during production/injection (temperature surveys, noise logs, radioactive tracer surveys, spinner logs)
    a. Detection of channel behind casing
    b. Flow profiles in injectors/producers

9.5 Well System Analysis

Well-system analysis (inflow and outflow) must be performed to check whether the well is producing at the expected rate or not.

9.5.1 Analysis of well sickness

Inflow restrictions:
Most oil and gas wells may become sick due to wellbore plugging, perforation restrictions, or formation damage. Therefore, removing or bypassing these inflow restrictions will increase production. Such damage may be indicated by production well tests, pressure buildup and drawdown tests, comparison with offset wells, and analysis of production history.

9.5.1.1 Wellbore plugging

Flow restriction in the wellbore is often due to debris (from the formation, e.g., sand) or scale. Both can reduce production drastically. Flowing and gas-lifted wells can be easily checked for debris by using a slick wireline unit, provided the tubing is open-ended. Where there is debris across the formation, a bailer or coiled tubing unit are commonly used for cleanout. For wells, on A/L modes like SRP or ESP, it may be necessary or economical to pull the tubing before cleanout.

Inorganic scale in the wellbore and perforation is a common problem that can easily reduce production. The first step is to determine the type of scale. Sodium chloride salt can be easily dissolved with hydrochloric or acetic acid. Large deposits of gypsum are more difficult to remove chemically and may best be drilled out. Barium and strontium chloride must normally be mechanically removed.

9.5.1.2 Perforation restrictions

Perforation plugging or just inadequate perforations can cause low production rate. In sandstones, normally at least four 0.5" multi-phased perforations per foot that penetrate the formation of more than six inches are required to reduce the skin value to a minimum. High-rate gas wells may require more shots per foot or larger perforation diameters to avoid "turbulent" flow restrictions.

Plugged perforations are very common. Considerable effort should be made on completion to ensure that the perforations are open and undamaged. Loss of dirty water into the formation that may plug both the perforations and the formation pores is to be avoided.

If plugged perforations are suspected, various steps can be taken. Ball sealers can be used to break down the perforations. In carbonate zones, a simple acid job may do the job. In sandstones, a staged HF–HCl acid treatment may be beneficial. Often the simplest procedure for short intervals is to re-perforate the zone. In some areas, it may be worthwhile to wash the perforations or to surge back the formation. In a few cases, it may even be beneficial to fracture the formation (a "short fat" fracture), which should clean out perforations and bypass near-wellbore damage.

9.5.1.3 Formation damage

Formation damage may be defined as any impairment of well productivity or injectivity due to plugging within the perforations; formation pores adjacent to the wellbore, or in fractures communicating with the wellbore. Contact with a foreign fluid, which is not compatible to the formation fluid, is the basic cause of formation damage. This foreign fluid may be drilling mud, completion, workover fluid, stimulation, or well-treatment fluid. The problem is to determine the degree of well damage (skin), causes of damage, and, finally, proper approaches to alleviate this damage.

9.5.2 Classifications of damage mechanism

The mechanism of formation damage can generally be classified by how it decreases production.

- Reduced absolute permeability of formation – caused by plugging of pore channels by induced or inherent particles.
- Reduced relative permeability to oil – results from an increase in water saturation or oil wetting of the rock.
- Increased viscosity of reservoir fluid, which results from emulsion of high-viscosity treating fluids.

Some of the key components of formation damage are discussed below.

9.5.2.1 Paraffin or asphaltene plugging

Asphaltenes are crude oil's heaviest and polar fractions. Asphaltene deposition can occur in formation pore spaces and is usually associated with maltenes, which suspend the asphaltenes. Once the maltenes as destabilizing substance for asphaltenes are removed in contact with acid, CO_2, or aliphatic solvents, asphaltene deposition occurs.

Asphaltene deposition is usually identified by:

- Change in wettability profile
- Sudden decline after successful acid job, etc.

Like asphaltene, paraffin deposition is also triggered by instabilities like pressure drop due to damage and restrictions, changes in temperature due to injection, hot oiling practices, etc., wax in the tubing/ casing in wellbore can be reamed out or removed by steam, a solvent, hot water, or hot oil. The preferred method of removing wax from the wellbore or formation is to squeeze an effective solvent into the formation (very low-rate low-pressure), followed by a 24-hour shut-in period. Normally, hot oil or hot water should not be used to remove wax from perforations, wellbore, or formation because when a portion of the melted wax gets cooled sufficiently and solidified, it forms a very hard plug, which at times becomes difficult to be removed. Wax deposited in formation pores or perforations is usually not soluble in reservoir oils. Crude oil analysis can identify potentially favorable conditions for precipitating paraffin/ asphaltene problems.

9.5.2.2 Emulsion blocks

They can be identified by step-rate decline in production plots. Emulsion damage may be alleviated with surfactants. If an emulsion block exists, the average well permeability as determined by injectivity tests will be much higher than the average permeability determined from production tests. This "check valve" effect provides a reliable way to predict emulsion blocks. Increasing or decreasing producing rates will not appreciably change the water percentages in an emulsion-blocked well.

For sandstone wells, an HF–HCl acid-surfactant treatment is preferred to remove emulsion damage. For matrix damage in carbonates, the usual approach is to bypass damage with acid. Emulsions formed during fracture acidizing of carbonates may be broken by pumping a 2% or 3% surfactant solution into the fracture.

Water blocking is a temporary shift of relative permeability in favor of water as the mobile fluid. Under these conditions, oil production will decrease, and the water percentage will increase. Circulating or killing the well with water usually causes water blocking. Water invading the pore spaces increases water saturation; thereby, relative permeability to water is increased. As the well is producing, water percentage usually declines by itself with time, because water saturation around the wellbore gets reduced. Water blocking can usually be prevented by adding surfactants to water used in well workover, well killing, or well circulating operations.

9.5.2.3 Fine particles (extraneous or inherent)

They are subjected to pore channel velocity. When the velocity exceeds the critical value, fine particles move and subsequently lead to pore plugging. Particles, however, tend to move with their wetting fluid, i.e., water-wet particles move with the water phase or oil-wet particles with the oil phase. With low water saturation (low relative permeability to water), water movement is slow and water-wet particles do not move or plug the pores. If the wellbore water saturation is high and the well is "cleaned up" too rapidly, particle plugging can happen. Where particle plugging is suspected, a matrix treatment such as sandstone acidizing or even merely injecting a non-damaging fluid to move particles back away from the wellbore may be successful.

9.5.3 Outflow restrictions

Outflow restrictions increase the backpressure, limit drawdown, and reduce production. A complete outflow analysis, i.e., tubing intake analysis (TIC) should be made on completion and should be reviewed periodically. The tubing size needs to be carefully selected. Tubing diameter smaller than optimum will cause extra friction loss, whereas larger than optimum will increase slippage and may cause the oil well to load and die. On a pumping well, smaller diameter tubing than the needed may cause friction, which will increase the lift energy requirement. Inflow and outflow performance curves are needed to find the proper tubing size. Any restriction in the tubing string will cause extra backpressure on the formation. The first stage separator operating pressure must be overcome by flowing wells.

9.6 Reservoir Problems

9.6.1 Low reservoir permeability

Low reservoir permeability may be an overall reservoir characteristic, or it may be limited to a specific area. If low permeability has been proved as a cause of limited production in a particular reservoir, it should be considered along with other possible causes of low productivity. Characteristically, in a low permeability reservoir, well productivity declines rapidly as fluids near the wellbore are produced. If available geologic and reservoir data do not readily prove low reservoir permeability, production tests and pressure build-up tests may aid in differentiating between low permeability and formation damage. Many low permeability reservoirs have been successfully stimulated for productivity increase.

9.6.2 Low reservoir pressure

Reservoir pressure history should be documented by periodic subsurface measurements. The next step is to consider the dominant reservoir drives in a particular reservoir and how it is associated with the real or apparent well problem being investigated.

For a partially depleted low-pressure oil reservoir, an effective artificial lift system should be planned. Pressure maintenance usually through water injection is usually the best possible approach to increase production rate and oil or gas recovery from partially depleted reservoirs.

9.6.3 Water production problems

During oil and gas production, a certain amount of water production is expected in the initial phase of life of the reservoir or well. However, when water production is considered excessive or a problem with increasing quantity of produced water, associated costs like lifting, processing, and disposal become high or uneconomical.

Water problems may result from:

- Natural water drive or water flood, aggravated by fingering or coning.
- Extraneous sources including casing leaks or poor cementing.
- Fracturing or acidizing into adjacent water zones or to below the oil water contact.

The best well completions and the best production practices can delay but not stop this water production. Temperature surveys of the bottomhole including the producing interval, run before hydraulic fracturing, often give a clue as to whether subsequent water production is due to primary cement failures, fracturing or acidizing into water, casing leaks, or water encroachment.

Water encroachment is complicated by stratified or layered permeability. Fingering is differential water encroachment through zones of high permeability. These zones will usually be watered out first. In stratified zones, a large volume of water is often produced before oil or gas is depleted from the remaining zones. Steps must be taken to shut off the high water-producing zones.

Some of the *common water production mechanisms* are:

9.6.3.1 Channeling through high-permeability path

This is defined as water movement in the more permeable zone, which is called "high-permeability streaks." This is more common in heterogeneous limestone and sandstone reservoirs.

9.6.3.2 Edge water encroachment

It is common in reservoirs with edge water drive.

9.6.3.3 Water coning

Coning in oil or gas wells is defined as vertical upward movement of water in a producing formation due to the large pressure drawdown. Cones for more than a few feet are rare. When the water cut trend follows the produced liquid rate, it is the time to suspect coning at play.

9.6.3.4 Water channeling

Vertical movement of water through a faulty primary cement job is often erroneously interpreted as "water coning." If coning is the problem, increasing production rate will usually increase the percentage of water produced and shutting in the well for one to three months and then the well to put back on production with reduced rate will decrease water cut. However, if it is otherwise and if cement/casing or cement/formation bond is poor, channeling behind casing can be the likely cause of water. These channels can occur at any time in the life of the well but are often observed immediately after a stimulation treatment or completion.

9.7 Methods to Reduce Water Production

9.7.1 Cement/gel squeezing or block cement/gel job

If production logs indicate flow behind the casing from a water zone, this communication can usually be stopped with low-pressure low-fluid-loss cement squeeze. The well can then be re-perforated in the desired interval. Gel squeeze, followed by cement, has been tried out with reasonable success. Though it conventionally used to be a rig-based job, the recent jobs have been more of rig-less, with the use of coiled tubing. In some cases, it would be proper to go for a block cementation job/gel job.

9.7.2 Formation of barrier or sealants

A matrix squeeze near the oil water contact using sodium silicate has been successful in some areas with the precaution that producing formation is to be kept out of sodium silicate reach. This is effective in case of water coning.

9.7.3 Plug back

For open hole and perforated casing completions where the lowest zones are producing water, a plug back in the form of bridge plug provides a result.

9.7.4 Straddle packer application

If the water-producing zone is above the oil/gas producing zones, straddle packer with sliding sleeve can be used to isolate the water-producing layer.

9.7.5 Re-completion

In cases of water coning, it is usual to plug back the existing perforations and re-complete as high above the oil water contact as possible. If the partial barrier to vertical flow exists, plug back becomes effective.

9.7.6 Stimulation/re-perforation

Stimulating a well to reduce the pressure drawdown can be helpful, especially if the well is damaged. Alternatively, re-perforation is tried to reduce drawdown.

9.7.7 High volume artificial lift installation

In high-permeability reservoirs where all zones appear to be producing with a high water cut and squeeze cementing is ineffective, it may be practical to produce the wells by high volume pumping or gas lift with all zones open.

9.7.8 Injection profile modification

Sometimes, it is easier to control water production in the producers by shutting off the high water intake intervals in water injectors. This can be done with gel/cement squeeze.

9.7.9 Well planning

In a steeply dipping reservoir, it is usually more profitable to produce wells high on the structure to avoid water coning until a high percentage of the reservoir oil or gas has been recovered.

9.7.10 Gas problems in oil wells

In a dissolved gas drive reservoir, gas saturation increases as oil withdrawal continues and reservoir pressure declines. When this gas is released from the oil, gas flows to the wellbore and with further decline in pressure, the gas tends to become the dominant mobile fluid. If there is no barrier to vertical flow in a reservoir with a gas cap, a decline in reservoir pressure may allow gas to expand into the oil-producing interval. With high drawdown at the wellbore, gas coning may occur in wells. In stratified reservoirs, premature fingering of gas may occur with high-pressure drawdown at the wellbore with gas usually flowing through high-permeability zones. Gas flow from zones above or below the oil zone may be due to (1) casing leaks, (2) poor cement job, or (3) natural or induced fractures communicating with gas zones. If gas flow is due to a channel behind the casing, the channel can be shut off with a low-pressure and low-fluid-loss squeeze cement/gel job. If a reservoir has a large gas cap, oil wells located low on the structure with low GOR should be produced in order to reduce gas production and to conserve reservoir energy.

9.7.11 High-viscosity oil

Viscosities greater than 200 centipoises often cause serious flowing and pumping problems. Production from wells producing viscous oil may be increased by decreasing viscosity of oil in the tubing and flow line with heat. Thermal stimulation (like steam injection) can be done in high-viscosity oil reservoirs. Many shallow wells with high-viscosity oil are being successfully produced by thermal stimulation. The tubing, casing, cement job, and packer installation are to be designed to withstand high temperature steam.

9.7.12 Sand control

Wells that sand up or produce more than 0.1% sand are often good candidates for sand control. Sand presents problems in artificial lifting, bean cutting, tubing erosion, etc. Gravel packing is the best approach to control sand in long-zone single completions. Gravel should be sized properly in relation to formation sand. Good placement techniques are essential to a successful job. Well should be put on production immediately after gravel packing, starting with a low rate of flow and increasing gradually to higher rate.

9.8 Mechanical Failures in Wells

Several different types of mechanical failures can cause loss of production and/ or increase costs in well operation. Some of the more common problems are:

- Primary cement channeling
- Casing, tubing, and packer leaks
- Artificial lift equipment malfunction
- Failure of well equipment – wellheads, subsurface valves, sliding sleeves, etc.

In locating casing leaks, water analyses are useful in differentiating between casing leaks and normal water encroachment. Temperature and noise logs and other production log are often beneficial in locating casing leaks.

Wellbore communication in conventional multiple completions can usually be detected by packer leakage tests, by production logging, by abrupt changes in producing characteristics, or by observing shut-in pressure.

9.8.1 Terminal well sickness

If the well is in such a condition that nothing can be done to revive it, or all the recoverable oil has been produced from the well (which may happen at the end of well/field life), the well has to be plugged and abandoned. Due consideration is to be given to all options before plugging a well. The possibility of using the well for an injector, effluent water disposal, as an observation well, or for providing cathodic protection to other wells needs to be looked into. If, however, a well is no longer economical to produce or has no use as a service well, then the immediate step is to plug and abandon the well. In most cases, a cement plug is placed across the entire open hole or all of the perforations. A cement retainer set above the zone and the formation squeezed with cement is often preferred as permanent abandonment. Cement plugs should also be placed across any zones where the casing has been severed or is in danger of external corrosion. It is necessary to protect the freshwater sands near the surface.

10

Water and Gas Shutoff

10.1 Why Water Control?

To reduce operating expenses

- Reduced pumping costs (lifting and re-injection)
- Reduced oil/water separation costs
- Reduced platform size/equipment costs
- Reduced corrosion, scale, and sand-production treatment costs
- Reduced environmental damage/liability

To increase hydrocarbon production:

- Increased oil production rate by reducing fluid levels and downhole pressure
- Improved reservoir sweep efficiency
- Increased economic life of the reservoir and ultimate recovery
- Reduced formation damage

10.2 The Scenario

Most of the oil fields of ONGC are in the mature stage of production. For every barrel of oil, the industry is producing an average of three barrels of water. Water production cost (lifting, transportation, processing and disposal, scaling and corrosion, and sand production, other than the intangible loss due to formation damage) has exceeded 25% of total production cost and is in increasing trend. Many wells are ceased and abandoned for high water cut production, a substantial number of which can be revived by proper

Table 10.1 The materials and techniques.

Chemical and physical plugging agents	Mechanical and well techniques
Cement, sand, calcium carbonate	Packers, bridge plugs, patches
Gels, resins	Well abandonment, infill drilling
Foams, emulsions, particulates, precipitates, micro-organisms	Pattern flow control
Polymer/mobility-control fluids	Horizontal wells

water control measures. In the international scenario, the situation is no less grim. On average in the United States, more than seven barrels of water are produced for each barrel of oil. Worldwide, an average of three barrels of water are produced for each barrel of oil. The annual cost of disposing of this water is estimated to be 5–10 billion dollars in the US and 40 billion dollars worldwide.

Since for more than two decades we have failed to discover any large reservoir, the situation demands an aggressive water control measure by applying the developed and proven technologies in high water producing wells on a mass scale.

10.3 The Objective

1. Creating awareness about the severity of the problem.
2. To equip with the technical know-how about water prevention mechanism.

10.4 The Philosophy

The philosophy of controlling excess water production problems is to attack the easiest problems first and diagnosis of water production problems should begin with information already at hand. A listing of water production problems is given below, which are then discussed in detail along with diagrams (wherever possible), and the possible remedial measures are discussed. Although a broad range of water-shutoff technologies are considered, the major focus is given when and where chemical gels can be effectively applied for water shutoff.

Many different materials and methods can be used to attack excess water production problems (Table 10.1). Generally, these methods can be categorized as chemical or mechanical.

10.5 The Categories

Category A: Easiest Problems – Needs Conventional treatment

1. Casing leaks without flow restrictions (medium to large holes)
2. Flow behind pipe without flow restrictions (no primary cement)
3. Flow behind pipe with flow restrictions (narrow channels)

Category B: Intermediate Difficulty – Needs Chemical treatment

4. Unfractured wells (injectors or producers) with effective barriers to cross-flow
5. Casing leaks with flow restrictions (pinhole leaks)
6. "Two-dimensional coning" through a hydraulic fracture from an aquifer
7. Natural fracture system leading to an aquifer

Category C: Moderate Difficulty – Needs Chemical treatment

8. Faults or fractures crossing a deviated or horizontal well
9. Single fracture causing channeling between wells
10. Natural fracture system allowing channeling between wells

Category D: Most Difficult Problems – No foolproof solution

11. Three-dimensional coning
12. Channeling through strata (no fractures), with cross-flow

Each of these problems requires a different approach to find the optimum solution. Therefore, to achieve a high success rate when treating water production problems, the nature of the problem must first be correctly identified.

Four problem categories are listed above in the general order of increasing treatment difficulty. Within each category, the order of listing is only roughly related to the degree of treatment difficulty.

10.5.1 Category A

Conventional treatments, which are normally used, include the application of water shutoff techniques that are generally well established, utilize materials with high mechanical strength, and function in or very near the wellbore. Examples include:

- Portland cement
- Mechanical tubing patches
- Bridge plugs
- Straddle packers
- Wellbore sand plugs

The first three listings (Category A, Problems 1–3) are the easiest problems and their successful treatment has generally been regarded as relatively straightforward. Of course, individual circumstances can be found within any of these problem types that are quite difficult to treat successfully. For example, for Problem Type 3, impermeable barriers may separate water and hydrocarbon zones.

However, if many water and oil zones are intermingled within a short distance, it may not be practical to shut off water zones without simultaneously shutting off some oil zones. The ranking of water production problems in Table 10.1 is based on conceptual considerations and issues related to the ease of treating each type of problem. It is experienced that operational and practical issues can make even the easiest problems very difficult to solve in practice.

10.5.2 Categories B and C

The intermediate problems (Categories B and C, Problems 4–10) are caused by linear-flow features (e.g., fractures, fracture-like structures, narrow channels behind pipe, or vug pathways). Problems 4–7 normally are best solved using gelants – i.e., the fluid gel formulation before significant cross-linking occurs. Problems 8–10 are best solved using preformed or partially formed gels (i.e., cross-linking products that will not flow into or damage porous rock.

10.5.3 Category D

In contrast, the last two problems (Problems 11 and 12) are difficult with no easy, low-cost solution. Certainly, much work remains to optimize the treatment of these problem types. However, substantial theoretical, laboratory, and field progress has been made in recent years toward solving these problems – especially using gels.

10.6 Identification

Identification of the excess water production problem should be performed before attempting a water shutoff treatment.

Several reasons exist for the inadequate diagnosis of excess water production problems.

- First, operators often do not feel that they have the time or money to perform the diagnosis, especially on marginal wells with high water cuts.
- Second, uncertainty exists about which diagnostic methods should be applied first. Perhaps 30 different diagnostic methods could be used. In the absence of a cost-effective methodology for diagnosing water production problems, many operators opt to perform no diagnosis.
- Finally, an incorrect belief exists that a “magic-bullet” method exists that will solve many or all types of water production problems.

10.7 Question – Is There Really A Problem???

The field engineer should recognize that two distinct types of water production exist.

- The first type, usually occurring later in the life of a waterflood, is water that is co-produced with oil as part of the oil’s fractional flow characteristics in reservoir porous rock. If production of this water is reduced, oil production will be reduced correspondingly. We call this *good water*.
- The second type of water production directly competes with oil production. This water usually flows to the wellbore by a path separate from that for oil christened as *bad water*.

In these latter cases, reduced water production can often lead to greater pressure drawdowns and increased oil production rates. Obviously, the second type of water production should be the target of water shutoff treatments but often confusion takes place between *good* and *bad* water.

Secondly the field engineer should ask: do significant volumes of mobile oil remain in the pattern or in the vicinity of the well of interest?

Three types of observations are commonly used to make this assessment.

- First, it may be noticed that certain well(s) exhibit a sudden increase in water cut.

- Second, a well or pattern of wells may be noted as producing significantly higher water/oil ratios (WORs) than other similar patterns.
- Third, plots of fluid production versus time may show an abrupt increase in WOR at a certain point.
- Results from reservoir simulation studies constitute a fourth, less common method to analyze water production problems in large reservoirs.

10.7.1 Key to success

- Distinguishing between the above two types of water production
- Successfully diagnosing the water production problem
- Successfully implementing and designing water shutoff treatments

10.7.2 Methods of diagnosis

Some of the common methods used to diagnose these problems include

1. leak tests/casing integrity tests (e.g., hydro testing);
2. temperature surveys;
3. flow profiling tools (e.g., radiotracer flow logs, spinner surveys, production logging tools);
4. cement bond logs;
5. borehole tele-viewers;
6. noise logs;
7. core and log analyses;
8. pulse tests/pressure transient analysis;
9. Inter-well tracer studies.

10.7.3 Problems and solutions

Let us now discuss the problem cases listed in page 10-2 & 3 of this chapter individually and try to find out what are the options available to tackle them.

10.7.3.1 Casing leaks (SI No. 1 and 4)

The difference between Problems 1 and 4 is simply a matter of aperture size of the casing leak and size of the flow channel behind the casing leak. Problem 1, involving casing leaks *without* flow restrictions, is where the leak is occurring through a large aperture breach in the piping (greater than roughly 1/8 inches) and a large flow conduit (greater than roughly 1/16 inches) behind the leak. Problem 4, involving casing leaks *with* flow restrictions, is where the leak is occurring through a small aperture breach (e.g., "pinhole" and tread leaks) in the piping (less than roughly 1/8 inches) and a small flow conduit (less than roughly 1/16 inches) behind the leak.

The most common methods to repair casing leaks (i.e., for Problem 1) involve either cement or mechanical patches. However, these methods have generally not been very successful when treating small casing leaks, such as "pinhole" or thread leaks (Problem 4). In particular, cement has difficulty penetrating through small leaks. With luck, cement may lodge in and plug the leak, but small mechanical shocks often easily dislodge the cement plug. *Gel treatments can be more successful for these applications.*

Appropriately designed gelants flow easily through the small casing leaks and some distance into the formation surrounding the leak. Thus, the gel treatment is directed at stopping flow in the porous rock around the vicinity of the casing leak, rather than solely attempting to permanently plug the casing leak itself.

Since the objective is to achieve total water shutoff from the leak and since small gel volumes are often used for this application, the gel plug should be relatively strong and must have a very low permeability.

10.7.3.2 Flow behind pipe (SI Nos. 2 and 5)

The difference between Problems 2 and 5 is simply a matter of aperture size of the flow channel behind the pipe. Problem 2, involving flow behind pipe *without* flow restrictions, is where the fluid flow is occurring through a large aperture flow conduit behind the pipe. Problem 5, involving flow behind pipe *with* flow restrictions, is where the flow behind pipe is occurring through a small aperture flow conduit. This problem often results from cement shrinkage during its curing during the well's completion.

Problems with unrestricted flow behind pipe are usually treated with cement. Cement can perform extremely well for this type of application if the channel to be plugged is not too narrow. When narrow channels are encountered, cement often cannot be placed effectively through small or constricted flow paths. Gels provide a better solution for this case, since they can flow or extrude readily through narrow constrictions.

10.7.3.3 Unfractured wells with effective barriers to cross-flow (SI No. 3)

Often, when radial flow exists around a well (i.e., fractures are not important), impermeable barriers separate hydrocarbon-bearing strata from a zone that is responsible for excess water production. When the water zone is located at the bottom of the well, cement or sand plugs are used most commonly to stop water production.

When the water zone is located above an oil zone, historically the most common water shutoff method is cement squeeze or mechanical packers. However, gelant injection has been used frequently to treat these problems with greater success because gelants can flow into porous rock, whereas cements and particulate blocking agents are filtered out at the rock surface.

When treating radial flow problems using gels or similar blocking agents, hydrocarbon zones *must* be protected during gelant placement. Otherwise, the blocking agent will probably damage the hydrocarbon zones.

10.7.3.4 2-D Coning: Hydraulically fractured production wells (SI No. 6)

When production wells are hydraulic fractured, the fracture often unintentionally breaks into water zones, causing substantially increased water production. Gelant treatments have significant potential to correct this problem. These treatments rely on the ability of these gels to be placed in the rock matrix adjacent to the fractures and to reduce permeability to water much more than that to hydrocarbon.

In these matrix rock treatments, gelants flow along the fracture and leak off a short, predictable distance into the matrix rock of all the zones (water, oil, and gas). Success for such a treatment requires that the gel reduce permeability to water much more than that to hydrocarbon in the treated matrix rock.

10.7.3.5 Faults or fractures crossing deviated or horizontal Wells (SI No. 8)

Deviated and horizontal wells are prone to intersect faults or fractures. If these faults or fractures connect to an aquifer, water production can jeopardize the well. Often, the completions of these wells severely limit the use of mechanical methods to control fluid entry. In contrast, gel treatments can provide a viable solution to this type of problem. However, conventional gelant treatments are not the desired form of remediation in this case. In a conventional gelant treatment, a fluid gelant solution is injected that flows down the well into the target fracture or fault and *also* leaks off into the porous rock around the wellbore and the fracture or fault. The resultant gel may plug or severely restrict water entry into the fracture or fault. Unfortunately, the gelant will *also* flow into the exposed hydrocarbon bearing rock all along the well during the placement process. Consequently after gelation, oil productivity can be damaged as much as water productivity. Alternatively, a formed gel can be pumped down the well and selectively placed in the fracture. The gel formulation may exist as an uncross-linked fluid at the wellhead, so long as significant gelation occurs before the gelant reaches the oil zone. Then, because formed gels do not enter or flow through porous rock, damage to oil productivity can be minimized. In contrast, the gel can extrude selectively into and plug the fracture or fault. When the well is returned to production, gel remaining in the wellbore can often flow back to the surface.

10.7.3.6 Naturally fractured reservoirs (Problems 7, 9, and 10)

Some of the most successful gel treatments were applied to reduce water and gas channeling in naturally fractured reservoirs. Preformed gel works best in case of naturally fractured producer treatment. Due to high viscosity and elasticity, the gelant does not intervene into oil matrix but preferentially seals the fractures. LMW ringing gel is the preferred material. During the injection well applications, large volumes (e.g., 10,000–40,000 bbls) of HMW elastic gel needs to be pumped. However, sizing of these treatments to date has been empirical – dictated primarily by perceived economic and operational limitations.

Gel treatments to reduce injector–producer channeling in naturally fractured reservoirs should be applied both in injection and production wells.

10.7.3.7 Three-dimensional coning (SI No. 11)

Gel treatments have an extremely low probability of success when applied toward three-dimensional coning problems occurring in unfractured matrix reservoir rock. When treating coning problems, a common misconception is that the gelant will only enter the water zones at the bottom of the well. In reality, this situation will occur only if the oil is extremely viscous and/or the aqueous gelant is injected at an extremely low rate (to exploit gravity during gelant placement). In the majority of field applications to date, the crude oils were not particularly viscous, and gelant injection rates were relatively high. Consequently, one must be concerned about damage that polymer or gel treatments cause to hydrocarbon-productive zones.

Even if a polymer or gel reduces k_w without affecting k_o, gel treatments have limited utility in treating 3D coning problems.

In contrast to the very limited potential of polymers and gels in successfully treating 3D coning, these treatments have much greater potential for successfully treating "two-dimensional coning" where vertical fractures cause water from an underlying aquifer to be sucked up into a well.

10.7.3.8 Injector–Producer channeling in unfractured reservoirs with cross-flow (SI No. 12)

Gelant and gel treatments are expected to be ineffective for treating injector–producer channeling in unfractured reservoirs where fluids can cross-flow between zones. It is recognized that near wellbore blocking agents are ineffective in these applications. Even if the blocking agent could be confined only to the high permeability channel, water quickly cross-flows around any relatively small plug. The only hope for blocking agents in these applications exists if a very large plug (i.e., plugging most of the channel) can be selectively placed only in the high permeability zone. Unfortunately, existing gelants enter and damage all open zones in accordance with the Darcy equation and basic reservoir engineering principles. Penetration and damage caused to the less-permeable zones is greater for viscous gelants than for low-viscosity fluids.

Also, penetration and damage caused to the less-permeable zones is greater when cross-flow can occur than when cross-flow cannot occur. Traditional polymer floods provide a more cost-effective and reliable solution for this type of problem.

10.8 Conclusion

1. When addressing excess water production problems, the easiest problems should be attacked first, and diagnosis of water production problems should begin with information already at hand.
2. Conventional methods (e.g., cement, mechanical devices) normally should be applied first to treat the easiest problems – i.e., casing leaks and flow behind pipe where cement can be placed effectively and unfractured wells where flow barriers separate water and hydrocarbon zones.
3. Gelant treatments normally are the best option for casing leaks and flow behind pipe with flow restrictions that prevent effective cement placement.
4. Both gelants and preformed gels can be applied to treat hydraulic or natural fractures that connect to an aquifer.

5. Treatments with preformed or partially formed gels normally are the best option for faults or fractures crossing a deviated or horizontal well, for a single fracture causing channeling between wells, or for a natural fracture system that allows channeling between wells.

6. Gel treatments should not be used to treat the most difficult problems – i.e., three-dimensional coning or channeling through strata with cross-flow without absolute confidence and understanding of the reservoir geology and geometry.

11

Introduction to Artificial Lifts

11.1 Background

In the extraordinary process of formation of oil and gas deep under the earth's crust, followed by their migration and accumulation as oil and gas reserve, a great amount of energy is stored in them. This energy is in the form of dissolved gas in oil, pressure of free gas, water, and overburden pressure. When a well is drilled to tap the oil and gas to the surface, it is a general phenomenon that oil and gas comes to the surface vigorously by virtue of energy stored in them. Over years of production, the decline of energy takes place. At one point of time, the existing energy is found insufficient to lift the adequate quantity of oil to the surface. From that time onwards, man-made effort is required and this is what is known as artificial lift. In other words, artificial lift is a supplement to natural energy for lifting well fluid to the surface.

Therefore, the flow of oil from the reservoir to the surface can be fundamentally dichotomized as self-flow period and artificial lift period. When a self-flowing oil well ceases to flow or is not able to deliver the required quantity to the surface, the additional energy is supplemented either by mechanical means or by injecting compressed gas.

11.1.1 Purpose of artificial lift

The purpose of an artificial lift is to create a steady low pressure or reduced pressure in the wellbore against the sand face, so as to allow the well fluid to come into the wellbore continuously. In this process, a steady stream of production to surface would result.

In other words, maintaining a required and steady low pressure against the sand face, which we call steady flowing bottomhole pressure, is the fundamental basis for the design of any artificial lift installation.

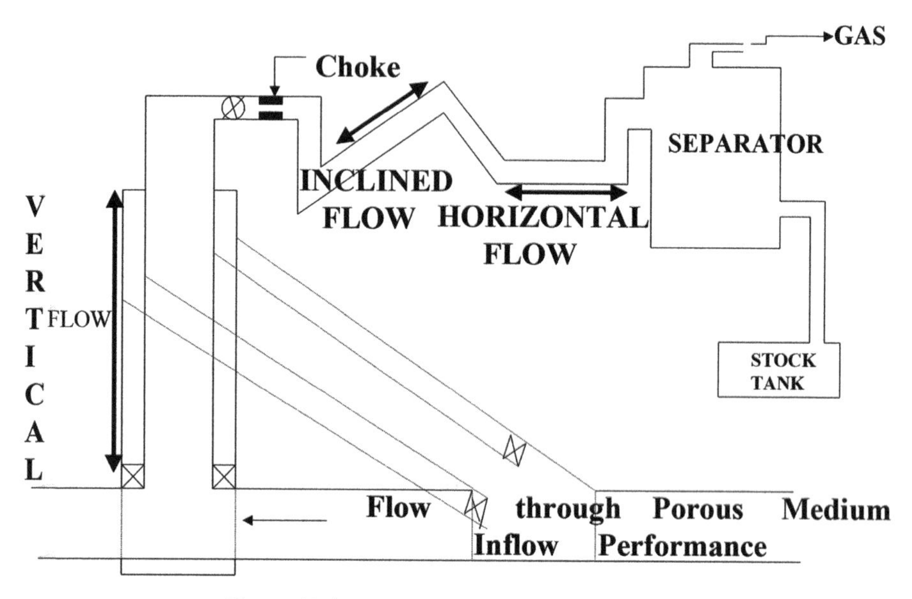

Figure 11.1 Representation of multiphase flow.

11.1.2 Path sectors influencing design of artificial lift system

Broadly four main sectors influence (Figure 11.1) the design and analysis of the artificial lift system. The first and second are the reservoir component from the periphery of drainage area to around the wellbore and then from around the wellbore to the wellbore that represents the wells' ability to give up fluids into the wellbore. The third component of flow path is the entire tubing in the vertical/inclined/horizontal path, which includes all systems like downhole artificial lift equipment, sub-surface safety valves, non-return valves, etc. The fourth component includes the surface flow path, which consists of length and diameter of flowline, valves, bends, wellhead, chokes, manifold, separator, etc.

Any change in the relevant parameters in any of the four sectors influences the parameters of other sectors. The required changes of parameters should be made till the flow gets steady.

For designing or functioning of artificial lift systems, it is important to ascertain two areas affecting production from an oil well. They are inflow performance relationship (IPR) and vertical lift performance (VLP) of the existing or envisaged well completion components, like tubing size, well deviation, restrictions, etc., of the well.

11.1.3 Productivity index and inflow performance relationship (IPR)

The productivity index (PI) of the well is generally denoted as "*J*." It is given by

$$J = \frac{Q}{P_r - P_{wf}}$$

where:

Q = Total quantity of fluid
P_r = Reservoir pressure
P_{wf} = Flowing bottomhole pressure in the wellbore against sand face

In fact, J is not a constant value, but it varies with the type of reservoir, type of drive mechanism, production rate, time of production, cumulative production, perforation density, skin, sand bridging, gas coning, infill wells on production, etc.

In order to define PI more correctly, the concept of inflow performance relationship (IPR) is introduced to define the liquid inflow in the wellbore. It is basically a straight line or a curve drawn in the two-dimensional plane, where X-axis is q, the flow rate and Y-axis is P_{wf}, flowing bottomhole pressure. Therefore, the concept that J is always a constant is not correct. PI here can be described as just a point on the IPR curve. The following are some of the typical IPRs being mainly influenced by different reservoir drive mechanisms.

11.1.4 IPR in Case of active water drive

Out of all types of reservoir drives, water drive is regarded as the strongest. However, the intensity differs in different types of water drive reservoirs. Some are moderately weak and some are strong, like the edge water drive is weaker than the bottom water drive. In bottom water drive, when the oil pool is underlain by a large aquifer of dynamic source, reservoir pressure is generally not mellowed at all with the advancing years of production – that is, the reservoir pressure practically remains constant and is not influenced by cumulative production. In this case, the IPR curve will simply be a straight line, i.e., the IPR curve will provide only one value of PI.

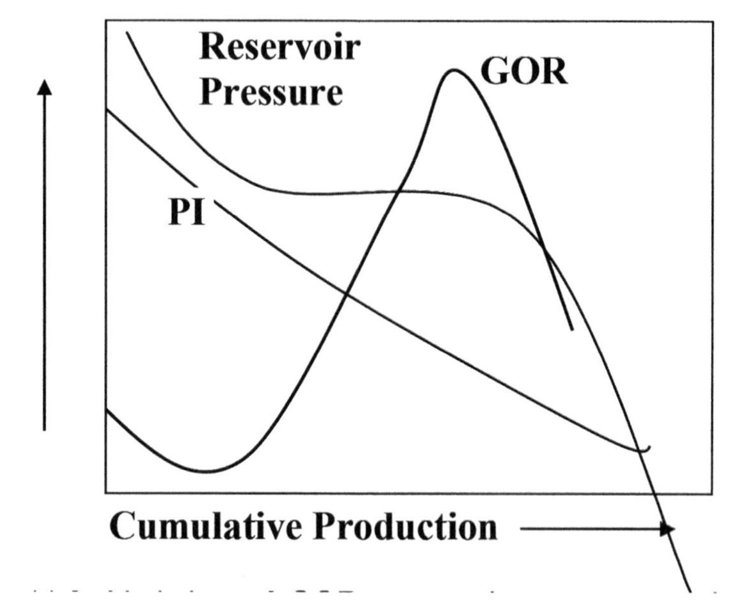

Figure 11.2 Variation of GOR, reservoir pressure, and PI with respect to cumulative production.

11.1.5 IPR in Case of solution gas drive

This type of drive is also called an internal gas drive or depletion drive. This is the least effective drive mechanism. If excessive drawdown is created, it results in an increase of permeability to gas and a corresponding decrease of permeability to liquid; thereby the ability of the well to deliver liquids is greatly reduced. Generally, the reservoir pressure for this type of reservoir declines at a very fast rate, and accordingly, it influences the pattern of the IPR curve (Figure 11.2).

11.1.6 IPR in Case of gas cap expansion drive

This drive mechanism is also called segregation drive because of the state of segregation of oil zone from gas zone, where oil zone is overlain by gas zone called gas cap. Also, as production continues, the gas cap swells and because of this the drive is also known as gas cap expansion drive. This type of reservoir drive mechanism is more effective than solution gas drive and less effective than water drive. Therefore, the profile of IPR curve for gas cap expansion drive lies somewhere in between those for solution gas drive and water drive (Figure 11.3).

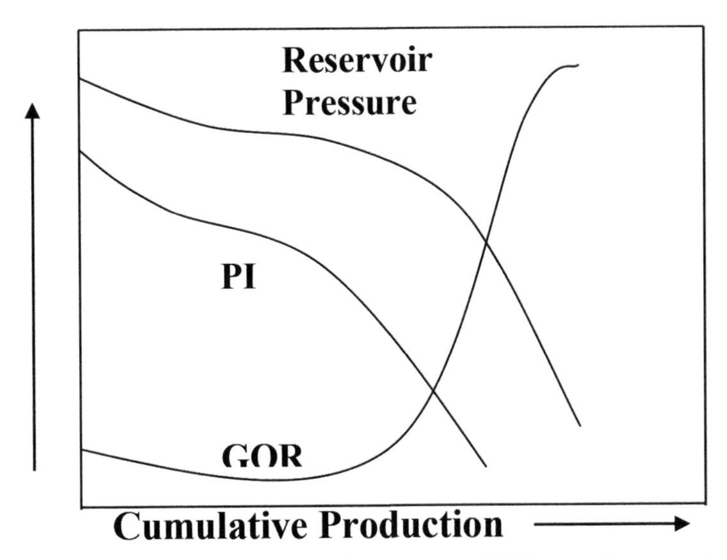

Figure 11.3 Variation of GOR, reservoir pressure, and PI for cumulative production.

11.2 IPR When P_r > Bubble Point Pressure (Saturation Pressure)

Up to a point B in the profile, AB is a straight line representing constant PI. At B, the gas separation starts in the reservoir. With more drawdown, i.e., by further dropping-in of bottomhole pressure, more and more gas will come out, and this affects the flow of liquid due to generation of more gas around the wellbore.

11.2.1 Vogel's work on IPR

A publication by Vogel in 1968 offered an extraordinary solution in determining the inflow performance curve for a solution gas drive reservoir for flow below the bubble point, gas cap drive reservoir, or any other types of reservoir having reservoir pressure below bubble point pressure. Vogel's performance curve is generated with the following equation, which is known as Vogel's equation.

$$\frac{q_o}{q_{\max}} = 1 - 0.2\left\{\frac{P_{wf}}{P_r}\right\} - 0.8\left\{\frac{P_{wf}}{P_r}\right\}^2$$

He plotted dimensionless IPRs in two-dimensional planes.

Here, *X*-axis represents $\frac{q_o}{q_{max}}$ and *Y*-axis represents $\frac{P_{wf}}{P_r}$ (both are dimensionless quantity).

The minimum and maximum values of $\frac{q_o}{q_{max}}$ and $\frac{P_{wf}}{P_r}$ in each case is 0 and 1.0.

When $\frac{P_{wf}}{P_r} = 1, \frac{q_o}{q_{max}} = 0$ and when $\frac{P_{wf}}{P_r} = 0, \frac{q_o}{q_{max}} = 1$.

11.3 Other Methods of Generating IPR

During the derivation of the equation, Vogel made the assumption that the flow efficiency is 1.00, indicating that there is no damage or improvement in the well. Standing expanded the Vogel's equation by introducing a comparison chart that includes flow efficiency values greater or less than one, which correspond to well damage or improvement, respectively. Fetkovich opined that oil well also behaves like gas wells so that IPR equation being used for gas well will also be applicable for oil wells. Detailed discussion of these methods is not in the scope of this training and can be found elsewhere in the literature.For all practical purposes, Vofel's IPR is reported to give reasonable results in case of oil wells.

11.4 Vertical Lift Performance (VLP)

A series of pressure drops occur along different flow paths of a well while producing to the surface. Pressure drop occurring in the completion string (tubing) constitutes the major portion of the total pressure drop. Generally, multiphase flow (oil, gas, and water) exists in the tubing. Estimation of this multiphase flow pressure drop depending on the phase fractions of different phases constitutes the vertical lift performance (VLP) of a flowing well.

Vertical multiphase flow pressure traverse is extremely important to select the completion string, predicting flow rates and design of artificial lift installation.

It is essentially the sum of three contributing factors, viz.:

- Static gradient or hydrostatic gradient.
- Friction pressure gradient or simply friction gradient.
- Acceleration pressure gradient or simply acceleration gradient.

The other factors like viscosity, surface tension, and density have also been included up to a certain specific limit.

Various flow regimes like bubble, slug, and mist flow occur depending on the velocities of different phases in the mixture.

The most important correlations for predicting pressure loss in vertical flow are:

1. Duns and Ros
2. Orkiszewski
3. Hagedorn and Brown
4. Winkler and Smith
5. Beggs and Brill
6. Govier and Aziz

These correlations are, in general, used judiciously for all pipe sizes and for any field.

All the correlations are based on certain common assumptions like

- Fluid must be free from emulsion.
- Fluid must be free from scale/paraffin build-up.
- Mashed or kinked joints should not exist in the tubing.
- Flow patterns should be relatively stable.
- No severe slugging should occur.
- Fluid (oil) should not be very viscous.

In order to have access to the multiphase correlation by the oil field design engineers, multiphase correlations as developed by different authors are available in two forms.

i) In the form of a set of pressure-depth working curves.

ii) In the form of computer solutions.

Both are very useful. A computer solution provides the design in no time. However, field engineers can acquire a fair idea when they apply working curves to solve problems.

There are several publications of multiphase flowing pressure curves viz. (1) Winkler and Smith curves in Gas Lift Manual of Camco, Inc., (2) Hagedorn and Brown curves in the book titled "Artificial Lift Methods" by Kermit E. Brown, Prentice Hall, Inc., and (3) U.S. Industries curves in Handbook of Gas lift, etc.

These correlations are useful for

i) Selecting tubing sizes.

ii) To predict when the well will cease to flow, i.e., when the well requires additional gas to be injected at some point in the tubing to make it flow at the desired rate.

iii) Designing of artificial lift systems.

iv) Determining flowing bottomhole pressures from the wellhead pressures and vice versa.

v) Predicting maximum flow rates possible.

As discussed earlier, the main function of artificial lifts is to supplement the reservoir energy so that oil flows to the surface. In flow performance relationship and vertical lift performance together will decide the quantum of energy to be supplemented by an artificial lift.

Now this can be achieved by two means:

1. Either by reducing the pressure drop in the tubing by lightening the liquid column by injecting gas at a pre-determined point. This is gas lift.
2. Or by directly imparting the required energy to the well fluid by means of a mechanical pump down hole.

   SRP, ESP, PCP, Hydraulic Lift, etc. will come in this category.

   Here pump basically bridges the pressure energy gap between IPR and VCP or TIC, so that required energy is available in the well fluid to overcome the pressure drops in the path ahead, so that it can flow to surface.

11.4.1 Planning and selection of lifts

Ideally, planning for an artificial lift must start before a field is developed or wells are drilled. Completion of well, i.e., casing and tubing sizes, must be envisaged to provide optimum production rates by lift at a future date. Following factors are to be considered when selecting the mode of lift.

1. Whether a group of wells or total field will require a lift
2. Productivity of the well

3. Location of well if isolated
4. Single or multiple completions
5. Offshore or onshore
6. Availability of high-pressure gas source
7. Availability of power
8. Weather conditions, extreme heat or cold, high winds, snow
9. Anticipated GOR or free gas
10. Straight or deviated/horizontal hole
11. Produced fluid properties, i.e., water cut, viscosities, H_2S, CO_2, paraffin, scale, temperature solids, etc.
12. Type of reservoir drive, depletion, water drive, or gas cap expansion
13. Technical ability of operating personnel

There may be more than one mode of artificial lift technically suitable in a well or group of wells. However, depending on economic considerations, suitable mode, which suits majority of the above-mentioned conditions, must be chosen.

Kermit Brown has indicated a broad range of rates for primary short-listing of suitable models:

< 100 blpd	–	Any except ESP
100–1000	–	Any
2000–10,000	–	Any except SRP
> 10,000	–	ESR or G/L

Similarly, he has categorized according to the depth of the lift:

< 8000 ft	–	Any
10,000–12,000 ft	–	Any except ESP (due to temp. limitations)
> 12,000 ft	–	Hydraulic pumping

12

Artificial Lift Methods

12.1 Various Modes of Artificial Lift

1. Gas lift
2. Sucker rod pump (SRP)
3. Electrical submersible pump (ESP)
4. Progressive cavity pump (PCP)
5. Jet pump

12.2 Selection Criteria for Artificial Lift Method

Selection of the appropriate and economical artificial lift method is imperative for the long-term profitability of most producing oil wells. With several factors influencing such a selection, it becomes a complex task. Artificial lift mode capabilities and the well productivity are required to be perfectly matched so that an efficient lift installation results. Design considerations must commence before drilling a well or a group of wells. To obtain optimum rates by artificial lift at some date, sufficient tubular clearances should be provided. Application of any of the lift techniques will also depend on whether a group of wells will be put on lift or only a single well needs artificial lift.

The type of lift may be influenced by whether the wells are conventional or multiple completions. Multiple completions present several problems. Often, here, the choice of lift method may be determined not by optimum design but by the physical limitations of the well. Producing location is yet another factor governing the choice of lift method. Capacity to withstand load and the limited space of offshore platform largely governs the type of the artificial lift system. The best artificial lift method for onshore may not be practical for offshore locations. Again, the choice of a proper lift method for marginal fields especially in the offshore is a difficult exercise. Severe weather conditions like extreme heat or cold, high winds, dust, or snow may limit the choice of lift. Corrosion again is very important in the selection of lift methods. Produced solids such as sand, salt, and formation fines along with paraffin asphaltene are also important factors for the selection of lift mode. Depth and temperature of producing zone and the type of hole deviation are important considerations while considering the application of a type of lift. Reservoir characteristics must be considered for the section of lift. For example, in a depletion drive reservoir, in the initial period of exploitation, high production is expected. An artificial lift may not be required at this stage. However, if it is decided to lower the lift before the start of production, the type of artificial lift and design considerations must be anticipated for operating the lift at a later date. A comparatively rapid decline in production with the rapid decline of reservoir pressure is one of the important characteristics of the depletion drive field for artificial lift selection and design.

In an active water drive reservoir, increasing water cut with the ongoing of production is anticipated. Logically therefore, in future, it requires larger volume of production to maintain the desired oil production. So, the type of lift must be considered in future production volumes as well as present volumes.

In a gas cap expansion reservoir, changing gas–oil ratios with oil production affects the type and size of artificial lift. More and more quantities of gas production take place with time. The increasing amount of gas gradually lowers the artificial lift efficiency. The choice of lift must take into account the anticipated maximum GOR/free gas during the life of the reservoir. Thus, the proper selection of any artificial lift mode depends upon several factors, as described above.

Some important aspects of the most commonly used modes of artificial lift are discussed below.

12.3 Gas Lift

12.3.1 Introduction

Gas lift term is a misnomer. In fact, liquid gets lifted with the aid of gas. High-pressure gas injected through annulus enters the tubing and it helps in lightening the tubing liquid column. This in turn reduces the flowing bottom-hole pressure against the sand face and results in attaining the desired liquid production from the wells. Before gas lift was introduced in oil industry as a very effective artificial mode of lift, a similar form was in vogue as early as in the eighteenth century. Water was being lifted with the help of air. Air was conveyed through tubing and water received on the surface through tubing–wellbore annulus. The same system of lifting, i.e., with the air, was adopted by oil industry in the beginning for lifting oil. It continued in this fashion up till around the mid 1920s. People started realizing the problems involved in the use of air as a lifting medium for oil, as mixing of air with hydrocarbon not only may form explosive mixture but also causes corrosion because of the presence of oxygen. So, from then onwards, compressed natural gas or high-pressure natural gas is being used in general to lift oil.

Early applications of gas lift adopted the simple "U"-tube or pinhole principle in producing oil from shallow wells. Then, with the advent of gas lift valves, the gas lift application could be extended to deeper wells.

The gas lift system is now broadly classified into two categories:

1. Continuous gas lift.
2. Intermittent gas lift.

12.3.2 Continuous gas lift

The basic principle underlying the natural flow and continuous gas lift is the same. The only difference between them is the source of gas. In the case of natural flow, gas comes into the wellbore either along with oil or in the dissolved condition in the oil, whereas, in the latter case, the gas is conveyed down the hole and is injected into the oil body. That is why continuous gas lift can be seen as an extension of the self-flow period of oil well.

The basic principle of continuous flow gas lift is to inject the gas in the oil body at some predetermined depth at a controlled rate to aerate the oil column above it, and as a result the density of oil column gets reduced to a point where a flowing bottomhole pressure for a desired rate of production is sufficient to lift the oil to the surface. Thus, oil is produced continuously from the well. It is also generally intended and the accepted practice that in the continuous gas lift, only one valve will be accomplishing the gas injection work and that this valve should be as deep as possible as per the available normal gas injection pressure. This valve is termed as "operating valve." The valves above it are used to unload the well to initiate the flow from the reservoir. Once the gas injection begins through the operating valve, the upper valves, termed as "unloading valves," are closed. In case there is disruption in gas injection, the well will be loaded. So, when gas lift is resumed, the well is required to be unloaded with unloading valves.

12.3.3 Intermittent gas lift

In intermittent gas lift, sufficient volume of gas at the available injection pressure is injected as quickly as possible into the tubing under a liquid column and then the gas injection is stopped. The volume of gas expands, and in the process, it displaces the oil on to the surface. So, the assistance of flowing bottomhole pressure is not required when gas displaces oil. Static bottomhole pressure, flowing bottomhole pressure, and productivity index of the well govern the fluid accumulation in the tubing.

In this system, a pause or idle period is provided, when no gas injection takes place. In this period, the well is allowed to build up the level of liquid, which depends upon the reservoir pressure and PI of the well. Then again, the next gas injection cycle is initiated to produce oil. In this manner, as the name suggests, intermittent gas lift works on the principle of intermittent injection in a regular cycle. It is to be noted that in the cycle, injection time should be as short as possible, so that a large volume of gas can be injected quickly underneath the oil slug. As a result, oil slug above the point of gas injection will acquire the terminal velocity (maximum velocity) within the shortest possible time, which would minimize the liquid fall back in the tubing string. Less fluid fallback will not only increase production but also help reduce the paraffin accumulation problem in the tubing, if oil is paraffinic in nature. In light of the above discussion, it can be comprehended that continuous gas lift system should be employed when well has a moderate to high reservoir pressure and PI. An intermittent gas lift system should be deployed when the well has a poor PI and low reservoir pressure. Thus, intermittent gas lift provides comparatively much lower volume of oil production than that of continuous gas lift.

12.3.4 Gas lift valve mechanics

A gas lift valve is analogous to a downhole pressure regulator. The surface areas of the gas lift valve are exposed to tubing and casing pressures. So, in response to casing or tubing pressure, the gas lift valve opens, which allows injection gas to enter the production string to lift fluid to the surface.

In the course of improvement of the gas lift system, several types of gas lift valves were developed. Probably the differential type of valve was a very early development and this type of valve was very prevalent before World War II. Advent of metallic bellow for making the gas lift valve has revolutionalized the gas lift system. The bellow operated nitrogen pressure loaded gas lift valve is the most common type of gas lift valves being used by oil industries.

In ONGC oil fields, whether it is in offshore or onshore, casing pressure operated, nitrogen loaded, unbalanced type gas lift valve is only being used.

The components include a valve dome for charging, bellows, a valve stem with a ball tip, a valve port or seat, and a reverse flow check valve.

12.3.5 Valve types

A. Casing-pressure/injection-pressure-operated unbalanced nitrogen-charged valves

As the name implies, the casing-pressure-operated gas lift valves operate predominantly with the injection pressure in the casing. So, the larger surface of opening and closing mechanism, i.e., the bellows area, is directly exposed to the casing pressure. That is, the casing pressure acts on the bellows and tubing pressure on the downstream side of the seat.

B. Tubing-pressure-operated unbalanced nitrogen charged bellows

As the name implies, the fluid-operated gas lift valves operate predominantly with the pressure of tubing. So, its larger surface of opening and closing mechanism, i.e., the bellows area, is directly exposed to tubing and not the casing pressure. That is, the tubing pressure acts on the bellows and the casing pressure on the downstream side of the seat.

12.3.6 Reverse flow check valve

A reverse flow check valve is either coupled with the gas lift valve or in-built with the gas lift valve. Its function is to prevent the backflow of fluids from the tubing to the casing. The backflow of fluids from the tubing to annulus is not desirable because:

1. The backflow of fluid has to be stopped during the setting of hydraulic packer with gas lift valves in the tubing string.
2. It may damage the gas lift valve seats.
3. It may result in accumulation of sand, etc., above the packer, making the servicing of well with workover difficult.

12.3.7 Gas lift mandrel

Gas lift mandrel is the port of tubing string. It houses the gas lift valve and the check valve.

There are two general types of mandrels in use – one for conventional or for fixed valve and the other is for wireline retrievable valves. In the conventional mandrel, gas lift and check valves are fitted on to the exterior side of the mandrel with the valve attachment lugs.

The mandrel for wireline retrievable valve is of a different type. The gas lift valve is housed inside instead of being on to the outside. The outer shape of mandrel's tubing body looks oval shaped with its eccentric end having box tubing connections. It has a pocket welded inside it in the eccentric portion, which is intended to house the gas lift valve. The pocket has drilled holes to connect the pocket bore with the tubing, i.e., with the mandrel and separate drilled holes to connect with the inside of the tubing. The eccentric form of the mandrel is required to ease the wireline job for the selective setting and retrieval of the gas lift valve.

12.4 Sucker Rod Pump

12.4.1 Introduction

Sucker rod pump, abbreviated as SRP, is a very old technique in the oil industry for lifting of crude oil from the wells, and, in fact, it is the most widely used mode of artificial lift system in the present-day scenario. SRP operated by beam pumping unit is more versatile and more common among other types of operating SRPs.

Although the sucker rod pumping system operation appears very simple, resembling a simple reciprocating tube well pump, in actual field practice, it has been found to be a very complex one owing to very deep installation of pump, lifting of a mixture of oil, gas, and water, which we technically term as multiphase fluid and other several factors, like rod/tubing-stretch/contraction, fluid viscosity, speed of pumping unit, length of stroke, etc. The various factors, which contribute to the complexity of the pumping, must be thoroughly studied by the design engineer and therefore he needs to be very familiar with the distinguishing features and the complex function of sucker rod pumping system.

Therefore, a superficial knowledge on the subject of sucker rod pump is not enough to understand operational complexities of pump and to make the running of the sucker rod pump efficiently.

It is, therefore, a necessity for a sucker rod pump engineer to have an in-depth knowledge of the total SRP system. Once an engineer knows fully the significance of the elementary principles of the pumping system, he will then make himself/herself familiar with the complex functioning and distinguishing features of each part of the system and as such he/she will be in a position to operate the pumping system in a fool proof manner. As a straight forward and simple strategy, let the whole pumping system be presented under three broad units, namely:

1. Surface unit.
2. Sub-surface sucker rod pump.
3. Sucker rods.

It is important, in brief, to visualize the motion of each of the units before they are described in detail and how they tie them together into a unique pumping system. With the help of a prime mover, say an electric motor of comparatively low rpm (like 720 rpm), a rotating motion is generated. This rotating motion is then passed on to the surface unit by the V-belt transmission system. It effects a reduction in rpm. Thereafter, the onward rotating motion is further reduced to about 1:29 with the help of a double reduction gear box of the pumping unit. This very low rotary motion (say 6 rpm) is then converted with the help of different components of pumping unit to linear motion at the polished rod.

This linear reciprocating motion is then transmitted to sub-surface sucker rod pump through the sucker rods, which is the linkage of the surface unit and sub-surface pump. In this way, a sucker rod pump operates and lifts well fluids to the surface from the well.

12.4.2 Pumping units

Pumping unit cum prime mover at the surface converts the rotary motion of the prime mover into the reciprocating/vertical motion with the help of several link arrangements. Majority of this pumping operation, worldwide, utilizes walking beam pumping unit and so it is named as beam pumping unit.

The structural parts of a conventional beam-pumping unit are as follows:

1. Walking beam
2. Horse head
3. Saddle bearing
4. Equalizer bearing
5. Equalizer
6. Pitman arm
7. Wrist pin or crank pin bearing

8. Crank
9. Counterweight
10. Crankshaft
11. Double reduction gearbox
12. Unit sheave
13. Sampson post
14. Ladder
15. Bridle (wireline hanger)
16. Carrier bar
17. Electric motor
18. Motor sheave or motor pulley
19. V-belt
20. Belt cover
21. Brake, its link and handle
22. Reducer gearbox
23. Pumping unit base
24. Motor base
25. Grouting nuts and bolts

When these are assembled, horsehead end of the walking beam hangs over the X-mass tree of the well, as such, the polished rod clamp on the polished rod is rested on to the carrier bar that is rigidly attached with the wireline hanger (briddle), which in turn is attached with the horsehead. The horse head that has a curvature-shaped surface and flexible hanger (i.e. briddle) together ensure that the polished rod is made to move in a vertical direction only. The other end of the surface unit has the prime mover, the pulley of which is connected to the gearbox reducer pulley with the help of a v-belt.

12.4.3 Sub-surface pump

It is a sub-surface reciprocating pump, actuated by the up and down motion of sucker rods, which is a connecting link between the surface unit and sub-surface pump. Its feature resembles a reciprocating tube-well water pump. It has five main components:

- Barrel
- Plunger
- Standing valve
- Travelling valve
- Pump seat or nipple

A conventional pump consists of a fixed "barrel" and a moving plunger, with a "standing valve" fitted at the barrel end and a "travelling valve" fitted at the plunger end. The word "travelling" implies that the valve moves or travels along with the plunger. The standing valve is fixed with the stationary barrel, hence the word "standing." Both the standing and travelling valves are unidirectional, implying, both allow fluid to pass through them in the upward direction only. The fluid will not pass through them in the downward direction. The pump seat or nipple seals the annular area between the pump barrel and tubing and thus prevents the pumped out fluid falling back into the pump intake again. The pump seat is also having other features, which makes the pump to get locked in that depth.

12.4.4 Types of sub-surface sucker rod pumps

The sub-surface sucker rod pumps are mainly categorized in two principal groups.

a. Insert (or rod) pump.
b. Tubing pump.

In the insert or rod pump, the barrel, plunger, traveling, and standing valve are the integral parts of the entire sub-surface assembly and are run as a unit on the sucker rod string.

In the tubing pump, the working barrel is run as part of the tubing and is placed at the desired depth. The standing valve is then dropped into the well followed by running in plunger along with sucker rod strings and is placed inside barrel.

Insert (or rod) pump is the conventional choice and is more commonly used. As a general rule, tubing pumps are used where greater liquid volumes are required to be pumped out. Tubing pumps are especially useful for pumping from inclined wells.

12.4.5 Sucker rod string

Sucker rod string, in fact, is the vital link between the sub-surface pump and the pumping unit. These sucker rods are available as per API in three different lengths – 25', 30', and 35'. These are connected to each other up to the depth of the pump. These are solid steel bars with forged upset ends with threads on it. API has standardized these solid steel sucker rods. The diameter of the rod body ranges from 1/2″ to 1 1/8″ with 1/8″ increments. Usually the rod body of diameters 5/8″, 3/4″, 7/8″, and 1″ are very common. At each end of the sucker rod, there is a short square section just before the sucker rod pin thread, which facilitates the use of sucker rod tongs for connecting two sucker rods. These sucker rods are generally available with one coupling fitted at one end.

12.5 Electrical Submersible Pump

12.5.1 Introduction

The electrical submersible pump (ESP) is basically a high volume mode of lift system. The minimum capacity of ESP is known to be around 200 bpd and the maximum capacity is as high as 90,000 bpd. ESP, in some situations, can provide the maximum possible drawdown by bringing the annulus level to the top of the perforations.

The ESP is extremely suitable for a very low viscosity liquid. This pump is also used to pump high viscosity fluids and can operate in gassy wells and high temperature wells.

The prime mover of the electrical submersible pump is the downhole motor coupled directly with the pump. The motor rotates at 3475–3500 rpm for 60-Hz power and 2900–2915 rpm for 50-Hz power.

Under normal operating conditions, the operating life of ESP can be expected from 1 to 3 years, with some units operating even over five years. With the recent improvement of ESP metallurgy and cable technology, some manufacturers claim that ESP run life is even more than five years under normal operating conditions. One of the main reasons of failure of ESP is the breakdown of insulation at the downhole in the cable, cable joint, motor, etc.

12.5.2 Applications

If we hark back a few decades from now, we find that ESP had application in lifting water from water well and thereafter ESP was used to produce an oil well with high water cut. Perhaps the first version of ESP was brought out in the name of REDA. The full form of REDA is "R" stands for Roto, "E" for Electro, "D" for Dynamo, and "A" for Arutunoff, after the name of a Russian Scientist who had first patented the pump for lifting water from under the ice-covered Alaska region.

ESP is currently producing many offshore and onshore wells, especially where wells are high producers. In ONGC, ESPs were tried in three wells of western offshore, but their life periods were short mainly due to different types of downhole electrical faults either by itself or engineered by the mechanical malfunction in the pump assembly. Currently, no offshore well in ONGC is operating on ESP. However, a good number of wells of ONGC's Assam field are being produced with ESP. Companies like M/s. REDA, Centrilift, TRICO, etc., manufacture electrical submersible pumps.

12.5.3 Surface and sub-surface components of electrical submersible pumps

Electrical submersible pumps consist of many types of equipment and their allied parts. The equipment can be broadly segregated as surface and downhole components.

The downhole components are as follows:

1. Electric motor
2. Protector
3. Pump intake/gas separator
4. Multistage centrifugal pump
5. Pressure sensing instrument (PSI)
6. Pothead extension power cable
7. Power cable
8. Centralizers
9. Cable bands
10. Check valve
11. Bleeder valve (drain valve)
12. Pump top substitute connection
13. Lower pigtail

Surface components are as follows:

1. Wellhead
2. Mini mandrel
3. Upper pig tail
4. Surface cable
5. Junction box
6. Booster
7. Switch board
8. Power transformer

12.5.4 Standard performance curves

The standard performance curves are the most important graphs for ESP design. For every type of ESP, in its dynamic flow condition, standard performance charts are drawn. The abscissa (horizontal axis) indicates the capacity of pumping in bbls/day or m^3/day and the ordinate (vertical axis) indicates liquid head to be generated, BHP (brake horse power) and efficiency of ESP. For a small type of pump with different rpm, the standard performance chart will be different. So for every pump performance, chart rpm is mentioned.

The head capacity is plotted with the head either in feet or in meters. For a simplistic approach, fresh water of density 1 gm/cc has been used to generate the performance curve by the pump manufacturing companies.

Also, the performance curve is plotted either with 100 stages of pump or with single stage; as such, some companies prefer the former one and some the latter.

In the pump performance curve, at very low rate or almost zero rate, the head capacity to be developed by the pump is maximum and as the pumping volume increases, the head capacity decreases and at one point of pumping, the head capacity is zero. It means there will not be any lifting of liquid in the tubing beyond that volume. Keeping an eye on the pump efficiency, every manufacturer has drawn a maximum and minimum range in each performance curve; as such, all ESPs are supposed to operate within this range. The space between the maximum and minimum lines is called the recommended range.

13

Oil and Gas Processing

13.1 Introduction

The produced well fluid is generally a mixture of oil, salt, water, and natural gas. Separation of gas and liquid is the first processing step. The separated liquid is further subjected to dehydration and desalting to remove salt and water. The pure oil is metered and dispatched to refinery. The water removed from oil known as effluent water is treated to meet the environmental system disposal requirements and is then disposed of. The gas separated from oil is further treated for dehydration, liquid recovery/LPG production and finally sent to consumers.

Field processing of crude oil involves three objectives:

1. Separation of crude oil from free and emulsified water or brine and entrained solids (mainly sand).
2. Stabilization of crude oil (removal of free or dissolved gases to the extent that it is safe for transportation).
3. Removal of impurities from the crude oil and any separated gases to meet the sales/transport/reinjection specification requirement.

The selection, design, and the operations of the processes used to separate oil from water and gas depend on the well fluid properties. The collection and processing of well fluids is carried out in group gathering stations (GGSs) on onshore and on process platforms in offshore.

The gathering system consists primarily of pipes, valves, and fittings necessary to connect the wellhead to the separation equipment. The gathering system may contain one or more pipelines with branches to each well or it may have separate flowline for each well, which are connected to a group header or test header. The processing facilities at the gathering station separate the liquid from the gas and process them to meet their sales requirements.

13.2 Separation

The separation of well fluids into gas and liquid components is the first step in the processing of well fluids and it is carried out in the pressure vessel called separator.

A separator can be referred in the following ways:

1. Oil and gas separator
2. Stage separator
3. Trap
4. Knock-out vessel/knock-out drum
5. Flash chamber/vessel
6. Scrubber

Oil and gas separator, stage separator, or trap is used interchangeably to refer to the oil and gas separator. These vessels are normally used near the wellhead, manifold, or tank battery to separate fluids produced into the oil and gas or liquid and gas. Knock-out drum may be used to remove only water from the well fluid or (oil + water) liquid from the gas.

13.2.1 Flash chamber/vessel

It normally refers to the conventional oil and gas separator operated at low separator with the liquid from the high pressure being flashed into it. It is normally a second or third stage of separation with the liquid discharged from the flash chamber to storage.

13.2.2 Filtration method

The oil–water mixture can be separated by filtration using the porous media.

13.2.3 Principles of separation

The physical separation of gas and liquids (oil and gas) based on three principles:

1. Momentum change
2. Gravity settling
3. Coalescing

Any separator may employ one or more of these principles, but the underlying factor is that the phases should be immiscible or have different densities for separation.

- Momentum change
  Fluid phases with different densities will have a different momentum. If two-phase stream changes direction sharply, greater momentum will not allow the particles of the heavier phase to turn as rapidly as the lighter fluid, so the separation occurs.
- Gravity settling
  Liquid droplets will settle out of gas phase if the gravitational force acting in the droplet is greater than the force of the gas flowing around the droplet.
- Coalescing
  Very small droplets such as fog or mist cannot be separated by gravity. These droplets can coalesce to form larger sized drops that will settle by gravity.

13.2.4 Section of the separator

Regardless of the shape or size, the separator consists of the following four major sections:

- Primary separator section
- Secondary separator section
- Coalescing section
- Liquid collection section
- Primary separator section

The primary separator section is used to separate the bulk portion of the free liquid in the inlet stream. The inlet component used to carry out the bulk separator is the diverter. These devices take advantage of the inertial effects of the centrifugal force or an abrupt change in the direction or momentum. Normally inlet baffles are used for this purpose.

13.2.5 The secondary section or gravity separation

The secondary section, also known as gravity separation, is specifically intended to harness the power of gravity to improve the separation of the droplets that are trapped within. It comprises a section of the container where the gas flows at a comparatively slow speed with minimal disturbance. Several internal components utilized for this purpose include wave breakers and defoaming plates.

13.2.5.1 Coalescing section

The coalescing section utilizes a coalescer or mist extractor, which consists of a knitted mesh pad or a series of vanes. Very small droplets cannot be separated by gravity alone. In this section, they are made to impinge on the surface where they coalesce to form the larger droplets, which settle, by gravity.

13.2.5.2 Sump or liquid collection section

The sump or liquid collection section acts as a receiver for all the liquid removed from the gas in the primary, secondary, and coalescing sections. Depending on the requirement, the liquid section should have a certain amount of surge volume, for degassing or slug catching over the minimum liquid volume required for control operations.

13.3 Classification of Separators

Separators can be broadly classified as:

1. Two-phase separators – used to separate gas and liquid.
2. Three-phase separator – Used to separate gas from liquid and water.

Separator configuration:

- Vertical separators
- Horizontal separators
- Spherical separators

13.3.1 Vertical separators

Vertical separators are usually used when the gas-to-liquid ratio is low or the total gas volume is low. In the vertical separator (Figure 13.1), the fluid enters the vessel striking the baffle plate, which initiates the primary separation. Liquid removed by the inlet baffle falls to the bottom of the vessel. The gas moves upward usually passing through the mist extractor to remove the suspended and then the dry gas flows out.

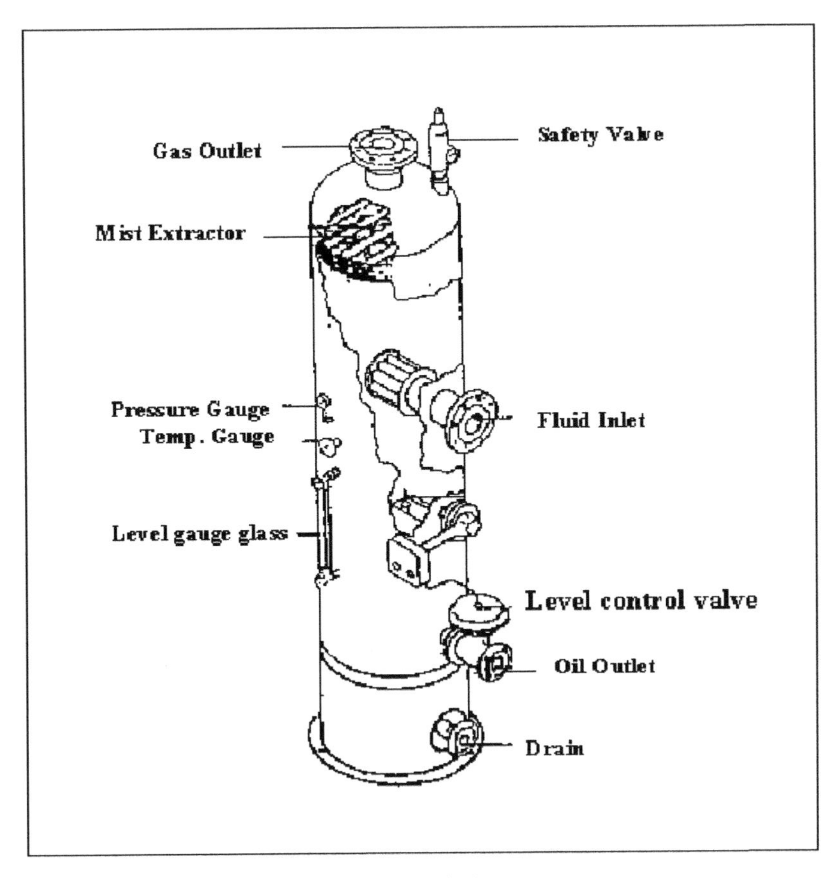

Figure 13.1 Vertical separator.

Liquid removed by the mist extractor is coalesced into larger droplets, which fall through the gas to the liquid section at the bottom of the vessel. The liquid collected is removed through a level control valve.

13.3.2 Horizontal separator

Horizontal separator (Figure 13.2) is used where the large volumes of fluids and large quantities of gas are present with the liquid. The greater liquid surface area in the separator provides optimum conditions for the releasing entrapped gas. The liquid, which has been separated from the gas, moves along the bottom of the separator to the liquid outlet.

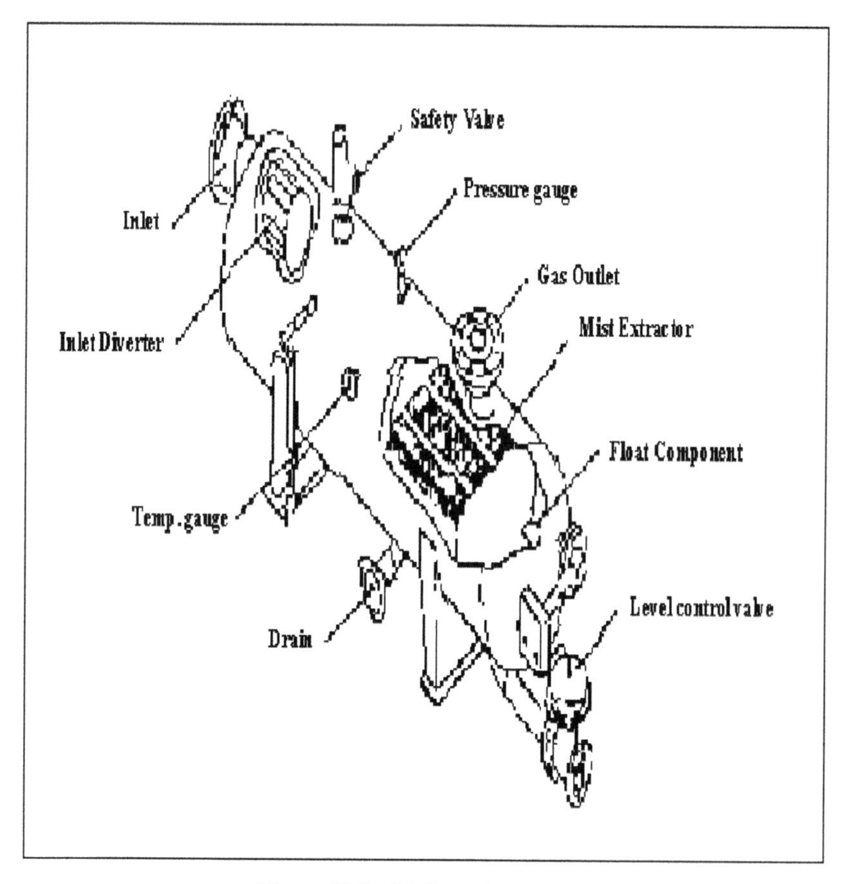

Figure 13.2 Horizontal separator.

13.3.3 Spherical separator

This is used where extremely large volumes of gas form extremely small volumes of liquid. It is mainly used as scrubber and seldom used at well site as an oil and gas separator. These are occasionally used as high-pressure separators.

13.4 Dehydration and Desalting of Oil

The presence of water in the crude oil presents the single largest problem to the producers. Difficulties experienced due to the water are:

1. Marketability of the oil is adversely affected.
2. Corrosion problems are enhanced.

3. Production costs are increased.
4. Frequently results in reduced net oil production.
5. Higher pressure drop in the pipeline.

13.4.1 Condition of water in petroleum

Water may be carried mechanically with the oil in either or both of the two forms.

1. Free water: The microscopic droplets remaining in suspension due to viscosity of oil and severe agitation, which settles out when left undisturbed.
2. Emulsified water: In this, the water is in the state of emulsion in which the droplets of microscopic size are more or less in permanent suspension. This water does not separate when left standstill. The petroleum emulsion contains the electrified charge. As the charge increases, the stability increases, and the reluctance of the globules from coalescing may be due to this phenomena.

13.4.2 Dehydration of crude oil

Dehydration of crude can be categorized as:

1. Removal of free water
2. Removal of emulsified oil

1. **Removal of free water**
   Free water can be separated from liquid hydrocarbons by giving sufficient settling time and absence of turbulence without the aid of heat. This can be achieved by:
   a) Free water knock-out
   b) Three-phase separator

2. **Emulsified water**
   There are three different methods of treating the emulsion, which are widely used in the industry.
   a) Mechanical:

   This can be further divided into :

   • Gravity settling

- Thermal treating
- Centrifuge method
- Filtration method

b) Electrical method

c) Chemical method

A) Mechanical methods

1. Gravity settling

In this process, the suspended water settles at the bottom due to higher specific gravity. It is necessarily carried out in large tanks to provide sufficient residence time for settling of water

B) Heat treatment

Sometimes the thermal treatment is an effective means of accomplishing dehydration and at times it is used in combination with other processes. Heat brings out separation due to the following reasons:

i) reduces the viscosity of oil;

ii) changes the interfacial tension;

iii) higher temperature changes the water into vapor.

Heating of crude oil results in the loss of lighter hydrocarbons. It is not effective in dehydrating crude oil completely unless the crude emulsion is instable or partially demulsified. It is therefore used in combination with other methods like chemical and electrical.

C) Electrical method

When high AC potential is applied, the dispersed water droplets coalesce into larger droplets, which settle down due to the action of gravity.

D) Chemical method

Organic and inorganic chemicals are used for demulsifying the crude oil. It is important that the chemical is tested in laboratory before any field application is done.

None of the methods described earlier are complete in itself and in practice a combination of methods are normally used to dehydrate the crude oil. In ONGC, heater treaters are most commonly used in onshore fields for dehydration.

13.5 Heater Treater

In the heater treater, thermal, chemical, and electrical methods are used for the removal of water from the crude oil. The heater treater is a horizontal cylindrical vessel. The heater treater can be divided into the following sections:

i) Inlet degassing section

ii) Heating section

iii) Differential oil control chamber

iv) Coalescing section

i) Inlet degassing section:

Oil mixed with demulsifying chemical enters the heater treater through degassing section, above the fire tubes. Free gas is liberated from the flow stream and is equalized across the entire degassing section. The degassing section is separated from the heating section by baffles. The fluid travels downward from the degassing area and enters the heating section under the fire tubes through multiple oil distributors.

ii) Heating section

The heating part comprises a fire tube that is bent at a 180° angle. Gas serves as the primary fuel source for heating purposes. The weir is used to maintain a steady liquid level. The oil is introduced into this area from the lower part of the degassing chamber and flows through the heater located at the bottom, where it undergoes a cleaning process. The water and solids pass are removed from the oil stream at no cost. The water level is regulated via a level control valve. The oil and the water that is carried along with it rise upwards from the surrounding areas, passing through the fire tubes until they reach the necessary temperature. Elevating the temperature of the oil results in the liberation of some gases. Subsequently, this gas combines with the unconfined gas from the inlet portion and is expelled via a pressure control valve.

iii) Differential oil control chamber

The heated fluid transfers from the heating section over affixed weir into a differential oil control chamber, which contains a liquid level float. The fluid travel downward to near the bottom of differential oil control chamber where the opening to the coalescing section is located.

iv) Coalescing section (electrical chamber)
The heater treater employs a high voltage potential across the electrodes to facilitate the coalescence of water droplets during the last stage of processing. The electrodes are suspended from insulated hangers in the upper part of the vessel. The ground electrode is equipped with sturdy steel hangers to provide a secure connection with the steel of the treater. A high-voltage transformer is installed externally to supply electricity to the electrodes. The transformer typically use a primary voltage of 240 V to generate a secondary voltage of 16,500 V. The high-voltage secondary is linked to the electrodes. When the oil and water that is carried along with it come into touch with the electric field in the grid region, the process of water coalescence occurs. The water descends back into the water reservoir while the purified oil ascends to the surface, where it is expelled via the clean oil control valve.

13.5.1 Demulsifier dosing

The demulsifier is mixed in the oil stream before it enters the heater treater. Chemical is injected using the dozing pump or by gas pressure.

13.5.2 Crude oil desalting

The water content of crude oil can be brought to the acceptable limit by electrostatic treatment; however, the salinity of oil might be still higher than the acceptable limit. Desalting operation is carried out to reduce the salinity. It consists of the flowing steps:

- Addition of dilution (fresh) water to the crude
- Mixing of dilution water with crude
- Second-stage dehydration to separate crude and brine phase

13.5.3 Pour point depressant

A large number of crude have very high pour points below which the ambient temperature goes and pumping of these becomes difficult. Organic polymers are available, which when mixed with crude reduces the pour point of the crude and enables it to flow at lower temperatures. Such chemicals are called pour point depressants.

13.6 Storage Tanks

Once the incoming fluid has been separated into oil, water, and gas, these must be either stored or transported. Gas is normally not stored because of the expensive storage facilities. Oil and water are stored in tanks for limited time.

Welded steel tanks are used extensively for the storage of crude oil. Various sizes of tanks are used depending on the storage requirements. The most common oil storage tank is the vertical upright cylindrical tank made of steel sheet.

13.7 Pumps and Compressors

Pumps are used at different purposes in the oilfield. Large pumps are provided to move the oil through piping from point to point during its gathering, processing, and pipeline transport. Crude oil is stabilized by stage separation to or near the atmospheric pressure for transportation and storage. Accordingly, main line oil pumps are required for pipeline transport. Centrifugal, reciprocating, and metering pumps are commonly used in crude oil handling, chemical injection, and other applications.

For the transportation of the crude oil, both centrifugal pumps and reciprocating pumps can be used. The selection of the pump depends on the properties of the crude oil, its quantity, and the process parameters.

Once the gas has reached the process facilities and its pressure is reduced in separation or other vessels, the pressure must be raised again to move it from one place to another.

Gas compressors are machines that raise the pressure for its transportation through pipeline. Centrifugal gas compressors and positive displacement (reciprocating type) gas compressors are used depending upon the quantity and other process conditions.

13.8 Metering of Oil and Gas

The measurement of crude oil is normally carried out by using any of the following meters:

1. Positive displacement meter
2. Turbine meter
3. Orifice meter
4. Ultrasonic meters
5. Mass flow meters

13.8.1 Positive displacement (PD) meter

As the name suggests, the PD meters are unique because they mechanically isolate and pass a known volume of liquid with every revolution. The trapped volume is defined by sliding vanes, oval gears, reciprocating pistons, etc. One type of sliding vane meter is the lease automatic custody transfer (LACT) unit, which is the traditional standard for measuring crude oil. In this meter, the measuring element measures the volumetric flow by separating the flow into segments or measuring chambers of known volume and counting the segments.

13.8.2 Turbine meters

Turbine meters operate under the principle of a free running rotor (coaxially mounted on the pipe centerline) with an angular speed proportional to the actuating fluid velocity. Inline turbine meters are the standard method for custody transfer of high volume, low viscosity crude. Other uses include natural gas measurement and water.

13.8.3 Orifice meter

An orifice meter consists of an orifice plate (thin plate having a circular hole located centrally) installed in the circular pipe with the pressure taps, upstream and downstream. The pressure difference measured by the pressure taps is used to calculate the flow rate of the fluid.

Ultrasonic meters and Coriolis mass flow meters are the recent development. The ultrasonic flow meter exploits the difference in the transit time when an ultrasonic beam travel with and against the beam. The uses include natural gas measurement in large pipelines.

13.9 Wastewater Treatment

Proper disposal of wastewater is essential. To prevent the contamination of surface and groundwater, the wastewater is not permitted to be discharged without proper treatment. The treatment consists of reducing the oil and sediment content within the permissible limits. Effluent treatment plants are installed to meet the effluent water disposal requirements.

14

Frontier Technologies

14.1 Coalbed Methane

14.1.1 Introduction

The technology of drilling coalbed methane reservoirs has evolved significantly since the first coal wells were drilled in the 1950s. Coal has several unique properties that must be considered during planning, drilling, completing, and producing a coalbed methane well. The variety in this unique reservoir and geologic characteristics has resulted in a corresponding diversity in the problems and practices of drilling coalbed methane wells. These are largely related to the different conditions existing for shallow coals and deeper coal seams. Generally, the primary drilling problems encountered when drilling coalbed methane wells are (1) excess water flows, (2) gas kicks caused by overpressure coal seams, (3) wellbore stability/coal sloughing, and (4) formation damage.

Coal reservoirs are not homogeneous and within the same area can have different reservoir and geologic characteristics. In addition, the reservoir is typically not a single coal seam but rather numerous thin coal seams interbedded with shale and sand. Reservoir characteristics such as permeability, pressure, saturation, gas content, relative permeability, and geologic characteristics such as coal rank, thickness, and natural fractures determine the productivity potential of the reservoir. However, the drilling completion and production techniques affect the deliverability, recovery effectiveness, and ultimate recoverable gas reserves. Although coal seam reservoirs are unique, properly applied conventional drilling technology is successfully being used to develop the resource.

14.1.2 Drilling considerations

Identifying and understanding the various geological and reservoir parameters of coal will have a significant impact upon the optimum design of the drilling, completion, stimulation, and production operations. The appropriate completion technique (hydraulic fracturing or open-hole cavity) depends upon the specific reservoir characteristics, and each technique requires a different drilling procedure. After completion, a coal reservoir typically produces large volumes of water requiring dewatering to reduce reservoir pressure allowing gas to desorb from the reservoir. Therefore, the wellbore configuration and completion technique must also conform to these special production considerations.

It is critically important that the initial pilot exploration well(s) is designed and implemented to include coring, logging, and drill-stem testing operations to collect all of the necessary geologic and reservoir date. Typically, pilot exploration wells are drilled using conventional techniques, with the primary objective of maintaining wellbore stability and pressure control by drilling with mud. The wellbore is typically cased and cemented across the coal steam interval(s), which is later perforated, tested, and fracture stimulated. Once the specific reservoir and geologic characteristics have been determined for the coal seam, optimization of drilling and completion of subsequent development wells can be initiated.

Much of the data required to determine the productivity of a coalbed methane well can be collected only during the drilling operations from coring, drill-steam testing, and open-hole logging. Therefore, a pilot exploration program consisting of a specifically designed well(s) to collect as many reservoirs and geologic data as possible should be implemented. This data includes reservoir thickness, fluid volumes in-place, and fluid/rock properties affecting flow.

One technique that has been utilized is the use of high-speed slim-hole drilling and coring technologies. The slim-hole system uses high-speed, small-diameter drag bits with a high-precision automated bit advance system to continuously core from surface to total depth. On site, geological and geophysical core analysis can be performed in addition to well tests for the determination of permeability.

14.1.3 Reservoir damage during drilling operations

Numerous studies have documented the swelling characteristics of coals by liquids and gases. While cleats have a very low porosity (1%–2%), they are responsible for the permeability of the coal and the flow path for the methane gas contained in the matrix. Therefore, even a slight swelling of the coal matrix by the sorption of fluids can lead to a relatively large reduction in cleat porosity and permeability. Research has shown that the sorption of liquids by coal, and the consequent swelling of the matrix and reduction in permeability, is highly irreversible.

Various drilling and completion techniques are currently being used to prevent or reduce formation damage. One technique that is commonly used is to drill the coal seam under balanced pressure (less than reservoir pressure) using air/mist, aerated mud, or formation water. This prevents the drilling fluid, chemical additives, and drilling solids from being injected into and plugging the cleat system of the coal. The coal is then completed open hole without casing or cement to prevent reservoir damage caused by the cement. Although this open-hole technique is successful in some areas, it is not practical in all coal basins, particularly in coal basins, which contain numerous thin coal seams and where multiple seam completions are desired. Therefore, drilling and completion practices vary significantly due to the variety of coal's unique reservoir and geologic characteristics.

14.2 Drilling Techniques

1. Cased hole with hydraulic fracture stimulation
2. Open hole
3. Open hole with cavity
4. Horizontal drain hole drilling

14.2.1 Cased hole with hydraulic fracture stimulation

Drilling is carried out with casing shoe set below the last coal seam. All the seams are behind casing. Casing is cemented and all the subsequent operations related to well tests, fracturing, etc., are carried out through casing for each seam. Typically, water-based drilling fluid is used.

14.2.2 Open hole

In this case, drilling is carried out through the coal seams interval and then casing is set above the entire coal interval, thereby completing the coal seams interval in an open hole.

14.2.3 Open hole with cavity

In this case, drilling is carried out and casing is set above the coal seam interval. Then the coal seam interval is drilled under balanced pressure with formation water, air, mist or foam. A cavity is then created in the coals naturally by the coal to slough into the open-hole wellbore. Some of the first CBM wells were drilled and completed in open hole in San Juan basin in USA. The open-hole section used to be under-reamed to prevent and remove suspected reservoir damage. However, these wells faced problems related to coal fines accumulation in wellbore. Gravel packs installed were also not useful as they use to get plugged very often. The cavity completion is an improvement upon this old technique.

14.2.4 Horizontal drain hole drilling

Horizontal drilling into coal seams is not a new idea. In-seam horizontal drilling that involves drilling into a coal seam at the coal face in underground mining operations has been used by coal mine operators for a number of years. It has been observed that horizontal wells drilled in coal mines perpendicular to the face cleat direction produce greater volumes of gas than horizontal wells drilled in any other direction.

Horizontal drain hole drilling, which involves drilling a horizontal well from a vertical wellbore, is also not a new idea. This type of horizontal drilling technology has been recently used to develop various naturally fractured reservoirs. Several horizontal drain holes have also been successfully placed in coal seams. Drilling horizontal wells in coal reservoirs requires special attention to wellbore stability, hole cleaning, and directional control. One of the primary advantages of a horizontal drain hole is that it can be oriented perpendicular to the maximum permeability direction and substantially increase production in areas having sufficient permeability anisotropy. Horizontal drain hole drilling experience in several western basins has indicated that gas production is no better than fracture-stimulated wells in the same area.

14.3 Reservoir Engineering Aspects of Coalbed Methane

The reservoir engineering aspects of classical coalbed methane production involve the physics of desorption, diffusion, and two-phase Darcy flow of gas and water. Generally speaking, early-time flow rates are controlled by the flow capacity of the coalbed and its ability to produce water since the coal cleats (natural fractures) are 100% water saturated at initial conditions. As a result, initial gas flow rates will be zero or very low. Once gas saturation in the cleats begins to increase, gas flow rates will peak and will then be controlled by the gas desorption rate. The micropore surface area upon which the gas is adsorbed is large, on the order of 1 million square feet per pound of coal. The generally accepted theory of gas storage in coal is that the gas is adsorbed in a monolayer on the micropore surfaces and the amount adsorbed is dependent upon the coal rank, temperature, and pressure. The coalbed reservoir is both the source of gas generation and the vessel in which the gas is stored. At similar depths and pressures, coalbeds contain from two to four times the amount of gas contained in a conventional gas reservoir.

14.3.1 Adsorption isotherm

Combining a monolayer adsorption assumption with the shape of the sorption isotherm has led most geologists and engineers to accept the theory of Langmuir as the operative sorption theory for methane on coal.

Mavor et al. (1990) describe the proper measurement techniques for developing the Langmuir isotherm. The measurement is made in the laboratory on crushed coal particles of approximately 100 meshes. Crushing to this degree minimizes the time required for the adsorption process. The sample is placed in a Boyle's law porosimeter, elevated to reservoir temperature, saturated to the equilibrium moisture content of the coal, and degassed. Beginning at low pressure and utilizing the gas expelled from a volumetric chamber of known size, the gas is then injected under pressure into the chamber containing the coal particles. After allowing time for equilibrium to occur, the pressure is re-measured, and by difference in the initial and equilibrium pressures, the amount of adsorbed gas can be computed from the known injection volume. An isotherm like the one shown in Figure 14.1 will be developed from a repetition of this procedure. This second-order curve is described by the following equation:

$$V = V_L \times P/(P + P_L).$$

The coal at reservoir pressure is determined from the adsorption isotherm. However, the fact that the canister gas contents are often lower than the isotherm values is a persistent problem in coalbed methane work. If the two values of gas content are truly different, the ramifications can be considerable.

A re-examination of the isotherm illustrates this so-called gas undersaturation inconsistency. Coal B is capable of holding more methane at initial reservoir conditions than the canister tests indicate. They postulated there mechanisms that could explain this situation. The simplest mechanism is that another gas(es) is occupying a portion of the sorption surface, thereby making it unavailable for methane molecules. A second explanation is that the coal is gas-wet and that due to the difference in capillary pressure, the pressures of the wetting and non-wetting phases could account for the gas desorption pressure being less than the hydrostatic pressure of the water in the cleats. A third possible explanation is that pulverization of the coal may add surface area to the coal in the laboratory isotherm measurement that cannot be accessed *in situ* due to small pore throats. We will dwell no further upon these observations; rather, our purpose is to understand how this inconsistency bears on gas recovery.

First, we need to understand the meaning of the Langmuir constants. The Langmuir volume V_L is the volume of gas adsorbed as the pore pressure approaches infinity, i.e., V_L is the asymptotic value of volume as pressure increases. The Langmuir pressure P_L is the pressure at which volume $V_L/2$ of adsorbed gas occurs. The shape or degree of curvature of the isotherm is important as well. The more nonlinear the curve, the more pressure drawdown work must be done to desorb an incremental volume of gas from the isotherm. Once the pressure approaches the sharp concave downward portion of the curve, more gas is desorbed for each pressure decrement of drawdown. This means that we would like to find those coals that have nearly linear (or only slightly nonlinear) isotherms. Such coals would provide a more even distribution of gas production over the entire range of pressure drawdown. When these desorption mechanics are combined with the effects of pressure drawdown away from the wellbore, we begin to understand why more linear isotherms are preferable. The effects of highly nonlinear isotherms are exacerbated in low-permeability coalbeds where lack of pressure drawdown as a function of radial distance becomes extreme. In such cases, the coalbed reservoir in the near-wellbore vicinity is experiencing any significant gas desorption. Finally, if one-half of all the desorbed volume is held below pressure P_L, then we must hope that P_L will be reasonably high, say several hundred pounds per square inch. Otherwise, we will not be able to lower the flowing bottom-hole pressure of the pumped well to values far enough below P_L to desorb the remaining $V_L/2$ of gas from the coal.

We can now revisit the undersaturated condition discussed in the preceding paragraph. Examination shows Coal A on the isotherm to be gas saturated at 13.1×10^6 N/m^2 (900 psia) of pressure. Coal B illustrates how an undersaturated condition would appear. Pressure drawdown to 6.2×10^6 N/m^2 (900 psia), the so-called desorption pressure P_D, would have to be accomplished before any gas could be desorbed from Coal B.

For gas-saturated reservoirs, the initial pressure and the desorption pressure are coincident. Because the degree of curvature of the isotherm is small at higher pressures, the amount of drawdown required to reach P_D can be considerable. Since the gas contained by the sorption isotherm above P_D is not available *in situ*, we cannot expect to recover it. So the operating range for gas recovery is from P_D, down to the absolute minimum flowing bottom-hole pressure attainable for the well and is maximized only at the wellbore itself.

The magnitude and shape of the pressure drawdown profile across the radius of drainage is dictated by the absolute permeability, the gas-water relative permeability, the porosity, and the two-phase flow as predicted by Darcy's law. The more radially distant a point is from the wellbore, the lower the drawdown pressure, and consequently the farther out on the isotherm (to the right) the reservoir pressure resides. Therefore, higher incremental volumes of gas are being desorbed nearer the wellbore with little or no gas (depending now on the gas saturation condition of the coalbed) being desorbed from points in the reservoir near the outer limit of the radius of drainage.

Because of multiphase flow, no analytic solutions for the equations for gas-water flow in porous media exist; therefore, reservoir simulation becomes necessary if one wants to investigate the relationship between coalbed properties and rates of gas recovery. Experience has shown that 50%–60% cumulative gas recovery is possible from saturated coalbeds having absolute permeabilities of 49.35×10^{-6} cm^2 (5 md) or greater. Gas recovery may be somewhat lower in the undersaturated case (Young et al., 1992). The time required to obtain these recoveries depends upon the absolute permeability, the relative permeability relationship, and the porosity of the coal. Higher porosities indicate that more water must be pumped before significant gas production can be developed. This fact directly impacts operating costs since in many cases produced water cannot be disposed of cheaply at the surface.

However, there is a note of caution: production of natural gas from conventional dry gas reservoirs can be commercially accomplished at permeability levels much less than 49.35×10^{-6} cm^2 (5 md) when massive hydraulic stimulations are performed. In the case of coalbeds, even large stimulations may not completely solve the problem. This is because a classic coalbed well is first a water well and only later, when dewatered, becomes a gas, and this metamorphosis cannot be ignored. Water has a much higher viscosity than gas and requires much higher absolute permeabilities so that water removal early on can be accomplished. This is precisely the reason many coalbed methane wells are initially stimulated with hydraulic fractures. When considering the time and amount of dewatering necessary to initiate economic rates of gas production, the cost of dewatering may be prohibitive.

14.4 Coalbed Methane Well Stimulations

14.4.1 Introduction

Coalbed methane wells require stimulation to effectively connect the wellbore to the reservoir. The stimulation of CBM wells is almost a necessity because unstimulated flows of gas and water are often very low. This may reflect high positive skin factors of 3–30, presumably due to mud invasion, cement damage, or only a few effective perforations. Hydraulic fracturing is the most common form of stimulation. However, alternative "stimulation" techniques, such as the open-hole cavity completion, have proved to be remarkably successful in coalbeds in parts of the San Juan basin.. A variety of stimulations have been tried, and these are summarized as follows:

1. Gel fracture treatments. These stimulations are conducted through casing perforations in coal seams. High fracture conductivities are achieved by using 12/20 mesh sand to concentrations of 10 ppg.
2. Water fracture treatments. In order to avoid gel damage to the formation, fracturing treatments have been conducted using water as fracture fluid, plus 12/20 sand to concentrations of a few ppg. In some cases, gas production is found to be greater than offset wells with gel fracture treatments, and the water fractures are cheaper by half.
3. *Sand-less water fracture treatment.* In the Black Warrior basin, water fracture treatments have been performed without sand, using ball sealers to open up more seams. Although their gas production may not be as good as wells fractured with water and sand, they are substantially cheaper, and when used to refracture gel-fractured wells, they generally improve gas production.

A slotted-casing completion utilizes a tubing string, a jetting tool, and surface equipment to pump water and sand at high pressure down the tubing to abrade and cut the casing and cement. If desired, slotting allows access in the anticipated vertical fracture direction, which can facilitate permeability measurement and reduce fracturing pressure. Perforations are relatively simple, fast, of low cost, and effective. Perforations are frequently used with one to six shots per foot. Use of 15%–28% hydrochloric acid at the beginning of fracture treatments can be effective at cleaning up perforations and reducing fracture pressures (Lambert et al., 1987). Perforated wells have the least problem with proppant flowback during production.

14.5 Hydraulic Fracture Stimulations

14.5.1 Background

Hydraulic fracturing is used for four primary reasons: (1) to bypass near-wellbore formation damage; (2) to more effectively connect the wellbore to the natural fracture system of the coal reservoir; (3) to stimulate production and thereby accelerate dewatering (to increase the rate of gas desorption); (4) to widely distribute the near-wellbore pressure drop (and reduce fines production).

A cased-hole completion ensures that multiple coal seams can be selectively stimulated. Fracture stimulation treatments in coal generally use a water-based fluid and sand as the proppant.

Coalbed fracturing does differ in some respects from fracturing of conventional formations. Some of the most important differences are that (1) the coalbeds often exist in the form of multiple seams, (2) damage to the formation caused by fracturing gel can be very serious, and (3) fracturing treatments are often accompanied by abnormally high treating pressures, and possibly T-shaped fractures.

14.5.2 Cross-linked gel fracture treatments

HPG cross-linked with borate has been the standard fracture treatment in the first two basins in the US, the northern San Juan basin, and the Black Warrior basin. The typical job involved 30 lb/1000 gal pumped at 30–60 bpm with 12–20 mesh sand, although some use 20/40, with high proppant concentrations to achieve high fracture conductivity.

In most cases, the sand mesh size was 20/40 or 16/30 total sand loading fell in the range of 100,000–200,000 lb, and maximum proppant concentration was 3 ppg or less. The optimum pad percentage has been reported to be between 38% and 50%. Many of these wells have been reported to be good producers.

14.5.3 Foam fracture treatments

Foam fractures are traditionally used in formations that are under pressure, or are sensitive to liquid damage, or to reduce leak off. However, they are more expensive than cross-linked gel treatments. Foam was not common in wells drilled in the overpressure part of the basins. However, it has been found that in areas that are under pressure and undersaturated, nitrogen foam treatments are being tried in several places. In areas that are normally pressured or a little less, there has not been great usage of foam fractures. The conclusion was that a simple water fracture treatment was more effective than nitrogen foam treatments or linear gel treatments. In another project, it was observed that there was no systematic difference in gas production between roam-fractured and water-fractured wells.

14.5.4 Water fracture treatments

The term, as used here, includes "slick" water as well as plain water for the fracturing fluid. In the early part of development in the Black Warrior basin, water fractures constituted 50% of the stimulations (they subsequently were replaced by cross-linked gel-fracture treatments). It has been found by some of the companies that water fractures were more effective than gel or foam fractures, at least in areas of the reservoir where the permeability was high (–20 md). They are also much cheaper – only about half the cost.

In a typical water fracture design, the proppant concentrations were very low (< 1 ppg), with total sand load around 70,000 lb and pump rates of 35–40 bpm.

The conclusion appears to be that the loss in production due to ineffective seam propping is compensated for by the gain in production by avoiding formation damage.

14.5.5 Sand-less fracture treatments

The basic objective and reasons for sand-less fracture treatments is that (1) they are cheaper, and (2) they avoid proppant flowback.

Sand flowback is a problem encountered in CBM wells. It is more severe during the initial time after fracturing and tends to improve with time elapsed since the fracture treatment. The success of sand-less water fracture treatments should be more likely where *in situ* stresses are comparatively low (i.e., shallow depths) and the induced fracture remains open due to self-propping.

The results have been varying from place to place. In some cases, there has not been any improvement over conventional fracturing. However, in a few cases , improved production rates have been reported over original gel fracture treatments by two times, which reflects the likely damage of coal seams due to gel.

14.6 Coalbed Methane Production

14.6.1 Introduction

Coalbed methane has been produced in commercial quantities in the United States since 1981 and has attached worldwide attention as a potential source of cost-competitive gas. Coal is different from other gas reservoirs in three primary ways: (1) gas is stored in the adsorbed state on the surface of the coal; (2) before gas can be produced in significant quantities, the average reservoir pressure must be reduced; and (3) water is usually present in the reservoir and is normally co-produced with the gas. These unique reservoir characteristics require low wellhead pressure (to maximize gas desorption), separation of gas and water at the surface, compression of gas to delivery pressure, and procedures to handle and dispose of produced water. Standard oil and gas production practices must be tailored to the unique characteristics of coalbed methane reservoirs to facilitate commercial production.

14.6.2 Requirement of low production pressure

The Schraufnagel study conducted in 1990 presents two adsorption isotherm curves that quantify the adsorption of the Mary Lee coal found in Alabama. Gas production from the Mary Lee coal only begins when the reservoir pressure drops below 366 psia (2523 kPa) due to its modest undersaturation. The amount of gas remaining in the coal at any location within the reservoir is directly influenced by the pressure within the cleat system. The difference between the remaining gas in the coal and the initial gas content indicates the percentage of gas that has been recovered, which is 50% in this instance.

14.6.3 Production equipment

14.6.3.1 Design of a production system

Surface facilities can account for more than half of the total project investment. Sequential development can allow for design improvements by taking advantage of lessons learned from reservoir characterization, well completions, and surface facility operation. The size of a given development varies from operator to operator and area to area. In each development, water and gas from each well site are transported to a single treating site serving water disposal, gas treatment, central compression, and sales.

Having water disposal facilities on a lower level than the majority of the wells allows gravity to drain the area, thereby reducing pumping costs. Due to low wellhead pressures, two-stage compression is often necessary. Satellite compressor facilities should be at higher elevations to minimize residual water in the gas lines and allow dispersion of compressor exhaust emissions.

14.6.3.2 Typical production well configuration

Successful design and operation of a coalbed gas production facility starts at the bottom of the well. Coalbed methane wells are often drilled through several gas-bearing formations that can be performed in several zones. To reduce reservoir pressure, the borehole is drilled to below the lowest producing zone to provide a sump. Water drains into this sump before being pumped to the surface.

A typical coalbed methane production well is configured with tubing placed inside production casing. These configurations have been adapted from oil field applications and have proven to be very effective. They provide an initial separation of gas and water in the wellbore, thus reducing the need for large surface separators, and allow only small amounts of back pressure down hole. Water is normally lifted by pumping through 2 3/8 or 2 7/8 in. diameter tubing string. Gas is produced up the annulus between the water-production tubing and the casing (4 1/2 or 5 1/2 in.).

Besides producing both gas and water, the wells and surface facilities need to handle solid materials (e.g., coal fines or stimulation sand). The well sump is used to collect solids. This minimizes debris entering pumps or any surface equipment. Additionally, a screen at the pump intake can reduce solids entering the production system but may get plugged with fines and require cleaning. Making slow changes to wellhead pressure during operations, and having sufficient clearance between the tubing and casing, can also reduce the occurrence of fines migration.

14.6.3.3 Dewatering pumps – Artificial lift

Dewatering systems have successfully been used for progressing cavity pumps, gas lift, and electric submersibles. Choosing the appropriate system depends on a number of factors including well depth, pressure, water rate, and gas rate. The systems used most commonly today are the sucker-rod and progressing cavity system work effectively in many situations.

14.6.3.4 Sucker-road pumps

Sucker-rod pumps are relatively simple and durable and require only minor routine maintenance. Special subsurface designs may be required in extremely gassy wells or wells that produce large amounts of sand and fines. A ring-type plunger pump that can tolerate solids, placed in the tubing at or below the lowest production zone has proven cost effective. Slightly undersizing the pump rings can increase tolerance for solids without significantly decreasing pump efficiency. The sucker-rod string commonly has 1 3/4 or 2 in. (45–50 mm) diameter plungers.

In the San Juan basin, pumping units of 160, 320, and 640 sizes (peak torque rating of the gear box) and 64–200 in. (1.625–5.08 m) stroke lengths are used. In the Black Warrior basin, because of shallower depths in the eastern part of the basin, units as small as 40 and 80 (peak torque rating) and 36–72 in. (0.915–1.83 m) stroke lengths are often used.

The pumping action is related to the motion provided by the surface equipment via the rod string. The surface pumping unit can have a conventional beam, air balance, or mechanical balance. The surface unit must be designed for anticipated depth, volume, and load of the well. The motion needed to operate these systems is supplied by a prime mover, usually an electrical or natural gas motor, connected to a gear box to reduce the motor speed.

Pump repair (workover intervals) depends on the amount of water pumped and the history of the well. Most commonly workovers are needed to replace failed rods and pumps worn by abrasion from produced solid material. More workovers are required early in a well's life, but frequency can be reduced to once a year after one to two years of production.

14.6.3.5 Progressing cavity pumps

The pumping action is created by the rotor turning clockwise in the stator. This action moves a series of cavities placed 180° apart and "progress" the fluid from the bottom suction end of the pump to the surface. A continuous seal between the rotor and stator keeps the fluid moving at a rate proportional to the rotational speed of the pump and the volume of the cavity.

Care should be taken so that the pump does not burn out when the well is pumped off. Improved elastomer systems have resulted in reduced pump wear from solids and entrained gas. If there is excessive wear, the entire down-hole assembly may have to be replaced.

14.6.3.6 Gas lift

The system consists of a series of gas lift mandrels or valves in a tubing string that are operated by gas injection from the surface. Down-hole valves should be wire line retrievable so that maintenance costs can be low and gas lift performance can be optimized.

14.6.3.7 Electric submersible pumps

Electric submersible pumps can move large volumes of fluid at both shallow and deep depths. A surface control unit regulates the speed of the down-hole motor and pump assembly using power supplied via cable clamped to the tubing string as it is run in the well. Fluid passes the motor, providing a cooling mechanism as it enters the intake of the centrifugal pump. Therefore, the pump must be placed above the producing zone and the well must never be pumped dry. Though this configuration may put additional back pressure on the coal compared to a pump placed below the coal, electric submersible pumps are primarily used when the wellbore cannot be dewatered by another system.

Variable speed drives increase the versatility of electrical submersible pumps. They have also been improved for better gas/liquid separation and solids handling.

14.6.4 Surface gas and water separators

14.6.4.1 Gathering system option

The purpose of a gathering system is twofold: (1) to transport gas from the wellhead to a central compression station at the highest economic advantage to the producer; and (2) to handle the water by product stream in a cost effective and environmentally acceptable manner.

14.6.4.2 Types of gas-gathering systems

Three types of gas-gathering system designs can be used for coalbed methane. First, treating and compression can occur at each well. Small-diameter, medium pressure piping then delivers the gas to a central compression facility.

In the second system, gas is transported from a small cluster of wells to a satellite compressor via low-pressure gathering lines. After initial treatment and compression, the medium pressure gas is transported to a central sales compressor.

In the third system, wellhead pressures are kept as low as possible – on the order of 5 psig (135 kPa) – and gas is transported via appropriately sized gathering lines to a central compressor facility.

14.6.4.3 Gas-gathering design considerations

Proper design of a gas-gathering system begins with an estimate of the gas rate, requires a definition of operating pressure limits, and must define how to handle moisture in the gas. Where there is limited production history or reservoir data, as well as high uncertainty or variability in gas flow rate, a tradeoff will have to be made between the higher cost and the increased capacity/reduced pressure drop of larger pipe. As it is important to have low wellhead pressure, the gain in reduced pressure drop and potential for extra flow capacity will often outweigh the increased cost of the next larger pipe size.

A critical low-pressure limit is the suction on the compressor. Proper line sizing can maintain the wellhead pressure and suction pressure in an acceptable range. For many low – and medium – pressure operations, polyethylene pipe is an excellent choice for gas lines. Where higher wellhead pressure is present, the gathering system may consist of epoxy-coated steel pipe.

Coalbed gas is usually saturated with water when it reaches the surface. Most of the water should be removed at the wellhead separator. However, condensation in the gathering line is inevitable because the temperature of the buried pipe is normally lower than the wellhead gas temperature. Condensation, which is more severe in winter than in summer, must be collected to reduce pressure losses. Collection points, called drips, are installed at low points along the pipeline to collect condensate. If large amounts of water are anticipated in the gas-gathering system, a slug catcher may have to be installed at the compressor station before the gas is dehydrated to pipeline specifications.

14.6.4.4 Water-gathering lines

Water lines need to be large enough to carry the water with minimal pressure drop. If there is excessive pressure drop, an auxiliary water pump may have to be installed at the well site to move the water through the lines and maintain low production pressures. If flowlines are too large, pipe costs are excessive and solid material or pools of water could accumulate in low spots. The lines should be installed on a level or uniform downward slope, away from the production wells, and be buried below the frost line.

High spots in the water lines should be avoided so that pumping costs can be minimized. Pockets of gas can collect in the gathering lines and prevent water flow. If there is a high point in the disposal line, a gas pressure relief valve may be required so that water is not suctioned back toward the wellhead. Besides releasing gas trapped at high points in the line, these valves also minimize surging caused by compression of gas pockets.

14.6.4.5 Flow measurement options

An awareness of operational issues related to flow measurement will improve measurement accuracy and field management. For example, gas production rates must be measured to accurately gauge gas sales. Measuring produced water rates is required for proper management of this stream. In addition, metering of individual wells for gas rate, water rate, pressure, and temperature is important for proper well management. In wells with multiple production horizons, an estimate of production by zone can also aid production decisions.

14.6.5 Water measurement systems

Accurate measurement of water production at the well site is critical for determining pump and well performance. Three commonly used measurement methods include a positive-displacement water meter, a turbine meter, and the bucket method.

A turbine meter is often installed at the outlet of a pump and provides a more accurate reading than a positive displacement meter. Intermittent flow, two-phase gas and water flow, and debris in the water can damage this meter or cause inaccuracies. Turbine meters are also used at holding ponds and water processing facilities to facilitate water handling. Both positive-displacement and turbine meters have increased accuracy at higher inlet pressure.

14.6.6 Gas measurement systems

Gas flow rates are usually measured at each well and at the central sales point. Primary measurement systems are orifice meters and turbine meters. Rotary or diaphragm meters (normal household gas meters) can also be used, particularly for measuring compressor fuel consumption.

The orifice meter or differential-pressure recorder measures gas flow rate by recording pressure upstream and downstream of an orifice plate inserted in the gas line. The pressure drop through the gas line is a function of flow rate, overall line pressure, temperature, and size of hole in the plate. Standardized procedures developed by the American Gas Association (AGA) are used to accurately calculate gas flow corrected to standard temperature and pressure (AGA, 1985). This is the most common method used to measure gas volumes at the point of sale.

One advantage of an orifice meter is that line pressure, temperature, and differential pressure are continually recorded on a circular chart, providing a permanent record of well gas production history. Few mechanical problems occur for vapor or particulates in the gas stream, and the system requires little maintenance. One disadvantage of orifice meters is that flow rates are subject to human interpretation and surges in production can make the charts difficult to read. However, once the charts are removed, usually weekly, optical scanners can provide an accurate estimate of well production.

The turbine meter measures flow rate and total flow. Gas flows over rotor blades and magnets located in the meter housing pick up signals that are linear to gas velocity. The area between the blades has a set volume; so the number of turns is related to the total volume of gas passing through the rotor. Pressure compensation corrects flow rates to standard conditions, and optional software is available to automate gas flow and total gas volume calculations. Many turbine meters have turn-down ratios of 10:1 or more so that wide variation in gas rate can be tolerated. A disadvantage of turbine meters is the multiplicity of moving parts that can make maintenance expensive.

In addition, turbine meters are more sensitive to liquid and entrained solid production than the orifice meter, which may result in more maintenance and less accuracy.

14.6.7 Measuring production by zone

Most coalbed methane wells produce from more than one reservoir horizon. Proper production management not only requires that gas and water production and wellhead pressure be known for each well but, if possible, some estimate be made of production of gas and water by zone. One method of determining production from multiple horizons is to produce a well until there is constant production from all zones. Then, a bridge plug is used to close off production from a lower zone, and production from the remaining zone proceeds at a new stabilized rate. This procedure may be appropriate in a research setting or a key well but will generally be unacceptable in a large operating field. Changes in gas or water composition have also been investigated as methods to estimate production from different zones.

14.7 Gas Treating and Compression

14.7.1 Gas composition

Typical composition requirements for the gas are shown in Table 4.

When the carbon dioxide concentration exceeds 3%, the impurity must be removed. Processes used to treat coalbed gas streams are the same as those used in other large-scale gas processing facilities. The gas composition of the coal reservoir in a new area must be determined early in the life of the project so that the economics of various gas processing options can be evaluated.

14.7.2 Dehydration

Initial water knockout and mist removal occurs at the well site. A filter screen on the compressor skid separator should be installed so that particulates in the gas stream do not enter the compressor or dehydration system and cause wear to the internal parts.

Skid-mounted glycol/gas contact towers with an integral scrubber and outlet mist pad are normally used for final water removal. Glycol absorption works best at low gas temperatures (765 to 70 °F; 18 to 21 °C) and higher operating pressure. Whether dehydration occurs after final compression, or after an intermediate level of compression, depends on the specific variables encountered in a given operation. Two types of tower internals (packing or trays) are used to assure adequate gas/glycol contact. After removing moisture from the gas stream, the glycol must be dehydrated. This is accomplished by separate regeneration equipment consisting of a boiler, circulation pump, and distillation column.

Fuel quality is important to maintain efficient engine operation; so water and any particulates need to be removed using a fuel filter. If natural gas is being used as the fuel, excess nitrogen, which may be present after a well has been re-stimulated, could cause the compressor to stall if the fuel heating value is reduced excessively.

14.8 Water Disposal

Because water is produced as a byproduct of coalbed gas production operations, water management can represent a large portion of daily operating costs of a coalbed methane project. Methods used to dispose of produced water and treatment cost depend on the amount of water produced and the water quality.

Water production will not begin at all wells in a field at the same time, nor will water be produced at a constant rate during a well's life. These temporal variations need to be considered in the permitting process and in the design of the water disposal system. Estimates of water production rates can be made based on the experience from nearby wells or by coupling permeability measurements into hydrologic or reservoir models.

14.8.1 Review of water treatment options

The most common methods for disposing of produced water are stream discharge and deep-well injection. Other methods include land application, evaporation, off-site commercial disposal, or reuse of water in hydraulic stimulations.

Development of a produced water management plan requires an understanding of the character of the produced water, the working range of available processes, the cost, and a knowledge of treatment constraints for existing legislation. If discharge limits become more stringent because of regulations, a mix of alternative processes may have to be put in place to cost-effectively manage the wastewater stream.

Process options vary from basin to basin and among operators in a given area. Injection wells are widely used in the San Juan basin and are more economical than evaporation when disposal volumes are less than 30,000 BWPD (4700 m^3/day) per facility

14.8.2 Injection wells

When gas fields are developed away from a continuously flowing river system and where water from coalbed methane wells is considered the same as a conventional oil and gas waste stream, other water disposal methods – such as deep-well injection – are required. A large cost component of water injection is the well. One injection well may be sufficient to dispose of water from 10 to 20 production wells. However, deep-well injection may be very costly.

Finding zones that can accept large volumes of water for an extended period is critical to the successful operation of an injection well. Candidate zones must exhibit significant porosity and permeability and show continuity of the reservoir. The volume of water injected depends on the reservoir characteristics and the maximum permitted injection pressure. Premature plugging or a rapid decline in injection rate at the maximum injection pressure can adversely affect the economics of the entire project.

Produced water must be properly treated before injection to minimize increases in injection pressure over time. Also, a sample of water from the injection zone should be compared with the produced water to ensure that there will be no solids precipitation. Filtration to remove micron-size particles is mandatory. Clay swelling caused by changes in water chemistry can result in reduced injectivity, and the presence of carbon dioxide and its potential impact on precipitate formation also needs to be reviewed. Sometimes scale or corrosion inhibitors or bactericides are injected with the water to assure that long-term injectivity is maintained.

14.9 Reverse Osmosis/Evaporation/Other Processes

14.9.1 Reverse osmosis

Reverse osmosis can be operated with no reject (i.e., 100% of the water is treated to a common reduced level) or, alternately, with a fraction of the original fluid rejected as a concentrated brine (on the order of 10%–25% of the original brine volume). Depending on the composition and initial TDS of the inlet salt solution, different separation membranes can be used. For TDS levels less than 20,000 mg/L, the processed stream will be less than 100 mg/L as the average TDS removal is of the order of 95%. However, some species, such as bicarbonate, may only have removal efficiencies in the 50%–80%

range. The principal process limitation for reverse osmosis is membrane fouling caused by suspended solids or oils that may be present at low levels in the produced water. The cost of reverse osmosis depends on the amount of reject, volume of fluid, and initial TDS. Reverse osmosis is relatively a very costly and sensitive process.

14.9.2 Evaporation

In sunny, arid environments, a retention pond with or without spray aeration can be used to evaporate water. Process costs are a function of the amount of water processed and are relatively low.

14.9.3 Other methods

In some instances, particularly for the areas with water scarcity or very high water cost, produced water can be reused in ongoing drilling and completion operations. Because of high water handling costs, this approach was recently used on a coalbed methane project in Pennsylvania (Hunt and Steele, 1992). Care should be taken, however, that the water is compatible with the quality control criteria for the process in which it will be used. If, for example, the water is to be used in a stimulation process, it is important that the residual breaker be neutralized in order to maintain the sand-carrying capacity of the stimulation fluid.

14.10 Production Operations / Reservoir Management

To optimize well field production, pressure and flow data on each well should be gathered on at least a daily basis. Well performance needs to be reviewed and operational changes made to optimize production. Radio transmission of data from individual well sites to a central office facility and computer production databases can keep daily operating costs down and assure that wells or facilities needing the greatest attention are cared for.

Cost-effective operations require that data be collected and organized in an orderly fashion and that operations personnel use time efficiently. Computer data acquisition routines can greatly simplify data management, and radio transmission of well site data to a central computer facility can greatly aid well management and increase efficiency.

14.10.1 Bringing a well on production

There are three operational stages in bringing a well on line. First, water is continually pumped up the tubing at the maximum capacity of the pump, but the annulus (casing) remains closed. During this stage, the water column in the well is replaced with a gas column with little change in surface pressure. If during this stage a vacuum is noted in the casing, the wellhead valve can be opened momentarily to equilibrate pressure. Over time as the water level in the wellbore lowers (and the reservoir pressure near the wellbore reduces), some gas is desorbed.

In the second state, the casing pressure is allowed to rise to some predefined level – say 200 psi (1400 kla) – and the annulus valve can be slowly throttled to initiate gas production. Over a period of perhaps a week, the pumper slowly lowers the casing pressure as the water level continues to decrease. Reducing the surface pressure allows the formation to produce additional water into the wellbore and dewater a region of the reservoir near the well. The objective is to have the surface pressure at its long-term operating pressure level (5–15 psig; 35–100 kPa) when the water level in the wellbore reaches the pump intake.

In the third and final stage of operation, the properties of the formation should control the rate of water and gas production, not wellhead pressures. The water level in the well should be below the lowest producing coal seam, and the annulus wellhead pressure should be as close to atmospheric conditions as the gathering system constraints limit. By first concentrating on removing water from the wellbore and then reducing pressure to the final operating pressure, a small area of reduced water saturation (or a "gas bubble") is formed near the well, and desorbed gas from the reservoir can flow unimpeded into the well.

This slow, careful procedure minimizes the flow of sand and coal fines into the well and may allow the formation of a small gas bubble in the formation near the wellbore. Slightly less gas will be sold during the initial weeks of production, and this procedure can be more manpower-intensive. There is often a degree of urgency to get a well on line. However, over the life of the well, this procedure may minimize workovers and prevent formation damage. This procedure will also achieve cumulative production over the first year as high as or higher than if the annulus remained wide open while the well was being pumped.

14.10.2 Reducing pump failure

The frequency of pump failure can be reduced by increasing down-hole separation to reduce gas production through the pump, using time clocks to place wells on intermittent pumping schedules (i.e., match pump capacity to well deliverability) and reducing back pressure on water discharge lines to the pump torque is reduced.

Knowing the daily performance changes of a well means that subtle changes in production behavior can be charted and appropriate modifications made. Operations data such as last workover date, pump speed, and pump failure are useful when making critical decisions regarding pumping efficiency, pump performance, and the frequency of pump replacements.

14.10.3 Reducing rod failure

Separation failure of the rod string can be reduced in beam pumps by adjusting the pump rate and stroke length to minimize stress. Rod failure can also be reduced if the proper grade and size of rod are selected and rods are made up to the proper torque when placed in the well (initially and after workovers). Time-delayed starting devices can reduce wear on rod parts due to power fluctuations, and failure due to solids or gas lock can be partially controlled by using proper down-hole separation devices. Also, minimizing the deviation of a well during drilling can reduce rubbing and other problems that could lead to rod wear.

14.10.4 Reducing solid material production problems

Modifications to the completion and operational protocols can reduce the occurrence of workovers caused by solid materials. For instance, by drilling a sump with a depth of 100-200 ft (30-60 m), any solid matter that may be generated during the production process can settle below the pump and be separated from the water.

Operational procedures can also affect solids production. Bringing a well on line slowly, without surging, minimizes flow of solids from the fracture into the wellbore and minimizes downtime. The size of the perforating charge can also influence future fines production; e.g., large charges may add metal debris or coal fines to the well. Given the choice of four large charges or eight smaller charges, eight smaller charges are preferable. Fines production can also be reduced by minimizing the time between perforating and stimulation.

14.10.5 Production management

The key to efficient coalbed production from a given field is to keep costs down and determine how to maintain gas production. The challenges lie in determining how to integrate the equipment and operating procedures to meet these objectives. The economics of coalbed methane production depends heavily on daily operating costs, which include maintenance/repair/materials, equipment rental, well servicing, chemicals and treating costs, pumping, and initial choice of production equipment.

Operating costs can vary substantially from project to project and during project life. Costs tend to be higher as facilities are brought online. The use of an automated database to track daily operations can reduce operating expenses and increase the efficiency of operations.

Every well has its unique performance record. It is essential to monitor the historical data to assess whether any action, particularly in terms of cost, is required to enhance the present gas production rate and sustain or improve the economic rate of return. Effective measurement equipment and surface amenities are essential for efficient well management. If the operator possesses a database, they can organize the daily and monthly production data into different completion kinds, geographic regions, or operational zones to modify production profiles.

In making production management decisions, it is important to understand why some wells produce poorly and other wells are good producers. Is poor production a result of poor reservoir conditions, a problem with the completion, or a result of an operational procedure? Is a good well producing at high rates solely due to good geology and reservoir conditions or are there specific operation differences present in these wells or this area that lead to increased gas production? For example, the back pressure may be lower because the configuration of the compressor facilities is different from those of a nearby area. Or, good wells might be "pumped off" while a significant column of water remains in the poorer producers. Does the individual pumper operating in a good area do something different that improves production? Are the workover frequencies of the good wells different from the poorer wells?

The overall integration of the surface facilities to the coal reservoir, the measurement of gas and water rates from each well, the geologic variability of the area, and the ongoing analysis of production data lead to operating decisions that can assure continued long-term cost-effective gas production from a coalbed methane production facility.

14.11 Deepwater Technologies

14.11.1 Introduction

14.11.1.1 What is deepwater?

The definition of deepwater in the offshore oil and gas industry varies from region to region and has changed over time. The Brazilian state-owned company Petrobras is one of the few organizations that have explicitly defined its usage of the term. It classifies fields occurring in water depths (WDs) of 400–1000 m as "deepwater" fields. Those lying below 1000 m are classified as "ultra-deep." For Shell and BP, in contrast, "deepwater" means beyond 500 m; for Exxon, beyond 400 m. For Oceaneering – one of the major global deepwater contractors – deepwater refers to depths more than 3000' (910 m), but it notes that for its client base in the Gulf of Mexico, the adjective can refer to WDs of 300–1200 m. Houston-based contractor Cal Dive, which maintains a strong presence in the gulf of Mexico where it has been active since 1975, has commented that for many years, WD 300 m (1000') was the standard definition. Still, nowadays WD 500 m (1640') has become more commonly used, and this is also the standard now adopted by Halliburton, another major global contractor. Infield Systems/Douglas-Westwood Report-III also has considered 500-m water depth as the deepwater definition. Water depth beyond 1500 m is considered as "ultra-deepwater."

The significance of deepwater for the offshore industry is based on three main characteristics:

i. **Relatively high rates of exploration success:** The industry's advance into deepwater has met with some remarkably high rates of exploration success. According to estimates by BP, some 120 deepwater exploration wells are currently being drilled each year, and their success rates are significantly better than in any other exploration play in the world.

ii. **Large reserve size:** To be candidates for development, deepwater prospects must hold reserve volumes sufficient to enable the operator to recover the high levels of investment.

iii. **High productivity:** Well-productivity is a major influence on project feasibility and is often used to determine priorities among a portfolio of potential developments.

14.11.1.2 Factors driving deepwater developments

With the ongoing depletion of shallow water reserves in most of the major offshore regions and the declining size of new prospects being found on the continental shelf, the main factors driving deepwater oil and gas developments are:

i. large reserve size and high productivity of deepwater fields;

ii. buoyant oil prices and the growth in global demand for hydrocarbons;

iii. reduced deepwater costs as a result of technological advances;

iv. reduced risk levels;

v. innovative structures and procedures within the industry;

vi. government support, especially through favorable fiscal policies.

14.11.2 Deepwater field development

14.11.2.1 Design parameters

A field development is influenced by a multitude of parameters that affect each of the production scenarios differently and play into the final selection of a production system and development philosophy. The parameters that generally have the most effect on the development of the field are:

- Recoverable reserves
- Water depth
- Production rate
- Reservoir structure
- Reservoir production characteristics/field life
- Environmental and geological conditions
- Existing infrastructure
- Economic criteria

All of these contribute to formulating the criteria and constraints with which one has to deal in developing the field. Table 14.1 shows a matrix of these and other parameters and their effect on the field development selection process. This matrix does not provide the solution to the various tradeoffs but highlights the *complexity of the design process*. The following paragraphs briefly describe how each of these parameters may affect the development scenario.

14.11.2.2 Recoverable reserves

The amount of recoverable reserves is a major consideration in assessing the worth of the discovery and is often used as an initial indicator of the field's economic feasibility. Figure 14.1 shows a graph that illustrates the relationship between recoverable reserves and the cost per barrel to produce the field as a stand-alone development. The same graph also shows the development cost of a subsea tieback to an existing host facility. It is evident that

Table 14.1 System selection matrix.

Key parameter	**Process platform**	**Processing equipment**	**Mooring system**	**Riser system**	**Well locations**	**Manifolding arrangement**	**Subsea flowlines**	**Export system**
Water depth	X		X	X		X	X	X
Reserves	X							
Field production	X	X		X		X	X	X
Field life	X		X	X		X	X	X
Number of wells	X			X	X	X	X	
Fluid properties		X		X	X	X	X	X
Reservoir area					X	X	X	
Reservoir depth					X			
Reservoir pressure		X		X		X	X	
Environment	X	X	X	X				X
Infrastructure	X							X

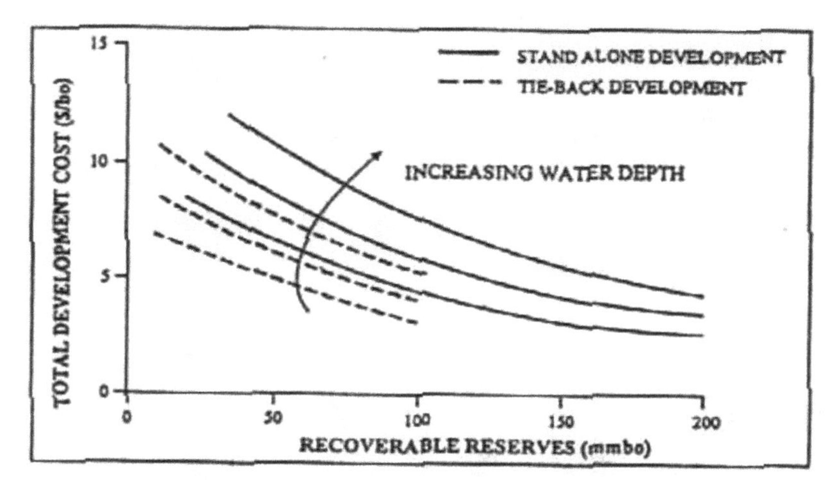

Figure 14.1 Relationship between recoverable reserves and development cost.

the tieback yields vastly better economics, upholding the feasibility of developing smaller reserves.

14.11.2.3 Water depth

Water depth is the other major factor in the selection of the development philosophy. Mooring systems, production risers, and installation considerations pose the main technical and economical limitations. Figure 1 also indicates the influence of water depth on the total development cost.

14.11.2.4 Production rate

Production rate is not so much a technical criteria as an economic threshold. Most economic models indicate that individual initial well rates of less than 1000 bopd are not economically viable in deepwater applications. Production estimates over field life are also critical in that rapid fall-off of rates (with low resultant ultimate reservoir recovery) can render a prospect uneconomical, even with high initial rates.

14.11.2.5 Reservoir structure

The structure of the reservoir will determine the number of wells required to produce the field. For instance, a highly fractured reservoir will require more drainage points than will a homogeneous reservoir. Low permeability coupled with low radial permeability will result in lower well rates and require more wells in the field. High vertical permeability may induce water coning or gas breakout early in field life. The configuration of the well system will affect the selection of surface facility.

14.11.2.6 Reservoir production characteristics

The main production characteristics that affect the system selection are:

a. Well pressure
b. Flowing well temperature
c. GOR
d. Productivity index (PI)
e. Fluid properties
f. Production functions (life of field gas, oil, and water production rates)

Reservoir pressure, for example, has a direct effect on well production capacity and on the practical offset distance of a subsea tieback system. In some circumstances, subsea pressure boosting could be used to offset flowline pressure drops associated with long flowline offsets and/or to augment low reservoir pressure late in the field production rate.

Flowing well temperature will affect the formation of hydrates and/or wax deposition. Wax can start to deposit inside wellbores or flowlines when fluid and flowline wall temperatures drop below the cloud point. If well production rates and/or reservoir temperatures are low, insulated well tubing may be required to assure that flowing fluid temperatures are high enough to avoid wax deposition in wellbores. In general, insulation would be used to limit wax deposition rate to levels that can be economically controlled by pigging. Dual flowlines will be required to allow flowline pigging from and to the host facility.

Well production rates can change dramatically (as much as a factor of 10 or more) throughout the production life of a reservoir. This leads to a difficult tradeoff between larger flow diameters to decrease initial-rate pressure drops and smaller flow diameters to improve low-rate flow performance (liquid holdup, flow stability) and fluid heat retention in late field life.

14.11.2.7 Environmental and geological conditions

Some of the environmental parameters to be considered are as follows:

- Weather conditions will mainly affect mooring system designs. For instance, it will affect the selection of turret mooring versus spread mooring for and FPSO. Also, a tradeoff may have to be made whether the FPSO should be permanently moored or whether a disconnectable or dynamically positioned system would reduce capital cost enough to offset the increased downtime associated with the periodic mooring disconnects.

- Soil conditions will affect the mooring and foundation design of the various surface facilities. Poor soil conditions will affect the cost and feasibility of the TLP's tendon foundations.
- Loop currents and other extreme current event will lead to high loads surface-piercing structures such as a SPAR. The size and cost to moor a SPAR in these conditions may negate the other benefits of SPARs.
- Extreme environmental conditions not only have a major impact on the design, hence the costs of offshore platforms also have considerable implications on installation and operating costs.
- A major factor influencing costs becomes the ability to be able to compress tasks into the short weather windows available in hostile environments. In the Atlantic Margin region, for example, there is a nominal four-month weather window between May and August, but even during this period, operations can be disrupted by the inclement climate.
- The costs resulting from unfavorable environmental conditions can be very significant; it is reported that 30% of the installation costs for BP Amoco's Foinaven development at a water depth of 490 m on the Atlantic Margin were weather-related.

14.11.3 Existing infrastructure

The existing infrastructure is one of the most important considerations in defining the preferred development method, and ultimately the economic performance of the development. As shown in Figure 1, the presence of an existing host facility will allow a field to be developed as a subsea tieback at a greatly reduced cost, compared to a full field development. Similarly, the absence of an accessible existing pipeline network will limit the export options and may in some cases sink the project. The ability to acquire export capacity and to pay acceptable tariffs is still another consideration even when infrastructure exists.

14.11.4 Economic criteria

To assess the economic performance of the system, different oil companies may use different criteria that include the following.

14.11.4.1 Minimum investment (capital risk)

It reduces the investment risk of a project. FPSO systems have an advantage in that they can be leased and booked to the project as operating cost. This does not necessarily reduce overall project cost but minimizes capital commitment to the project. Often this is the preferred choice for small fields with unproved reservoir performance or when political or tax conditions are unfavorable.

14.11.4.2 Early payback period

It reduces the length of time the project is "in the red." Low capital investment does not necessarily improve payback. For instance, multiple subsea well clusters will allow simultaneous drilling with two rigs, bringing the field to full production sooner. Another way to recoup costs sooner is by compressing the overall development schedule (the so-called Fast Track approach)

14.11.4.3 Maximum return on investment (ROI)

It is a commonly used yardstick to measure profitability. Typically, ROI improves with high initial production rates (early breakeven). "Incremental" or "marginal" ROI is often used to assess the worth of an option in sensitivity analyses.

14.11.4.4 High net present value (NPV)

It is primarily employed as a metric to assess the present value of all costs and earnings throughout the lifespan of the field, using a predetermined monetary discount rate. The relationship between NPV and ROI is close, as ROI is the discount rate at which NPV reaches a breakeven point of zero. They will offer compatible perspectives when comparing various field development scenarios for a specific field. Like ROI, NPV can be employed to assess the economic viability of various findings and make comparisons between them. In general, the sooner and more rapidly reserves are produced, the greater the Net Present Value (NPV). When employed as a benchmark for evaluating other domains, NPV does not quantify profitability. For example, a smaller field with a lower net present value (NPV) can yet have a better return on investment (ROI) and be more profitable.

14.12 Deepwater Production Facilities

Common facilities concepts in vogue to develop deepwater fields are given below.

14.12.1 Semi-submersible systems

Reservoir properties, operational considerations, and cost dictate whether the wells are placed underneath the floating production facility (FPF) or are offset from it. Placing the wells below the FPF, on a drilling template or clustered around a manifold, allows vertical access for drilling, completion, and maintenance from the production platform and avoids the high cost of contract rigs. In deepwater, the incremental cost associated with a heavy drilling or workover facility on the FPF can be significant. The advantage of the FPF is its diversity and adaptability to different functional requirements. This also allows more optimal placement of the wells and allows drilling from several locations at the same time to accelerate the development. On the other hand, it also increases the cost associated with flowlines and contract drilling. The key elements that need to be considered in configuring the FPF include the following.

14.12.1.1 Drilling and well maintenance requirement

An FPF can be designed for simultaneous production, drilling, or workover operations, provided that the subsea wells are located beneath the FPF. By drilling the subsea wells from the FPF, contract-drilling costs can thus be significantly reduced.

14.12.1.2 Well system configuration

A cluster arrangement generally offers the most field development versatility, since wells can be drilled before the placement of a manifold. For a template system, the template will have to be fabricated and installed before drilling can start. If the extent of the field, limited well reach, or the high cost of full drilling operations on the FPF favors placement of the wells remotely from the FPF, the wells can be drilled as individual satellites, tied back to the FPF through individual flowlines. If pigging of the flowlines is required to mitigate paraffin problems, two wells can be daisy-chained together, forming a continuous flow loop to the FPF. Satellite wells can also be manifolded to reduce the number of flowlines.

14.12.1.3 Newbuild versus conversion

The choice between new build and conversion will be dictated by total life-cycle field economics. A converted vessel can often be delivered to the field earlier and at less initial cost, thus improving cash flow, but may not have the required life expectancy or desired deck load capacity.

14.12.1.4 Production riser configuration

Several production riser concepts have been used. Flexible pipe risers are most commonly used. Current technology extends the use of flexible risers to 6000 ft, with diameters up to 14 in. The small footprint of the freestanding riser lends itself best for well systems that are placed below the FPF. Hybrid riser systems composed of a lower freestanding riser section and an upper flexible riser section are currently being developed in various forms to reduce cost. The steel catenary riser (SCR) has been used with more stable structures, such as TLP's and fixed platforms. Large deepwater FPF systems are likely to exhibit motion characteristics appropriate for an SCR.

14.12.1.5 Mooring system configuration

Current technology allows FPFs to remain moored in water depths more than 6000 ft. These mooring systems typically consist of chain-wire systems and contain submerged buoys to reduce the suspended weight and improve the mooring system performance and simplify installation. The complexity of these hybrid-type mooring systems and the associated cost for installation make the cost of deepwater mooring systems disproportionately higher.

14.12.2 Floating production storage and offloading (FPSO)

There are currently 66 FPSOs operating worldwide at water depth ranging from 20 to 1500. The FPSO provides integrated oil storage capacity, making it the system of choice for remote locations where there is no existing network of pipelines. Instead, the oil is periodically offloaded by shuttle tanker and carried to port.

The FPSO is also the most mobile of all the production systems, allowing it to be easily moved and reused in different locations. This is especially important when the field has a short life span or when local conditions do not favor the fabrication of more "permanent" facilities. The key elements that need to be considered in configuring the FPSO include:

14.12.2.1 Newbuild versus conversion

Most FPSOs in operation today are converted oil tankers. The cost of acquiring a used tanker and converting it to a production platform for a 5–10-year field life is found to be generally less expensive and faster than building a new vessel. During its 5–10-year field lifetime, the tanker needs an underwater inspection in lieu of dry-docking after a five-year service in order to remain in class. To keep an FPSO in service for much longer than 10 years without bringing it back to dry-dock for repairs requires a considerable amount of life extension work as part of the conversion. Building a new tanker or barge may be more economical in those cases.

14.12.2.2 Mooring system configuration

The two systems that are normally considered for deepwater permanent tanker mooring are the bow turret and central turret designs. In relatively mild areas, such as West Africa, a spread mooring system may be considered. A spread mooring may have a larger number of mooring lines than a turret mooring, and the cost of additional mooring lines will have to be compared to the cost of the turret system and tanker structural modifications.

Using thrusters to assist in station-keeping can reduce size and cost of deepwater mooring lines; alternatively a full DP (dynamic positioning) system can be used to eliminate the mooring system altogether. In either case, the operating cost of the DP system and the risk of potential downtime have to be weighed against the cost of a passive mooring system.

14.12.2.3 Well system configuration

The main drawback of the FPSO is that the larger vessel motions preclude surface wellheads. Remote subsea wells need to be used and a contract-drilling rig will have to drill, complete, and maintain the wells. The associated cost can be substantial. Systems have been proposed, whereby a drilling rig is placed on the FPSO. This so-called FDPSO would accommodate subsea wells placed directly underneath the vessel, thereby eliminating the need for contract rigs and long subsea tieback flowlines.

14.12.2.4 Production riser configuration

Flexible pipe risers have been most commonly used with FPSOs since they accommodate large vessel motions. Freestanding riser or hybrid riser systems are currently being developed in various forms and offer advantages when a larger number of individual production lines need to be brought up to the surface. The larger motions of the FPSO make it generally impractical to suspend SCRs directly from the vessel, although systems have been proposed whereby the SCR is suspended from a larger subsurface buoy.

14.12.3 Tension leg platforms (TLPs)

Currently there are 08 TLPs installed worldwide. TLPs are thought to be suitable for deepwater production in WDs down to 1500 m. A tension leg platform (TLP) is a floating structure held in place by vertical, tensioned tendons connected to the sea floor by pile secured templates. The piles to which the tendons are attached may be either directly driven into the seabed or drilled and grouted.

Although the TLP is a floating platform, the vertical tendon mooring system virtually eliminates platform heave motions. This eliminates the need for subsea completions with flexible riser systems and allows direct vertical access risers to surface wells for simpler well maintenance and facilitates the design of the process facilities. The risers are tensioned so that they can move vertically and flex laterally to follow the displacement of the TLP due to currents and wave action. The limited movement of the TLP also accommodates the use of simple steel catenary risers (SCRs) for use as export lines or to bring product aboard from satellite wells.

A penalty paid by the TLP design as opposed to a semi-submersible is the requirement for increased buoyancy to pretension the tendons. This results in a larger platform than is required to just support the deck loads. This translates to increased capital cost. However, this increased cost is partially offset by less expensive surface completions and conventional surface well maintenance. The key elements that need to be considered in configuring the TLP include the following.

14.12.3.1 Drilling and Well Maintenance Requirement

Similar to an FPF, a TLP can be designed for simultaneous production, drilling, or workover. By drilling the subsea wells from the TLP, the cost for pre-drilling the wells can be significantly reduced. The cost to acquire a deepwater drilling system and to provide deck space and very large TLP displacement is quite high and must therefore be compared to the cost of pre-drilling the subsea well with a contract rig instead. Full well maintenance capability is still required on the TLP, however.

14.12.3.2 Well system configuration

One of the main advantages of the TLP is that the wellheads can be located on deck, similar to a fixed platform. A high-pressure tieback riser connects the well casing to the surface where the well is completed with a surface tree.

14.12.3.3 Mooring system configuration

Of all the components of TLP, the mooring system is most affected by water depth. In 4000-ft depth, the tendon weight may exceed the combined steel weight of the hull and deck. Beyond 3000-ft water depth, the cost of the tendon system increases exponentially.

14.12.4 SPAR systems

Currently there are two SPARs in operation and two more are in an advance stage of design and implementation. The first installed SPAR began production in early 1997 for Oryx Energy in the Gulf of Mexico. The SPAR is a long vertical cylinder, moored the seabed with a taut wire mooring system. The deep draft of the SPAR creates very good motion characteristics with relatively small heave motions, which allow locating the trees at the surface. The high pressure risers are kept under tension by large cylindrical buoys and are free to slide up and down inside the moon pool. The SPAR also provides a large integrated oil storage capacity. This eliminates the need for a separate in-field storage tanker in locations where there is no existing export pipeline infrastructure.

The key elements that need to be considered in configuring the SPAR include the following.

14.12.4.1 Drilling and well maintenance requirement

Similar to an FPF, a SPAR can be designed for simultaneous production, drilling, or workover. The vertical load of a full drilling operation is substantial and will result in very large topside loads, which will have to be compensated by additional ballast, hull length, and displacement. The high cost of a full drilling operation on the SPAR should be compared to the added cost of pre-drilling the subsea wells with a contract rig instead.

14.12.4.2 Well system configuration

One of the main advantages of the SPAR is that the wells can be located on the deck, similar to a fixed platform and a TLP. A high-pressure tieback riser connects the well casing to the surface where the well is completed with a conventional surface tree.

14.12.4.3 Mooring system configuration

The very large frontal area of the SPAR hull will generate very high mooring loads, in particular in high current areas. The need to maintain near-vertical production risers that require a very stiff lateral mooring system composed of a large number of radial taut-wire mooring lines. The mooring system is complex, and the associated installation costs will form a large part of the overall cost of the system.

14.12.5 Subsea tieback systems

A subsea production system typically consists of a subsea drilling/production template or cluster manifold with subsea trees. The produced fluid from these wells is commingled and then flows back to a host surface production facility through one or more production flowlines. A control umbilical controls the valves on the trees and manifold. In addition, the subsea production system may contain facilities for gas lift, water injection, and gas injection and well testing. Subsea systems are typically used with any of the above deepwater production facilities, and can be placed within the reach of the surface facility for direct vertical access, or they can be remote and tied back though subsea flowlines.

In some circumstances, subsea pressure boosting could be used to offset flowline pressure drops associated with long flowline offsets and/or to augment low reservoir pressure late in the field production rate. Reducing flowing wellhead pressure can improve its production rate and/or increase the total recovery of reserves from the reservoir. There are several methods to provide pressure boost, including subsea positive displacement pumps, subsea rotodynamic pumps, and subsea separation combined with single-phase pumping of the separated phases. Alternatively, a small surface facility can be placed over or near the subsea system to serve as a pressure boosting station. Production is brought to the surface facility where multiphase boosting or partial separation and single phase boosting is accomplished. The product is then re-injected into a transport pipeline (or pipelines) and transport to the host process facility where full separation and treatment takes place. The in-field boosting platform can be a semi-submersible, a TLP, or a SPAR.

Without pressure boosting, remote subsea oil production tiebacks have been produced over distances in excess of 30 miles (Tarwhine – Australia). The Mensa subsea gas development, in production since early 1998, is tied back 68 miles to a shallow water host facility. With pressure boosting, it is expected that these distances can be increased significantly, provided hydrate and paraffin issues are also successfully addressed.

14.13 Subsea Technology

14.13.1 Subsea production

A subsea completion is one in which the wellbore terminates or is "completed" on the seabed, as distinct to a surface completion where the tubing carrying full well pressure continues up to the platform deck. The emergence of subsea technology has revolutionized the industry's offshore activities and it has developed at a remarkable pace in recent years.

The technology of subsea production was first introduced in the USA in the early 1960s and, a decade later, the operator Hamilton Brothers pioneered its use in the harsher waters of the North Sea, using subsea wells connected to a converted semi-submersible drilling rig to produce the UK's first offshore oil from the ArgyII field in 1975. For a variety of reasons, other operators were hesitant to follow this lead and significant growth in subsea production did not really begin until the 1980s, and was then centered largely in the North Sea and offshore Brazil. In the USA, despite the early pioneering work, there was little subsea activity until the discovery of deepwater reserves in the Gulf of Mexico, which have prompted a surge of development over the past five years. Recent advances in driverless technologies have further boosted the application of subsea systems.

14.13.2 The significance of subsea

Two parallel trends have become evident in offshore developments over the past decade, namely those toward subsea production and floating production systems. These trends tend, of course, to be mutually reinforcing and are consequent, to a large degree, on the industry's requirement for systems enabling the exploitation of prospects in deeper waters.

14.13.3 Why subsea?

Surface completions are cheaper to manufacture, easier to install, and far less troublesome to maintain than subsea ones, so operators will only opt for subsea development when other context-specific criteria make it demonstrably superior in terms of overall cost effectiveness. Such criteria include:

- Water depth (WD)
- Prevailing climatic and environmental conditions
- Reservoir size and reserve distribution
- Oil:gas ratios in the output stream
- Wellstream flow characteristics
- Well placement and numbers
- Topography at the proposed well site, and between the site and the host facility
- Availability of appropriate subsea field construction vessels
- Well maintenance requirements

- Distance to existing or planned infrastructure (especially pipelines)
- Treatment/disposal of non-production gas

Contexts, which particularly favor the adoption of subsea technology, include deepwaters, harsh environmental conditions, and fields, which are small or which have a scattered reserve distribution. In addition to the above criteria, one increasingly important consideration is whether a phased development – designed to achieve early production, which is then augmented by later stages of development – is planned.

14.13.4 Subsea development options

There are three main alternatives for the layout of subsea facilities:

i. **Satellite wells**: Subsea production was pioneered in the USA in the 1960s and further developed in the North Sea during the 1970s and offshore Brazil. Most of these early subsea systems were composed of satellite wells with individual control umbilicals and flowlines linking production back to an FPS or a platform. The subsea connection and maintenance work was carried out by divers; therefore, these systems were depth limited, and the individual flowlines and umbilicals were prohibitively expensive for long distances. With technological advances in production hardware and the development of remotely operated underwater vehicles (ROVs), the satellite concept has become a practical and effective development strategy for deepwater prospects, and has been adopted in many projects.

ii. **A. Integrated template/manifolds:** Manifolds are used to combine the flows from number of wells. Their complexity and cost vary according to field characteristics and maintenance requirements. Though costly in themselves, their viability revolves around the often considerable savings in flowline costs that they allow. A template can provide the base for multiple wellheads, manifold, and a protective structure. In the early 1970s, designs emerged for integrated subsea template/manifolds through which a dozen or more wells could be drilled and completed allowing the co-mingling of their production. In addition to increased complexity and associated costs, the integrated template/manifold has significant disadvantages that can offset the savings in flowlines and umbilicals:

    - Weight – Shell's 1982 South Cormorant UMC was 2200 tons, requiring a dedicated crane barge for installation.

- Most development drilling must wait until the template is in place.
- Increased drilling costs – a central location requires more deviated wells.
- Large templates generally require precise leveling and pilling to the seabed.
- Compared to individual satellites or clusters, access to the Xmas trees on templates is more restricted, and IMR operations are more problematic.
- Templates require a large early Capex outlay and render phased development impractical; financial exposure is therefore increased and the project has a slower cash flow.

However, large templates are still used in both shallow and deep waters.

ii. **B. Mini-template:** Mini-templates are smaller versions of the integrated template/manifolds, offering the same advantages of minimizing flowlines and umbilicals, plus the ability of being installed from an MSV (multi-purpose support vessel) or from the drilling rig itself and the wells can be pre-drilled. Disadvantages are:

- A centralized drilling location is required and hence extended reach drilling may be necessary.
- Unlike the cluster approach, mini-templates do not enable existing exploration wells to be tied-in without extra costs.

iii. Cluster developments: A simple approach appeared in the late 1980s involving a simple manifold together production from a cluster of individual satellite wells typically separated by 30 m and connected by short flowlines and control line "jumpers." Production can be commingled for transmission to the host platform, minimizing flowline costs. In some instances, two or more manifold clusters may be used on a single field. The well-manifold separation gives good access for ROVs and reduces the chances of damage to other wells during drilling. Control pods are usually ROV retrievable. The cluster system has the advantage of flexibility in terms of installation scheduling, allowing incremental expansion or contraction of the development as it progresses. Advantages of the integrated system are its ability to provide aggressive mechanical remediation techniques for removing hydrates and paraffin plugs and the fact that it is the only system by which down-hole tools can be deployed without the need for a rig or specialist intervention vessel. However, there are some disadvantages of cluster developments, namely:

- The costs associated with flowline and umbilical jumpers.
- The need for a centralized location, which, in turn, necessitates extended reach drilling requirements.

14.13.4.1 Xmas trees

On a subsea completion, the wellbore terminates in a wellhead on the seabed. The wellhead completes the vertical bore of the well and allows the connection of the flowline. A series of valves – the "Xmas tree" – is positioned on top of the tubing hanger to allow pressure control, well shut down, and re-entry for well workover.

i. **Horizontal trees:** Most Xmas trees that have been installed to date are dual bore versions, with the production stream routed inside an outer annulus. Horizontal trees are increasingly seen as viable alternatives to conventional dual bore trees, particularly for large diameter completions, or where frequent interventions and/or workovers are expected during the life of the well. Their main advantage is that, compared to conventional vertical trees, they allow easier access for intervention work by ROVs and obvert the need for equipment disassembly, thereby reducing downtime. The tree valves are located at 90° to the wellbore, which has the advantage of enabling workovers to be carried out with the tree still in place as a BOP can be landed on top of the horizontal tree. Other advantages include the simplification of tooling packages and installation procedures and a reduced overall tree height that simplifies the design of over trawl protection structures. Initially there were claims of considerable cost savings, but it has been reported that in practice the difference in whole life costs (Capex and Opex) between conventional and horizontal trees is minimal.

14.13.4.2 Control systems

Subsea control systems are coordinated by a master control system (MCS) located on the host facility. In addition to monitoring and controlling subsea equipment, the MCS also provides an interface with the main platform control systems, including process control and emergency shutdown procedures. The MCS and the subsea equipment are hooked up to an uninterruptible power supply.

i. **Control buoys**: The use of buoys to house control systems for remote subsea installations is designed to allow advanced communications technology to transmit control signals to and from a subsea production system instead of having to use expensive lengths of electrohydraulic umbilicals. Such umbilical-free subsea control systems are limited by the communication media available, and by the need to generate enough power locally to run the subsea equipment hydraulic valves and the control system itself.

14.13.4.3 Manifolds

A manifold is used to commingle production from a number of wells prior to routing it back to the host facility via a common flowline. The significant savings in flowline costs that manifolds allow have meant that they are fairly common features in deepwater developments, although their size can vary greatly depending on the project.

14.13.4.4 Templates

A template is a structure, which supports a number of subsea Xmas trees, housed within a protective frame. In addition to providing protection for the wellheads, a template offers obvious advantages for inspection, repair, and maintenance (IRM) operations since the wellheads are grouped together at a single location, and can therefore be serviced via a single manifold. Templates are only feasible when the reservoir can be exploited via a centralized drilling location.

14.13.4.5 Flowlines

The term "flowlines" refers to the lengths of pipe that carry the wellstream away from the subsea wellhead, or which deliver water or gas to subsea wellheads for injection purposes. It can be distinguished from the term "riser," which refers to the vertical section of the flowline stretching from the seabed to the production facility on the surface. Since the riser is very often merely an extension of the flowline, much of what is said here in relation to flowlines is also relevant to risers, but the two are dealt with separately for the sake of clarity. Flowlines and risers tend to be either flexibles or rigid steel. Flexible flowlines are typically ten times more expensive than conventional steel pipe, but they are cheaper to lay, more economical for shorter distances, and can be recovered for reuse. The diameter of subsea pipelines varies according to their function and to the operational conditions they will encounter, but flexible pipe sizes range from 12-inch to 16-inch diameter, and their steel counterparts can be produced with even larger diameters. In the design of subsea pipelines, there are three main issues of concern. These are: hydrostatic collapse resistance, thermal insulation properties, and weight.

14.13.4.6 Risers

The term "riser" refers to the vertical section of a subsea pipeline stretching between the seabed and the production facility on the surface. Because they are more exposed to environmental forces and must accommodate the motions of FPSs, risers are more problematic in design, engineering, and installation terms than the flowlines, which lie on the seabed.

14.13.4.7 Umbilicals

Umbilicals link subsea wells to control systems on offshore platforms or floating production vessels, and are also used to link production facilities. They usually contain a combination of thermoplastic hoses, electrical cables, fiber optics and/or metallic tubes, and are armored with steel wires prior to the extrusion of an outer thermoplastic sheath. A production control umbilical typically comprises electrical power and signal conductors, optical fibers and, in many cases, chemical injection lines as well. The cables and hydraulic lines may be bundled together or separated as two umbilicals, and may consist of both static and dynamic sections.

The functions of umbilicals are to provide:

- Hydraulic control
- Chemical injection
- Power and signal transmission

14.14 Multiphase Metering and Subsea Processing

14.14.1 Introduction

Multiphase technology holds real significance for the offshore industry generally as it attempts to monitor and handle variations in wellstream composition. It is of particular importance in deepwater developments where productivity is at a premium, flow assurance issues are intense, and the hydrocarbon often has to be transported over relatively long distances despite low reservoir drive. The potential for very considerable cost savings and productivity improvements on fields worldwide are driving some of the most intense R&D efforts presently underway in the industry.

14.14.2 Multiphase metering

On a single subsea well, the flow can be tested at the platform, and fluid composition data obtained by using test separators for oil, gas, and water and associated metering trains. However, where longer flow distances and multiple wells are involved, a manifold is often used to combine the flows into a single flowline for transmission back to the platform. In this case, a separate test line may be installed to monitor the production of individual wells, the alternative being to lose production by shutting them all down and testing one by one. However, the expense of providing separate test lines for each well is often prohibitive and consequently there has been a drive to develop a reliable way of metering multiphase well flows on the subsea manifold itself.

Thus, subsea metering systems eliminate the need for test lines from subsea wells or manifolds to the host platform, and for test separators and associated equipment on platform topsides, enabling operators to make savings reckoned in the million-dollar range. A study done in 1996 suggested that, in the case of a subsea development 10 km from the host platform, subsea multiphase metering might give a cost reduction of 62% over the use of separate test lines. In addition to this, the resulting improvements in system management might yield gains of 6%–9% in the value of the oil recovered.

In addition to such cost savings, subsea multiphase meters also promise to deliver more accurate well performance data since they record the status of the wellstream as it leaves the seabed, prior to any changes in product characteristics that may occur during flow to a surface-based meter. This enhanced accuracy could enable better reservoir management and tighter risk analysis parameters. These factors combine to guarantee a promising market for simple low-cost meters.

In principle, five measurements are normally required for multiphase metering: three velocities (oil, gas, and water fractions) and two phase fractions (the third phase fraction can then be deduced). For well test purposes, an accuracy of +/– 5%–10% is usually thought acceptable. There are currently at least 12 multiphase metering systems under development worldwide and at least three are commercially available.

14.14.3 Subsea processing

One of the main challenges in deepwater production is to ensure the flow of hydrocarbon from the seabed or "mudline" to the surface. In deepwater, the problems are exacerbated by the fact that the extreme environmental conditions tend to modify the properties of the different phases in a multiphase fluid such as the wellstream, and the risk of plugging – due to the formation of waxes and hydrates in the flowlines – is higher than in equivalent systems in shallower waters. Subsea processing is something of a portmanteau term, but in essence it involves technologies such as subsea separation, subsea metering of the production stream, subsea boosting to pump well products to the surface, and subsea power generation to drive all these processes.

Subsea processing is still in its infancy, and its applicability is highly dependent on individual field characteristics such as reservoir depth and pressure, gas:oil ratios, water cut, distance from host platform facilities, etc. However, the concept offers a wide range of highly significant benefits, principal among which are:

I. Increased productivity
II. Increased recovery
III. Improved flow assurance
IV. Longer tieback distances
V. Reduced topsides processing requirements

I. *Increased productivity* can be achieved in two main ways. Firstly via the use of multiphase pumping systems to supplement reservoir drive and secondly via the separation and disposal of produced water subsea rather than transporting it with the hydrocarbons back to the facility for treatment and disposal. In fields with early and significant water production, the subsea separation and disposal of produced water can be a major benefit since it would reduce the back pressure on the wells and allow a freer flow of hydrocarbons.

II. *Increased recovery* results from the reduced back pressure on the wells consequent on the pressure boosting and/or separation of water from the wellstream. By reducing this "hydrostatic head," reservoir drive is maximized and more reserves can be produced. This is a particularly important consideration in deepwater where very significant back pressure can be generated by the long risers required.

III. *Improved flow assurance* is derived from the addition of pressure via single or multiphase pumps and/or from the separation of wellstream components, either by removing water or by separating gaseous and liquid fractions. The removal of water from the production stream reduces the risk of hydrate formation, which is a particular concern in deepwaters. Liquid/gas separation eliminates the troublesome phenomenon of liquid "slugging" in flowlines.

IV. *Longer tieback distances* can be achieved as a result of the improvements in flow assurance and the addition of energy (pumping) to the wellstream enabling it to flow over longer distances. Subsea systems have produced using reservoir pressure over flowline distances of up to 50 km in length; however, this is most unusual and in practice few flowlines exceed 20 km. The attractions of increasing this distance are considerable and could lead to the exploitation of many reservoirs that are currently too small or too isolated for economic development.

V. *Reduced topside processing requirements* result from the fact that partial (two-phase) or possibly total (three- or even four-phase) separation has been completed subsea; so there is no need for bulky topside separators, with consequent space, weight, and cost benefits for the host platform. Subsea processing could thus serve to extend the economic life of existing platforms by reducing the topside additions associated with hosting a new satellite development. In the case of processing, which involved water separation, the topside gains would be enhanced by reductions in the size and number of flowlines required to drain the field and the elimination of the need for a dedicated water injection line running from the host platform.

14.14.3.1 Multiphase Pumping

The benefits of subsea multiphase pumping show considerable overlap with those of subsea separation mentioned above. In essence, multiphase pumping has the potential to dramatically increase the viability of many deepwater, remote, or marginal fields.

Subsea multiphase pumping offers the advantages shown below:

A step-change reduction in CAPEX:

- Boosting new, remote, or marginal field to an existing or central host facility.
- Reducing the total number of platforms and topside investment required.
- Affording a potential reduction in the number of wells required to produce a field.
- Accelerating production from existing wells and thereby deferring the need for later development phases

Accelerated production and improved economics:

- Higher production rates due to reduced wellhead back pressure.
- Increased financial viability of deepwater, remote, or marginal fields.

Flow assurance:

- Increased well productivity.
- Maintenance of pressure and flow rate regardless of gas volume fraction (GVF).

Multiphase pumping demands:

- A pump with the capability to operate under the changing well conditions experienced over the reservoir lifetime, with wellstream content ranging from 100% oil to 100% gas.
- A variable speed prime mover, able to supply perhaps 1 MW at the pump shaft. The options, depending on the field and its location, are variable speed electric motors, hydraulic motors, or turbines.
- A pump control system.
- Power connectors.
- Cooling and lubrication.
- Chemical injection to overcome any scaling effects.
- Hydrate control.

The various pump designs include:

- Centrifugal
- Twin screw
- Diaphragm pump
- Hydro-booster
- Reciprocating piston type.

14.15 Issues in Deepwater Development

14.15.1 Environment

For practical purposes, most of the world's deepwater regions can be categorized into benign environments or harsh environments. Specific examples are:

Table 14.2 Variations in key environmental parameters across different regions.

Maximum design values	Northern North Sea	Gulf of Mexico	Brazil
Wave height (meters)	31.0	12.3	8.4
Wave period (seconds)	14–18	12–15	11
Surface current (m/sec)	1.50	0.25	2.50
Bottom current (m/sec)	0.75	–	1.00
Wind (m/sec)	41	45	46

- Benign – Brazil, Gulf of Mexico, West Africa. Although severe weather conditions can be experienced, most are predictable.
- Harsh – West of Shetlands and Northern Norway. In such situations, the environmental conditions are often hostile and unpredictable.

Table 14.2 below illustrates the differences in key environmental parameters that exist between regions.

14.15.2 Deepwater installation

14.15.2.1 Installation problem

Extreme environmental conditions not only have a major impact on the design and hence the costs of offshore platforms but also have considerable implications on installation and operating costs.

Specific problems, which can be encountered, are listed below and illustrate the range of obstacles, which can hamper deepwater installation:

a. A descent time of up to 1 hour for ROVs to reach the seabed.

b. Effect of currents on ROVs, which could be pushed out of vertical by 250 m during decent.

c. Difficulties of station-keeping with ROVs.

d. Time taken to install buoyancy units on risers.

e. Vessel station-keeping using DP.

f. Strain-induced rotation of flexible flowlines resulting in tools not matching.

g. Reduction in soil strength characteristics because of the structural loading on soft sea beds.

h. Temperature and current changes throughout the water column.

Table 14.3 Subsea hardware – typical dimensions and weights.

Item	Type	Size	Weight
Xmas tree	Conventional	4 m × 4 m × 8 m	30 tons
	Horizontal	4 m × 4 m × 4 m	27 tons
Control module		1.5–2 cm (H) × 0.7–1 m (D)	1 ton
Template	Four-slot	15 m × 20 m × 6 m	400 tons
	HOST	5 m × 7 m	40–100 tons
Manifold	HOST	5.7 m × 5.3 m	21.5–35.5 tons
		Various	60–200 tons
	Kvaerner	Eight-slot	170 tons
Pipeline	Flexible	8" ID	120 kg/m in air
		14" ID	330 kg/m in air
	Rigid	10" OD	> 100 kg/m in air
Spool pieces	Various	variable	max 30 tons
Umbilical	SSIV	76 mm OD	10 kg/m in air
	Booster pump	146 mm OD	37 kg/m in air
	Various	Variable	2–40 kg/m in air
Multiphase meter	Various	1 m × 3 m × 2.5 m	1–2.5 tons
Multiphase pump	Framo	4 m × 0.8 m	8–20 tons
Subsea structures	Various	Variable	2–120 tons

i. The difficulties of late-season weather forecasting due to the long sea state hysteresis.

j. Uneven seabed terrain.

k. High hydrostatic pressures.

l. Very low temperatures.

m. Varying surface and subsurface current directions and velocities.

n. Turbidity caused by installation activities affecting visibility for ROV pilots.

By way of illustrating the range and complexity of offshore installation tasks, the table below lists the main items of subsea hardware that can be expected to feature in deepwater developments, giving approximate size and weight ranges for each (Table 14.3).

The larger items, such as Xmas trees and manifolds, tend to be installed by the drilling rig or a specialist workover or heavy-lift vessel. Others are installed by lighter construction vessels, which are often specially designed to install pipelines and umbilicals as well as the less bulky items of subsea equipment.

Table 14.4 Construction vessel day rates.

Vessel type	Approximate dayrate
Supply vessel	$3,200–$4,500
Anchor-handling tug	$4,800–$8,000
DP Class II survey vessel	$40,000–$50,000
Diving support vessels (DSVs)	$80,000–$100,000
DP Class II Derrick/laybarge	$120,000–$150,000
DP Class II pipelay/construction vessel	$150,000–$200,000
DP Class III pipelay/construction vessel	$200,000–$250,000

14.15.2.2 Installation costs

Day rates for installation vessels are project-specific and are susceptible to changes in oil prices. Approximate day rates for deepwater work are shown in Table 14.4.

14.15.2.3 Flowline installation

Several types of construction and installation methods may be used for the flowlines that carry production to a manifold for co-mingling with production from other wells, or which run from the well directly back to the platform. The choice of installation method will depend on considerations such as technical feasibility, risk analysis, and company specifications and economic factors such as laying rate and possible requirements for investment in equipment modification. Some of the methods used and the respective capabilities of each are outlined in the table below. of these, J-lay, S-lay, and reel are the most common (Table 14.5). The reel method has the latest lay-rate of the three, but it requires dedicated reel vessels.

14.15.2.4 Reel-lay

This is the fastest flowline installation method, but it is also the most expensive. The line is unwound from a reel carried by a specialist lay vessel, and paid out over the stern. Flexible flowlines are installed in this way, as are rigid steel lines. For rigid steel installation, lines up to 14-inch or 16-inch diameter are welded onshore and coiled onto a reel mounted on the lay vessel. The advantages of this method relative to the more conventional S-lay or the newer J-lay techniques are that it allows a higher standard of quality control (because welding is conducted onshore), the actual installation is faster, safer and less expensive, and is less prone to weather-related disruption.

14.15.2.5 J- or S-lay

Length of rigid steel pipe are welded on the lay vessel and deployed either vertically (J-lay) or horizontally (S-lay) over the stern. S-lay is the traditional

Table 14.5 Flowline installation methods.

Method	Description	Flowline type	Maximum diameter (inches)	Water depth limits (m)
Reel	Pipe wound on a reel	Rigid and flexible	16	1000
			12	1800
		Bundle	2 × 6	2000
J-lay	Welded lengths Vertically launched	Rigid	20	900
			16	1200
			12	1500
S-lay	Welded lengths, Horizontally launched	Rigid	24	1000
			20	1500
Tow	Onshore weld and tow-out	Rigid, flexible, and bundle		Tow-out limits: 6–10 km (mid water) 20–25 km (bottom tow)

method for large diameter lines. It is not usually considered economic for the small diameter flowlines used in deepwater development.

The J-lay technique is ideally suited for installation of deepwater pipelines, including steel catenary risers and pipe-in-pipe designs. By laying pipe at near-vertical angles, distance to the touchdown point is reduced and consequently the lay vessel's tensioning requirements are reduced (since shorter lengths of pipe are held suspended in the water). The J-lay technique allows pipelines to be laid to exacting routes, safely negotiating sea bottom hazards or in-field pipe networks.

14.15.2.6 Bundled

In this application, flowlines or risers are fabricated onshore and towed out to location, using either surface, bottom, or controlled depth towing (in some instances, small diameter bundles can be reel laid). One disadvantage of this approach is the risk of damaging the bundle during the tow-out operations.

14.15.3 Flowline Tie-in

Once the flowlines are in place, they must be hooked up to the wells and/or manifolds – an operation that can be very challenging, particularly in deepwaters and harsh environments. On subsea developments, one major consideration is the method by which the flowline is pulled in and connected to the manifold. The aim is to minimize the requirements for a large pull-in "porch" in order to restrict the manifold size. There are various methods of laying and tie-in, including:

Horizontal connection – this is the conventional method. Pipelines are deployed to the seabed and then pulled into place by a winch mounted on an ROV.

Vertical connection – this is now used extensively by Petrobras for its deepwater developments. The flowline is installed on the production adapter before the tree installation and the non-simultaneous operations reduce the logistics and planning constraints. While the design of vertical pipe joints tends to be more complex than for horizontal connections, operations on the seabed are simplified and the system allows the separation of the rig and lay-vessels operations.

Layaway – Petrobras also uses lay-away connection. Indeed it initially standardized on the lay-away method for diverless connections but found that the method is "technically great, but the logistics (in particular those relating to vessel availability) can be a problem."

Stab and hinge over – this is a concept popular in the Gulf of Mexico and becoming widespread in deepwater developments worldwide. The connection is vertically stabbed in, and then the flowline is hinged over to the horizontal as the lay vessel "lay-away."

14.16 Flow Assurance

Flow assurance refers to interventions that are designed to maximize productivity of both surface and subsea wells in all water depths. Because of the distances involved in deepwater developments and the temperature and pressure regimes they operate under, flow assurance is a crucial concern.

14.16.1 Slugging

"Slugging" is the partial separation of oil, gas, and liquid phases in the line, with disruptive slugs of gas or liquid being transmitted along the line. Slug flow can develop under normal operating conditions because of the hydrodynamic properties of the wellstreams, or it can be induced by undulating seabed topography causing liquids to pool in the dips along the line. The intermittent flow that slugging causes is characterized by sudden drops and surges in liquid and gas volumes and can be highly problematic for topsides processing facilities.

Server slugging can also occur at the base of production risers. This is similar to terrain-induced slugging and disrupts the vertical flow with blockages at the seabed as the liquid slug builds up until the pressure of the trapped gas is sufficient to force it up the riser to the topside facilities. To control this erratic and highly unpredictable flow, riser base gas lift is commonly applied.

14.16.2 Water and gas injection

Some fields make use of either gas or water injection to boost reservoir pressure. Such injection is essentially an addition of energy, which helps to maintain product flow out of the reservoir. As reservoirs become depleted, their initial flow rate declines, sometimes quite considerably, so gas and/or water injection is a fairly standard strategy adopted by operators to ensure continued productivity. In a typical application, produced water is separated and cleaned on the surface platform, and then injected into the lowest strata of the reservoir via separate injection wells, thus raising the level of the oil and sweeping it toward producing zones. Such reinjection, however, is not without problems and may lead to fracturing and reservoir plugging or increased scaling and souring in production fluids.

Gas may also be re-injected to raise the reservoir pressure and sometimes as a means of gas storage for later production. Gas injection requires a source of separated gas, compressors, flowlines to the wellhead and tubing down the well. One drawback of this technique is that the additional gas tends to increase slugging in the flowlines; so while oil output is improved by the reduction in back pressure on the well, gas lift can introduce limitations to tieback distances.

14.16.3 Chemical injection

Chemical injection to counter hydrate blockage, waxing, or scaling can, for multiple wells, involve separate lines to each wellhead, or be effected through a small subsea manifold and metering system. In some cases, the chemical injection lines are contained within the control umbilical.

14.16.4 Gas lift

Gas is also used to improve product flow up the wellbore to the surface – a process known as gas lift. The gas is routed down into the outer annulus of the wellbore and enters the wellstream through a series of gas inlets in the inner (production) conduit. This reduces the density and the hydrostatic head of the wellstream fluids, increasing recovery.

A similar application, known as riser-base gas lift, is also utilized – particularly on deepwater developments – to counter severe slugging at the base of the production riser. Pressurized gas is injected at the base of the production facility. To avoid the erosive effects of direct gas injection on the internal wall of the riser opposite the injection inlet, gas can be delivered indirectly, being injected into a sleeve wrapped around the riser and hence entering less forcefully via a series of inlets in the riser wall.

14.17 Hydrates and Waxing

14.17.1 Hydrates

At high pressures and temperatures of 20–30 °C, gas hydrates – ice-like compounds of natural gas and water can form in subsea pipeline and valves, posing a series of potential risks to production and well-control operations. Hydrate formation can plug up risers and the wellbore itself. Deepwater development is particularly prone to hydrates since their production lines tend to be longer and hence suffer more thermal loss than those on shallow water development.

Currently, there are two predominant approaches to preventing hydrate formation in flowlines:

- **Thermal:** Either by insulating flowlines or by applying heat to the lines using hot water pumped from the platform and/or direct electric heating.
- **Chemical injection:** For multiple wells, this can involve separate lines to each wellhead, or a small subsea manifold and metering system. In some cases, the chemical injection lines are contained within the control umbilical. Chemical inhibitors include methanol and glycols.
- New low-dose anti-hydrate compounds are being developed, including kinetic inhibitors – which slow the rate of hydrate formation and anti-agglomerates, which prevent hydrate plug formation.

14.17.2 Waxing

Some hydrocarbon liquids contain a large number of heavy components, which as the liquids cool during transport, tend to settle out as solid crystalline wax deposits inside the pipelines. These paraffin deposits clog and can eventually block flow through the lines. In deepwater, it may be impossible to clear the paraffin plug from a pipeline and it may become necessary to lay a new replacement line. This obviously is an extremely costly exercise; so it is imperative to minimize paraffin build-up.

As with hydrates, prevention of wax deposition and build-up typically focuses on retaining the heat of the wellstream using flowline insulation and heating technologies.

14.17.3 Pigging

Conventional pigging of subsea equipment employs the "round-trip" pigging technique, which requires the installation of dual flowlines. Under this system, pigs are launched from the platform down the auxiliary line running to the subsea equipment. They then return to the platform via the main flowline, cleaning off deposited solids as they proceed. Devices developed to run pigs in subsea systems include: dual-diameter scarper pigs (which can be run through pipes of different sizes) and subsea pig signalers, which use magnetic sensors to detect the pig's passage down the line. To prevent damage to flexible flowlines, pigging operations in flexibles rely on non-metallic scrapers and brushes, which will not harm the internal surface of the line.

14.17.4 Asphaltenes and scales

Other problems, which can arise in production lines, are those associated with the gradual deposition of scale and asphaltenes. Scale tends to become a problem toward the latter part of field life when water cut increases. Asphaltene build-up is a relatively new flow assurance challenge for deepwater development.

14.18 Subsea Reliability

The reliability of subsea equipment, which is located out of sight and often several kilometers away from manned installation, is critical for the cost-effective operations of deepwater projects.

The following problems are experienced with subsea hardware:

- complex metallurgy resulting in highly accelerated corrosion;
- hydrostatic pressure causing seawater ingress into control systems;
- poor pipeline connections;

- failure of hydraulic and electrical connections;
- umbilical tube failures;
- overall quality control.

14.18.1 Inspection, Repair, and Maintenance (IRM)

All subsea facilities will require routine inspection, occasional repair, and regular intervention for maintenance, data, or reservoir management. Intervention has been classified into four main types:

1. Services performed through flowlines or umbilicals (e.g., pigging or bullheading scale chemicals into a wellbore).
2. Services external to the wellbore, including ROV services and module change-outs.
3. Down-hole services performed through the wellbore. This includes all slick-line, electric-line, and coiled tubing work that can be performed without removing the production tubing.
4. Service requiring the removal of the production tubing. For these "workover" operations, a marine BOP is essential.

14.18.2 ROVs

Three factors that have conspired against the deep diving operations are:

- Health and safety – in countries such as Norway, concerns over diver health and safety effectively limited manned operations to shallow waters.
- Cost – although proven technically feasible, manned intervention beyond WD 300 m is prohibitively expensive.
- The increasing capabilities of remotely operated underwater vehicles (ROVs).

These factors have made ROVs the logical tool for deepwater intervention and ROVs are currently deployed on both fixed and floating platforms, as well as on support vessels, including DSVs, and pipelay vessels. Industry standardization has been crucial in the development of these tools.

There are a number of distinct types of unmanned underwater vehicles in operation, but the most important group are the "work-class" ROVs. These vehicles have in many instances become an enabling technology for deepwater production. Costing $2–5 million each, at least 446 are in operation worldwide.

There have been few really major advances in ROV design in recent years, most changes being more a case of continuing evolution together with the application of external technical development in computing and electronics.

14.18.3 Technology development

The tendency for operators to adopt a more "hands-off" stance toward field development can arguably be expected to have negative impacts on the generation of new technologies. As a result of their emerging preference for EPIC contracts, operators' links with the smaller supply companies, which have traditionally been the source of many elements central to new technologies in the offshore industry, are weakening. The latter find it more difficult to field test equipment and because of their lack of direct communication with the oil companies, their ability to innovate is restricted.

14.19 Deepwater R&D

The new technological development is vital if the offshore industry is to continue to find and develop deepwater reserves. In today's environment when technology development has become costly, it is necessary that resources be strategically shared between the operator, industry, and technology centers. The need led to the industry initiatives toward developing joint industry projects namely:

- Deepstar co-operative industry R&D program led by Texaco, known as Deepstar, USA.
- PROCAP 3000 strategic corporate R&D Program led by Petrobras, Brazil.
- Norwegian Deepwater Program.
- AMJIG (Atlantic Margin Joint Industry Group).
- SLOOP 2000 (simulation of long-term offshore oil field production).

These projects have been developed to tackle the deepwater challenges with various priorities based around the environment in which they have to operate. The number of technological developments made by those JIPs has in general benefited the industry.

14.19.1 Deepstar

The Deepstar co-operative industry R&D program is aimed at identifying and developing new technologies for Gulf of Mexico and worldwide deep-water projects.

Deepstar began in 1992 as a contracted study regarding the feasibility of using long offset subsea tiebacks to shallow water platforms to develop the Gulf of Mexico deepwater tracts. Texaco made the funding commitment and then sought partners to share the costs. Study results benefited from access to intellectual talent of 11 oil companies and the minerals management services (MMS), which joined in Phase-I.

14.19.2 Mission of deepstar

"An industry wide cooperative effort focused on identification and development of economically viable, cost – risk methods to produce hydrocarbons from Tracts up to 10,000 ft water depth."

14.19.3 Deepstar membership

Deepstar is open to industry participants in following two forms:

- Direct oil company participation in the form of participation fee and active involvement in Deepstar Technical Committee. E&P companies like Shell, BP, UNOCAL, Texaco, and TotalFina Elf are members of the consortium.
- Vendor participation in the form of voluntary contribution or technology sources. Companies like ABB, Shclumberger, Kvaernenor, and JP Kenny are such members.

14.19.4 Administration

Texaco, as the project organizer, has program administration and co-ordination responsibility for Deepstar work. Third parties required to perform technical work in support of various study activities are contracted through Texaco. It also has a technical advisor and any necessary project support staff for administrating the work progress.

The Senior Advisory Committee has representatives from each participating company and is charged with putting together a balanced program to fit the budget.

14.20 PROCAP/PROCAP-3000 Program

The PROCAP was set up back in 1986. The PROCAP-2000 program has been completed and was renamed as PROCAP-3000 as much of its work target aimed at enabling production from waters of up to 2000 m have been achieved.

The PROCAP 3000 program will give capability to produce from 3000-m water depth.

PROCAP-2000 is the Petrobras Technological Innovation Program on Deepwater Exploitation Systems. The aim is a step reduction in production costs, increasing productivity in deepwater fields while enabling oil production at water depth over 1000 m.

With a strong accent on floating production systems and related technologies such as subsea systems, This program undertook more than 100 interdisciplinary studies; the main result was the full technological capability through floating systems (based on semi submersibles) for production in water depths of up to 1000 m.

PROCAP-2000 was developed with the objective of enabling PETROBRAS to produce from field in ultra deepwater of 1000–20,000 m of water depth.

The overall commitment to the technology research budget is of the order of US$ 125 million, which is mainly in the prototype construction and installation. It estimates that it can cut development cost per barrel up to 30% through these technologies.

14.20.1 First PROCAP

The first PROCAP studied in detail the following important points:

- Semi-submersible weight reduction (production facilities and hull) – specific design and drilling to production conversion.
- Platform positioning and mooring systems.
- Subsea completion and subsea connection systems (pull in).
- Subsea production equipment such as subsea trees, wet and atmospheric manifolds, template-manifolds.
- Rigid and flexible risers and flowlines.
- Monitoring and remote control systems.
- Installation, maintenance, retrieval, and inspection of subsea equipment.

- In the innovation technology projects, the following alternatives were studied:
- Compliant towers.
- Tension leg platforms (TLP).
- Semi-submersible with dry completion.
- Subsea separation systems.
- Subsea multiphase pumping systems.

14.20.2 The PROCAP-2000

This program has been implemented to give continuity to the efforts of the first one to improve Brazilian technological skills in deepwaters. The program has got the following objectives:

1. To enable PETROBRAS to produce oil and gas from offshore field situated in ultra deepwater (1000–2000 m) aiming at incorporating the reserves located at these water depths.
2. To develop technological innovative projects aiming at reducing investment (CAPEX) and operational (OPEX) expenditures, as well as anticipating the production and enhancing the final recovery of oil and gas, in waters over 3000 m deep.

14.20.2.1 PROCAP-2000 strategies

The strategies chosen for PROCAP-2000 reflect not only today's level of oil process – expected to hold steady or climb only slightly in the near future – but also Brazil's current situation, particularly as a developing country. The program endeavors to optimize reliance on know-how and resources available locally and abroad, and reduce the costs of the development of deepwater technology. This strategy stresses:

a. Links with the Brazilian Technological Community (Sharing Efforts).
b. Links with the International Technological Community (complementation).
c. Focus on Essential Technologies (Selectivity).
d. Links with the Government funding agencies (Financial Resources).

14.20.2.2 PROCAP-2000 Portfolio

PETROBRAS had chosen 12 systemic projects to be developed throughout 5 years by PROCAP-2000 from 1992 to 1997. These projects represented the essential technologies for PETROBRAS to achieve the goals of that program. Once more, the aim was to get a steep reduction on production costs, increasing productivity at deepwater fields while enabling oil production in water depth over 1000 m.

- Stability in horizontal and highly deviated wells.
- Drilling highly deviated wells in unconsolidated sandstones and unstable shale.
- Kick and blowout control in deepwater wells.
- Electrical submersible pumps in subsea wells (ESP).

14.20.3 The PROCAP-3000

This program has been developed to meet the challenge of exploration and production of oil from depths as great as 3000 m. Another objective of the current program lies in funding technology solutions that will allow the company to reduce the cost of the oil extracted in deepwater. Initially, the program will be conducted through 19 systematic projects:

1. Extended reach wells in ultra-deepwater
2. Design of wells in ultra-deepwater
3. High rate production wells in ultra-deepwater
4. Drilling, well test evaluation and completion in ultra-deepwater
5. Light fluid and underbalanced drilling in ultra-deepwater
6. Intelligent completion in deepwater
7. Gas-lift optimization in ultra-deepwater
8. Subsea multiphase pumping system
9. Flow assurance in ultra-deepwater
10. Subsea equipment for ultra-deepwater
11. Unconventional subsea production systems
12. Subsea risers and pipelines (gathering, exportation, and control) for Albacora Leste and Marlim Leste Fields
13. Subsea risers and pipelines (gathering, exportation and control) for Roncador and Marlim Sul Fields

14. Subsea risers and pipelines (gathering, exportation and control) for 3,000 water depth
15. Subsea connection system for 3,000 water depth
16. Mooring systems for ultra-deepwater
17. Cost-effectiveness analysis for different hull concepts considering dry and wet completion
18. Dry completion stationary production units
19. Acquisition and treatment of geological, geophysical, geotechnical and oceanographic data for ultra-deepwater

The projects subdivide into 62 subprojects, encompassing all areas of offshore development; well technology, artificial lift, flow assurance, subsea equipment, non-conventional production system, subsea connection system, stationary production units, mooring systems, and last but not the least the acquisition and treatment of geological, geophysical, geotechnical, and oceanographic data for ultra-deepwater.

14.21 Deepwater Case Studies

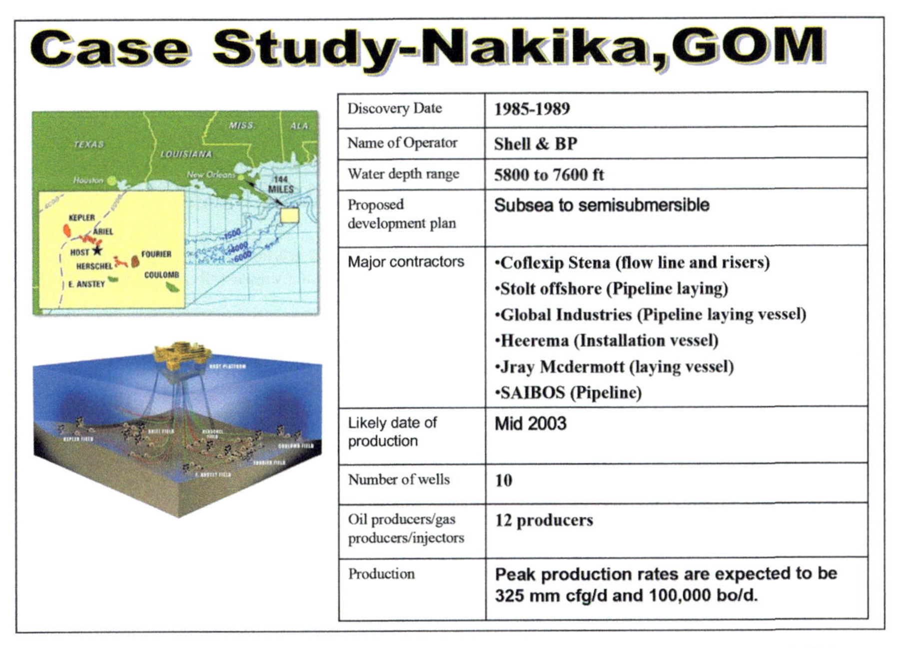

Case Study-Nakika,GOM

Discovery Date	1985-1989
Name of Operator	Shell & BP
Water depth range	5800 to 7600 ft
Proposed development plan	Subsea to semisubmersible
Major contractors	•Coflexip Stena (flow line and risers) •Stolt offshore (Pipeline laying) •Global Industries (Pipeline laying vessel) •Heerema (Installation vessel) •Jray Mcdermott (laying vessel) •SAIBOS (Pipeline)
Likely date of production	Mid 2003
Number of wells	10
Oil producers/gas producers/injectors	12 producers
Production	Peak production rates are expected to be 325 mm cfg/d and 100,000 bo/d.

Figure 14.2 Production and well data for a offshore well drilled in Nakika, GOM.

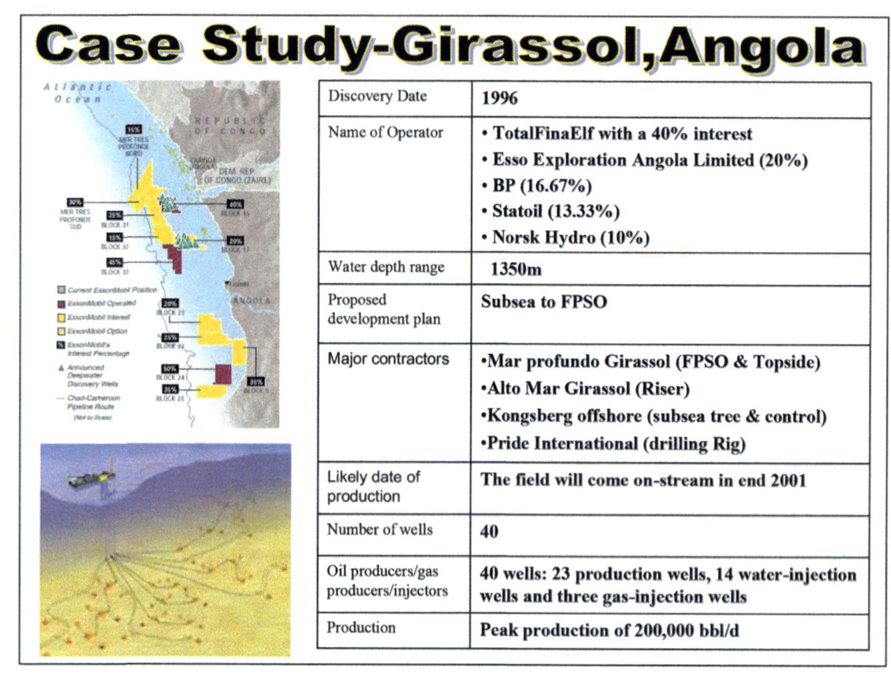

Case Study-Girassol,Angola

Discovery Date	1996
Name of Operator	• TotalFinaElf with a 40% interest • Esso Exploration Angola Limited (20%) • BP (16.67%) • Statoil (13.33%) • Norsk Hydro (10%)
Water depth range	1350m
Proposed development plan	Subsea to FPSO
Major contractors	•Mar profundo Girassol (FPSO & Topside) •Alto Mar Girassol (Riser) •Kongsberg offshore (subsea tree & control) •Pride International (drilling Rig)
Likely date of production	The field will come on-stream in end 2001
Number of wells	40
Oil producers/gas producers/injectors	40 wells: 23 production wells, 14 water-injection wells and three gas-injection wells
Production	Peak production of 200,000 bbl/d

Figure 14.3 Production and well data for an offshore well drilled in Girassol, Angola.

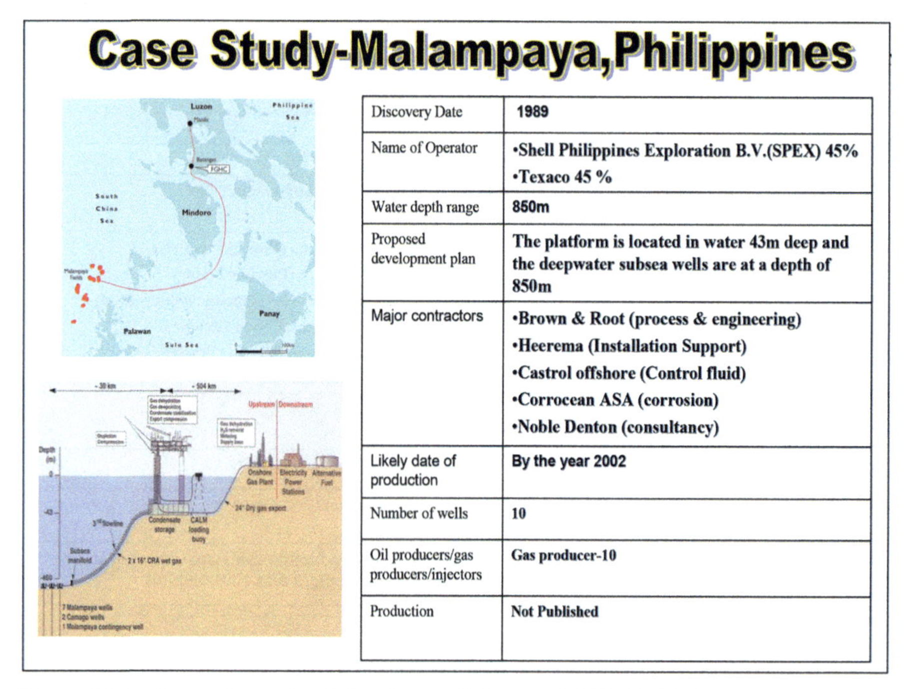

Case Study-Malampaya,Philippines

Discovery Date	1989
Name of Operator	•Shell Philippines Exploration B.V.(SPEX) 45% •Texaco 45 %
Water depth range	850m
Proposed development plan	The platform is located in water 43m deep and the deepwater subsea wells are at a depth of 850m
Major contractors	•Brown & Root (process & engineering) •Heerema (Installation Support) •Castrol offshore (Control fluid) •Corrocean ASA (corrosion) •Noble Denton (consultancy)
Likely date of production	By the year 2002
Number of wells	10
Oil producers/gas producers/injectors	Gas producer-10
Production	Not Published

Figure 14.4 Production and well data for an offshore well drilled in Malampaya, Philippines.

Case Study-Bonga,Nigeria

Discovery Date	1995
Name of Operator	•Esso (20%) •Nigeria Agip (12.5%) •Elf Petroleum Nigeria Limited (12.5%).
Water depth range	1245m
Proposed development plan	Subsea with FPSO
Major contractors	•Coflexip (Flowline) •ABB (EPC) •Emerson (Process management) •McDermott Fabricators (Fabrication)
Likely date of production	2003
Number of wells	36
Oil producers/gas producers/injectors	20 producers and 16 injectors
Production	225,000b/d

Figure 14.5 Production and well data for an offshore well drilled in Bonga, Nigeria.

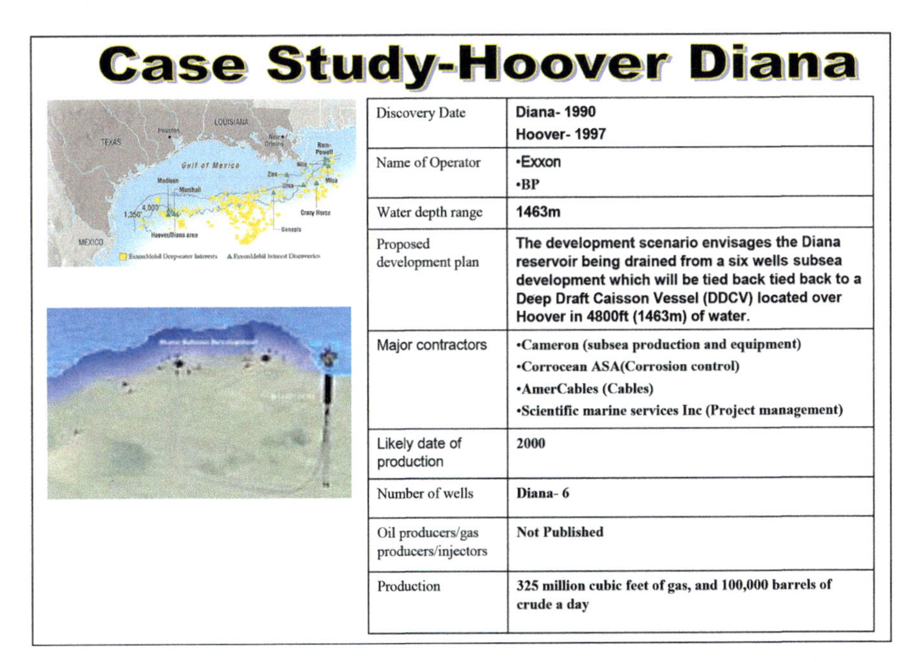

Case Study-Hoover Diana

Discovery Date	Diana- 1990 Hoover- 1997
Name of Operator	•Exxon •BP
Water depth range	1463m
Proposed development plan	The development scenario envisages the Diana reservoir being drained from a six wells subsea development which will be tied back tied back to a Deep Draft Caisson Vessel (DDCV) located over Hoover in 4800ft (1463m) of water.
Major contractors	•Cameron (subsea production and equipment) •Corrocean ASA(Corrosion control) •AmerCables (Cables) •Scientific marine services Inc (Project management)
Likely date of production	2000
Number of wells	Diana- 6
Oil producers/gas producers/injectors	Not Published
Production	325 million cubic feet of gas, and 100,000 barrels of crude a day

Figure 14.6 Production and well data for an offshore well drilled in Hoover, Diana.

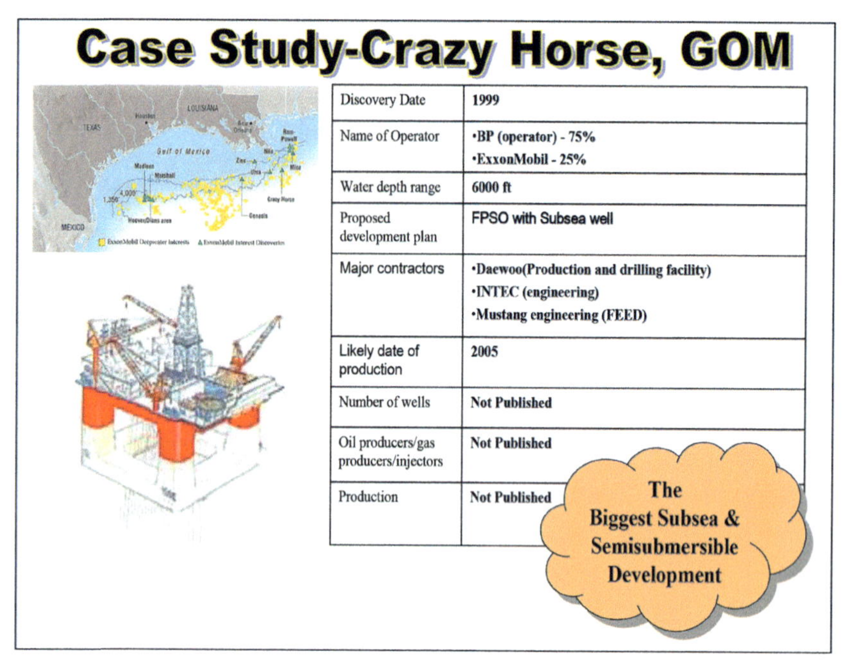

Case Study-Crazy Horse, GOM

Discovery Date	1999
Name of Operator	•BP (operator) - 75% •ExxonMobil - 25%
Water depth range	6000 ft
Proposed development plan	FPSO with Subsea well
Major contractors	•Daewoo(Production and drilling facility) •INTEC (engineering) •Mustang engineering (FEED)
Likely date of production	2005
Number of wells	Not Published
Oil producers/gas producers/injectors	Not Published
Production	Not Published

Figure 14.7 Production and well data for an offshore well drilled in Malampaya, Philippines.

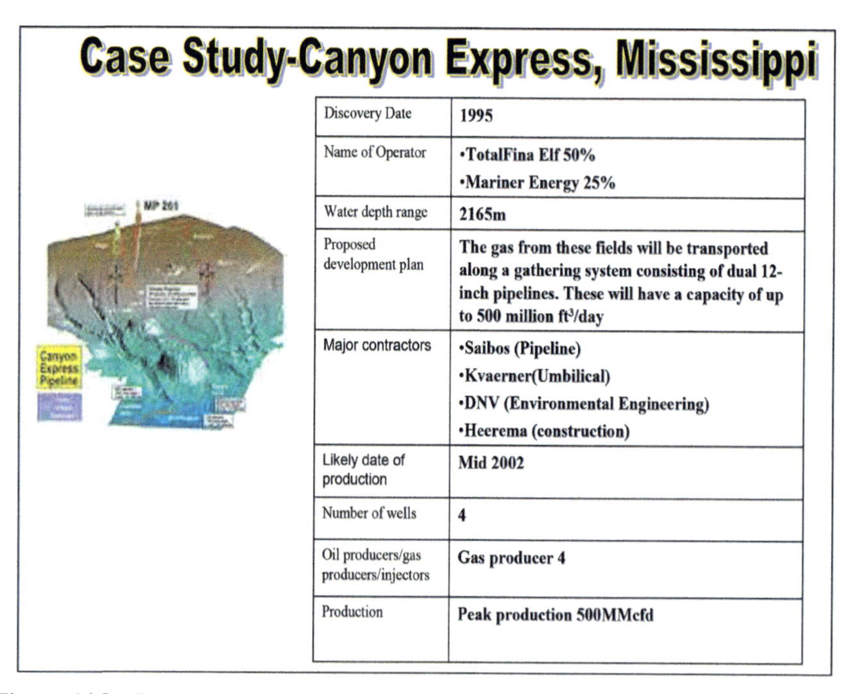

Case Study-Canyon Express, Mississippi

Discovery Date	1995
Name of Operator	•TotalFina Elf 50% •Mariner Energy 25%
Water depth range	2165m
Proposed development plan	The gas from these fields will be transported along a gathering system consisting of dual 12-inch pipelines. These will have a capacity of up to 500 million ft^3/day
Major contractors	•Saibos (Pipeline) •Kvaerner(Umbilical) •DNV (Environmental Engineering) •Heerema (construction)
Likely date of production	Mid 2002
Number of wells	4
Oil producers/gas producers/injectors	Gas producer 4
Production	Peak production 500MMcfd

Figure 14.8 Production and well data for an offshore well drilled in Canyon Express, Mississippi.

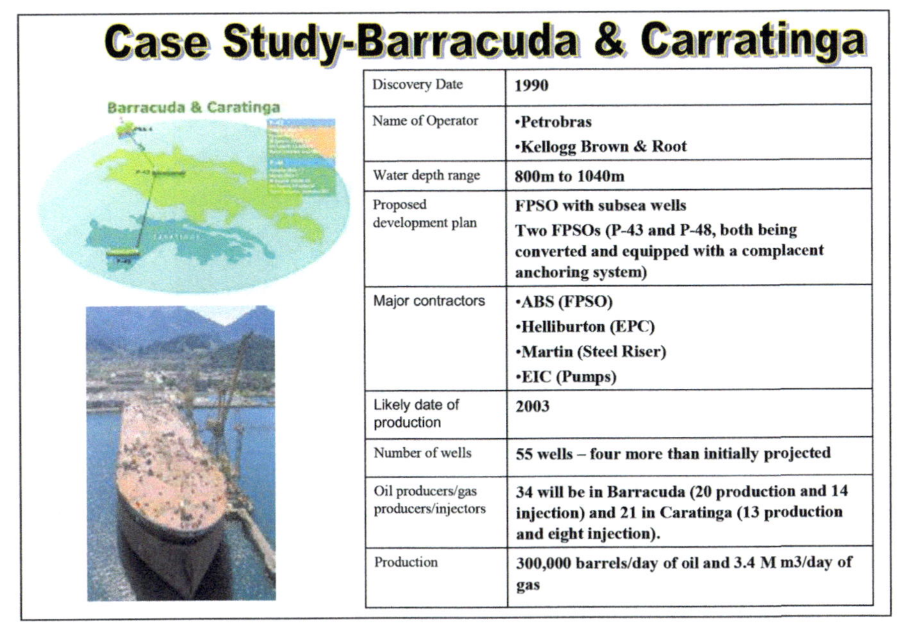

Discovery Date	1990
Name of Operator	•Petrobras •Kellogg Brown & Root
Water depth range	800m to 1040m
Proposed development plan	FPSO with subsea wells Two FPSOs (P-43 and P-48, both being converted and equipped with a complacent anchoring system)
Major contractors	•ABS (FPSO) •Helliburton (EPC) •Martin (Steel Riser) •EIC (Pumps)
Likely date of production	2003
Number of wells	55 wells – four more than initially projected
Oil producers/gas producers/injectors	34 will be in Barracuda (20 production and 14 injection) and 21 in Caratinga (13 production and eight injection).
Production	300,000 barrels/day of oil and 3.4 M m3/day of gas

Figure 14.9 Production and well data for an offshore well drilled in Barracuda, Carratinga.

14.22 Natural Gas Hydrates

14.22.1 Natural gas hydrates

- The depleting conventional oil and gas reservoirs, increasing supply demand gap and growing recognition of the need for increased supplies of cleaner fuel has necessitated the need to look for both conventional and unconventional hydrocarbon resources in harsher environments.
- One such unconventional resource is natural gas hydrates(Figure 14.10).

14.22.2 Introduction to natural gas hydrates

- *Natural methane hydrate*, an ice-like crystalline substance, is the most common form of a unique class of chemical compounds known as *clathrates*.
- Clathrates are characterized by a rigid, open network of bonded host molecules that enclose, without direct chemical bonding, appropriately sized guest molecules of another substance.

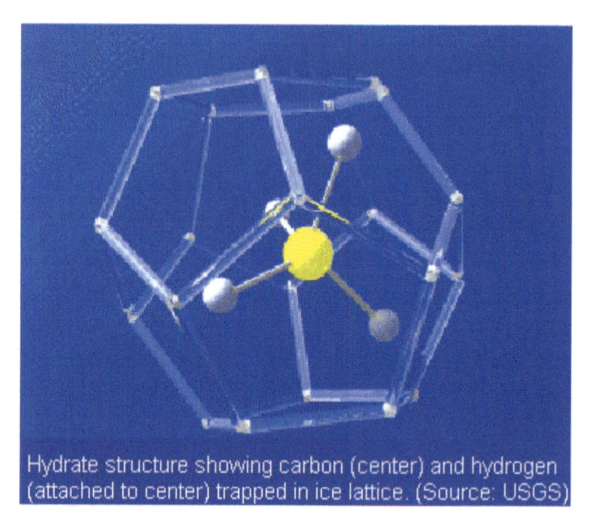

Figure 14.10 Gas Hydrates.

- **Hydrate as a lab curiosity:**
  - Clathrate compounds are known since 1810 when Humphrey Davy and Michael Faraday noticed a solid material forming at temperatures above natural freezing point of water with chlorine–water mixture.
- **Hydrate as an industrial nuisance:**
  - Hydrates were responsible for plugging gas pipelines, particularly those located in cold environments (E.G.Hammerschmidt, 1930).
- **Hydrate as a naturally occurring substance:**
  - In the late 1960s, the global view of clathrate science began to change dramatically when "solid natural gas" or methane hydrate was observed as a naturally occurring constituent of subsurface sediments in the giant gas fields of the western Siberia and also in the shallow sun permafrost sediments on the north slopes of Alaska.
- **Structure of hydrates**:
  - Hydrates are formed with differently-sized molecules. Each unit cell includes a fixed number of strongly bonded hydrogen–water molecules. Size and shape of the unit cell is determined by the size and the energy of the clathrate hydrate former guest molecules (Figure 14.11).
- **Structures I and II:** Two primary types of hydrate structures are known to exist commonly in nature, termed simply, structure I and structure II. These structures represent different arrangements of water molecules resulting in

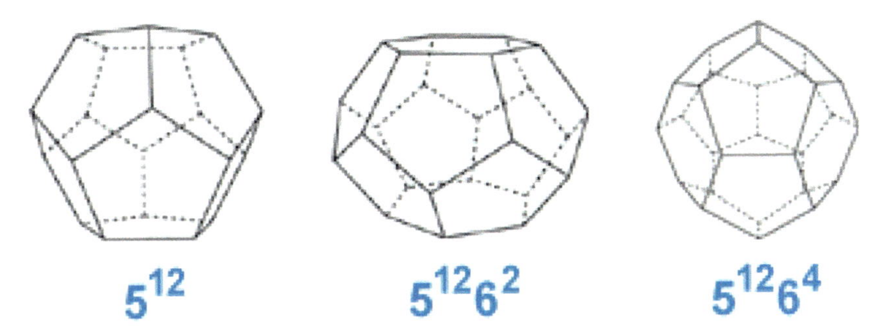

Figure 14.11 The three types of cavities present in structure I and II methane hydrates (Courtesy of Center for Gas Hydrate Research at Heriot-Watt University).

Table 14.6 Geometry of hydrate cavities for different hydrate sizes.

	Geometry of hydrate cavities						
	I	I	II	II	H	H	H
Cavity size	Small	Medium	Small	Large	Small	Small	Huge
Cavity shape	Round	Oblate	Round	Round	Round	Round	Oblate
Cavity description	5^{12}	$5^{12}6^{2}$	5^{12}	$5^{12}6^{4}$	5^{12}	$4^{3}5^{6}6^{3}$	$5^{12}6^{8}$
Number/unit cell	2	6	16	8	3	12	1
Average radius (A)	3.91	4.33	3.902	4.683	3.91	4.06	5.71
Rel. size of CH_4	88.6%	75.7%	88.9%	67.5%	88.6%		
Coordination no.	20	24	20	28	20	20	36

slightly different shapes, sizes, and assortments of cavities. Which structure forms depends on various aspects of the available guest gas.

- Methane preferentially forms structure I.
- A unit cell (the smallest repeatable element) of a structure I hydrate consists of 46 water molecules surrounding 2 small cavities and 6 medium-sized cavities.
- The unit cell of structure II hydrates consists of 136 water molecules creating 16 small cavities and 8 large cavities.
- Both structures I and II can be stabilized by filling at least 70% of the cavities by a single guest gas – and are therefore known as *simple hydrates* (Tabe 14.6).
- **Structure H:** The third hydrate structure (structure H) that requires the cooperation of two guest gases (one large and one small) to be stable. A unit cell of this new *double hydrate* consists of 34 water molecules producing 3 small cavities, 12 slightly larger cavities, and 1 relatively huge cavity (Figure 14.12).

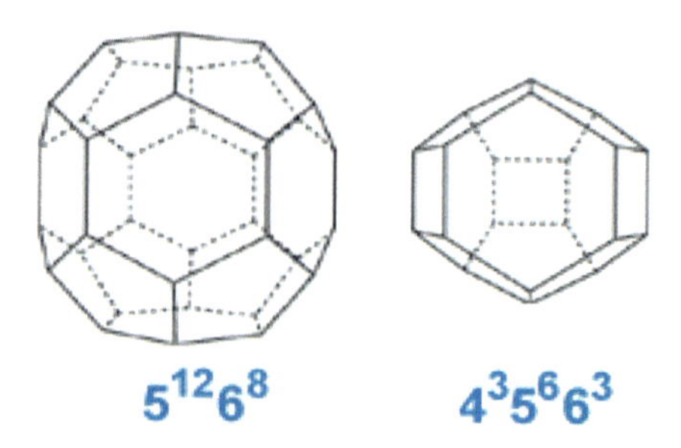

Figure 14.12 The two types of cavities unique to structure H methane hydrates (Courtesy of Center for Gas Hydrate Research at Heriot-Watt University).

Table 14.7 Hydrate structures and properties.

Structure and formers	Number of polyhedral cages in unit cell	Number of water molecules	Formula if all cages are filled[a]
Structure I. Small molecules such as methane, ethane, carbon dioxide, and hydrogen sulfide in large cages.	8 cages (2 large and 6 small cages)	46	8X.46 H_2O or X.5 ¾ H_2O, where X is the guest molecule
Structure II. Larger[c] molecules such as propane, and isobutene in the large cages.	24 cages (8 large and 6 small cages)	136	24X.136 H_2O or X.5 2/3 H_2O or 8X.136 H_2O or X.17 H_2O[b]
Structure H. Larger molecules but only in the presence of a smaller hydrate former, such as methane.	6 cages (1 large, 3 medium, and 2 small cages)	34	X.5 Y.34 H_2O where X is the large molecule and Y, the small[d]

[a] Hydrates are non-stoichiometric.
[b] When only the large cages are occupied.
[c] Examples of hydrate formers are 2-methylbutane, methyl cyclo-pentane, methyl cyclo-hexane, and cyclo-octane.
[d] Large molecules and small molecules occupy large and small cages, respectively.

It is this large cavity that allows structure H hydrates to incorporate large molecules (such as butane and larger hydrocarbons), given the presence of other smaller *help gases* to fill and support the remaining smaller cavities (Table 14.7). Structure H hydrates are rare, but are known to exist in the Gulf of Mexico, where supplies of thermogenically produced heavy hydrocarbons are common.

Physical properties of hydrates and ice:

Table 14.8 Summary of published values for properties of ice and pure gas hydrates modified from Davidson (1983).

Property	Ice	Hydrate
Dielectric constant at 273 K	94	~58
NMR rigid lattice second moment of H_2O protons (G2)	32	33 ± 2
Water molecule reorientation time at 273 K (μsec)	21	~10
Diffusional jump time of water molecules at 273 K (μsec)	2.7	> 200
Isothermal Young's modulus at 268 K (109 Pa)	9.5	~8.4
Speed of longitudinal sound at 273 K velocity (km/sec)	3.8	3.3
transit time (μsec/ft)	8.0	92
Velocity ratio V_p/V_s at 272 K	1.88	1.95
Poisson's ratio	0.33	~0.33
Bulk modulus (272 K)	8.8	5.6
Shear modulus (272 K)	3.9	2.4
Bulk density (gm/cm^3)	0.916	0.912
Adiabatic bulk compressibility at 273 K 10–11 Pa	12	~14
Thermal conductivity at 263 K (W/m^{-K})	2.23	0.49 ± 0.02

~ symbol = nearly equal to, approximately

Gas storage capacity:

- The structure of methane hydrate compresses methane molecules into a very dense and compact arrangement (Table 14.8).
- When dissociated at normal surface temperatures and pressures, a 1 ft^3 block of solid methane hydrate with 100% void occupancy by methane will release roughly 164 standard cubic feet of methane (however, occupancy typically ranges from 70% to 90%).
- This gives methane hydrate an energy content of roughly 184,000 btu/ft^3. This value lies in between the values for methane gas (1150 btu/ft^3) and liquefied natural gas (LNG: 430,000 btu/ft^3).

Necessary conditions for methane hydrate formation:

- Methane hydrate, much like ice, is a material very much tied to its environment – it requires very specific conditions to form and be stable. When these conditions are removed, it is very likely that it will quickly dissociate into water and methane gas (Figure 14.13).

Current understanding of natural methane hydrate indicates that the fundamental controls on hydrate formation and stability are:

1. Adequate supplies of water and methane

Figure 14.13 Methane actively dissociating from a hydrate mound.

2. Suitable temperatures and pressures
3. Geochemical conditions
4. Other controls, such as sediment types and textures, may also play an important role

14.22.3 Hydrate stability

14.22.3.1 Temperatures and pressures

Given adequate supplies of water and methane, Figure 14.14 shows the combination of temperatures and pressures (the phase boundary) that marks the transition from a system of co-existing free methane gas and water/ice solid methane hydrate. When conditions move to the left across the boundary, hydrate formation will occur. Moving to the right across the boundary results in the dissociation (akin to melting) of the hydrate structure and the release of free water and methane.

In general, a combination of low temperature and high pressure is needed to support methane hydrate formation.

- In addition to temperature and pressure, the composition of both the water and the gas is critically important for fine-tuning predictions of the stability of gas hydrates in specific settings.
- However, natural subsurface environments exhibit significant variations in formation water chemistry, and these changes shift the

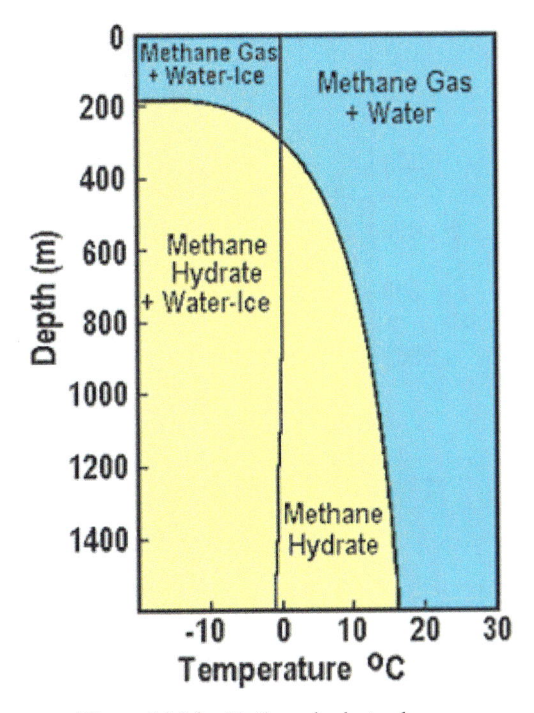

Figure 14.14 Methane hydrate phases.

pressure/temperature phase boundary (higher salinity restricts hydrate formation causing the phase boundary to shift to the left). Similarly, the presence of small amounts of other natural gases, such as carbon dioxide (CO_2), hydrogen sulfide (H_2S), and larger hydrocarbons such as ethane (C_2H_6), will increase the stability of the hydrate, shifting the curve to the right.

- The phase diagram in Figure 14.15 illustrates typical conditions in a region of arctic permafrost (with depth of permafrost assumed to be 600 m). The overlap of the phase boundary and temperature gradient indicates that the GHSZ should extend from a depth of approximately 200 m to slightly more than 1000 m. Both the permafrost thickness and pressure/temperature gradients can vary with locality, requiring specific phase diagram for predicting hydrate stability.
- The phase diagram in Figure 14.16 shows a typical situation on deep continental shelves. Temperature steadily decreases with water depth, reaching a minimum value near 0 °C at the ocean bottom. Below the sea floor, temperatures steadily increase. In this setting, the top of the GHSZ occurs at roughly 400 m – the base of the GHSZ is at 1500 m.

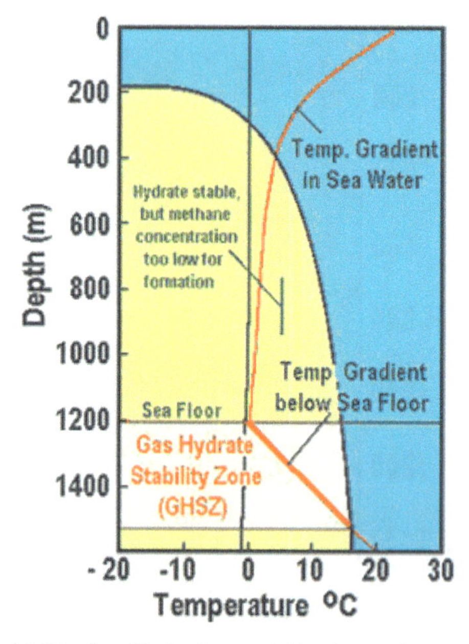

Figure 14.15 Specific hydrate stability for arctic permafrost.

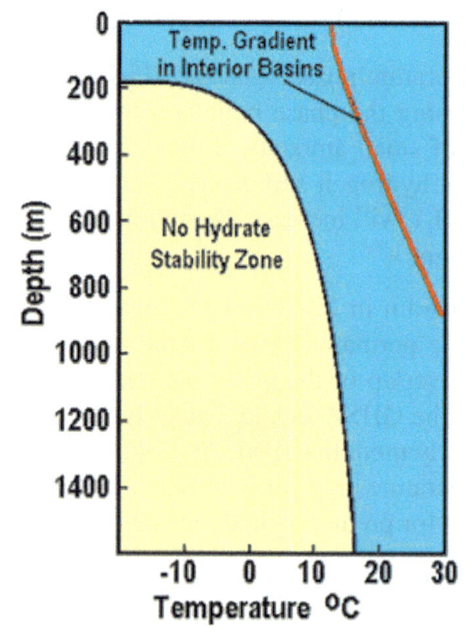

Figure 14.16 Typical occurrence of the gas hydrate stability zone on deepwater continental margins.

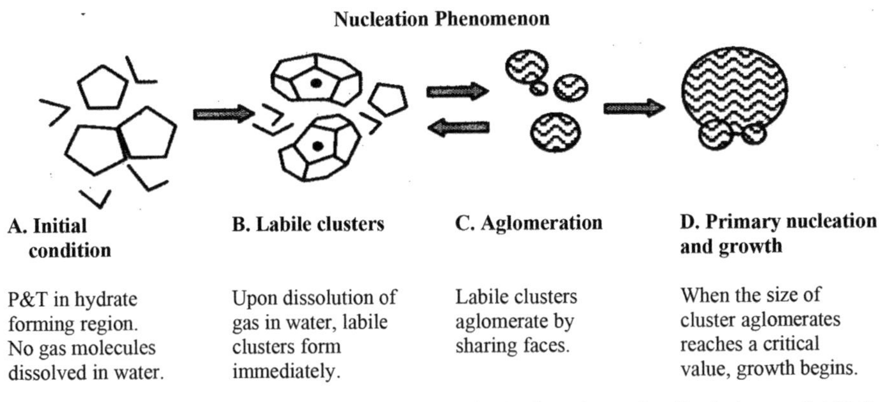

Figure 14.17 Autocatalytic reaction mechanism for hydrate formation (Lederhos et al. 1996).

However, hydrates are likely to accumulate in the sediments, which restrict them from floating up (they are less dense than water) and out of the stability zone.

- From the phase diagram, it appears that hydrates may accumulate in the ocean-bottom sediments where water depth exceeds ~400 m.
- However, very deep (abyssal) sediments are generally not thought to house hydrates in large quantities. The reason is that deep oceans lack both the high biologic productivity (necessary to produce the organic matter that is converted to methane) and rapid sedimentation rates (necessary to bury the organic matter) that support hydrate formation on the continental shelves.

14.22.3.2 Mechanism of hydrate formation

This process is complex and not well understood; several theories have been developed explaining the mechanisms of hydrate formation. Lederhos et al. (1996) proposed that gas hydrates form in an autocatalytic reaction mechanism, when water molecules cluster around natural gas molecules in a structure similar to the ones shown in Figure 14.17.

This attraction between neighboring guest molecules is termed "hydrophobic bonding," which can be described as an attraction between the apolar molecules inside the clusters [B]. Large and small clusters forming structures I and II are termed "Labiles" because they are easy to break down, but relatively long-lived. Labels can dissipate or grow to become hydrate unit cells or agglomerations of unit cells forming what is known as "Metastable Nuclei" [C]. Then, growth can continue until crystals are stable, indication the onset of secondary nucleation [D]. This process is illustrated in Figure 4. This theory implies reversibility of the process when the system is heated up.

14.22.3.3 Types of hydrate deposits

a. **Finely disseminated:** These are dispersed in the pore spaces of the rock and does not exceed the size of the pores of the carrier rock. This type was found in Orea and Mississippi Canyon areas of the Gulf of Mexico.

b. **Nodular hydrates:** These are up to 5cm in diameter and may occur such as those found in Green Canyon, Gulf of Mexico. The gas in this hydrate may be thermogenic in origin that migrated from some depth. Formation of hydrate nodules is accompanied by the destruction of the carrier rock and hydrates act as cementing material.

c. **Layered type:** Growth of hydrate nodules may result in the formation of lens-shaped hydrate accumulations extending up to several meters in length and tens of centimeters in thickness. These layers are separated by thin layers of sediments, such as cores recovered from Blake-Bahama ridge. It can occur both in permafrost and offshore.

d. **Massive hydrates:** This type was recovered from sites 570 of DSDP Leg 84 off the Middle America Trench. It may be as thick as 3–4 mts and contain 95% of hydrate and less than 5% sediments.

- Thermogenic hydrates are more mature and massive compared to biogenic, as greater supply of gas was available in faults and diapirs.
- Similarly, hydrate is typically found with uniform distribution, showing clear but subtle vertical trends of increasing or decreasing abundance. However, examples of heterogeneous distribution with zones of sparse or no hydrate interspersed with zones of high concentration are also common. The factors that control the ultimate type, distribution, and amount of hydrate are:

  1. The porosity and permeability.
  2. The degree of lithification of the enclosing medium.

- In general, there are two primary geologic/geographic environments for hydrate accumulation:

  1. Areas with deepwater in close proximity to land.
  2. Continents in polar regions.

14.22.3.4 Hydrate distribution worldwide

Distribution of organic carbon on earth:

- The worldwide amounts of carbon bound in gas hydrates are conservatively estimated to total twice the amount of carbon to be found in all known fossil fuels on Earth.
- The magnitude of the energy available for methane hydrates was initially quantified by the USGS in 1995. The USGS suggested that hydrate deposits entrap approximately 112,000 trillion cubic feet of methane gas.
- The estimate was refined in 1997 to a more conservative 200,000 trillion cubic feet. This estimate is significant when compared to the 1400 trillion cubic feet in the nation's conventional reserves. On a worldwide basis, it is estimated that methane hydrate reserves are 400 million trillion cubic feet compared with 5000 trillion feet in known gas reserves

In many areas, the existence of natural methane hydrate is inferred only from indirect evidence obtained through geophysical surveys or geochemical analyses of sediment samples. However, there are a growing number of localities where detailed information is being collected. Each of these localities, with their own unique geologic settings, is unveiling surprising information that questions the initial theories of hydrate formation and ultimately advances the general state of knowledge of natural gas hydrate.

Although hydrates associated with permafrost contain only a small fraction of the global methane hydrate resource, areas like the North Slope of *Alaska* provide excellent opportunities to study natural hydrates by combining the data gained from more than two decades of drilling of wells with relative ease of access. In 1998, the *Mackenzie River Delta of Canada's Northwest Territories* was the site of the world's first research well drilled specifically to study natural methane hydrate.

The *Messoyahka gas field of the West Siberia basin* is another well-known example of an arctic hydrate accumulation. Hydrate was inferred from well logging and other data during initial drilling of the field in 1964, and debate continues as to whether dissociation of the hydrate in response to production of deeper free gas zones has resulted in actual production of methane from hydrate.

Perhaps the best-known and most closely studied oceanic hydrate locality is the *Blake Ridge*, a large pile of deepwater sediment located off the eastern coast of North America. The Blake Ridge has been scanned and probed regularly since the first evidence of hydrate was collected there in the early

1970s. The Blake Ridge's uniform sediment makes it an ideal laboratory for fine-tuning the tools and techniques that will be used to study hydrate accumulations around the globe.

14.22.3.5 Hydrate mound on sea floor

A third prominent locality for hydrate occurrence is the deepwater *Gulf of Mexico*. Unlike the Atlantic shore, the Gulf is an area of significant production of conventional oil and gas. The hazards that unintentional hydrate dissociation pose for drill rigs, pipelines, and other equipment are a prime driver for focused study of hydrates in the Gulf. The significant geologic differences that exist between the Blake Ridge and the Gulf result in the presence of unique features, including visible mounds of hydrate directly on the sea floor. In recent years, scientists have visited the deep gulf in submersible vehicles to observe and sample the mounds. Among the many discoveries are unique chemosynthetic communities, including previously unknown species called ice worms, that derive sustenance not from the sun but directly from the methane slowly dissociating from the hydrate.

In 1999, *Japan's Nankai trough* region was the target of the first well drilled specifically to test the resource potential of oceanic hydrate. The geologic setting (a subduction zone), characterized by the close proximity of deepwater to the land and the resultant improved reservoir character of the sediment, may eventually result in the Nankai region being the host to the first attempts at commercial methane production from hydrate. Examination of a similar tectonic setting on the Pacific coast of North America has resulted in identification of a promising hydrate locality named "Hydrate Ridge," located offshore Oregon. Hydrate Ridge was recently ranked the most important marine hydrate site for scientific study, and therefore will be a prime focus of an upcoming leg of the Ocean Drilling Project.

14.22.4 Gas hydrates: Indian scenario

- Over the last few years, the economic development in India has been remarkable but unfortunately there has not been any significant breakthrough in the conventional hydrocarbon energy resource.
- There has been a constant decline in the reserves of conventional energy resources in India. However, significant efforts are being put in by Indian scientists and researchers to discover new fields to meet the ever-increasing demand of energy in India.

Table 14.9 The demand–supply balance based on the demand and supply projections till 2025 (Hydrocarbon vision 2025).

	2002	**2007**	**2012**	**2025***
SCENARIO-1				
Demand	117	166	216	322
Supply as a given scenario	70	58	45	36
Gap	**47**	**108**	**171**	**286**
Supply optimistic scenario	70	64	78	84
Gap	**47**	**102**	**138**	**238**
SCENARIO-2				
Demand	151	231	313	391
Supply as a given scenario	70	58	45	36
Gap	81	173	268	355
Supply optimistic scenario	70	64	78	84
Gap	81	167	235	307

* In the absence of supply projection for 2025, it is assumed that gas supply in 2025 is the same as those for 2020.

Some success has been achieved in this regard in the form of marginal fields, but that may not be sufficient. India still largely depends on conventional resources for its energy requirements and a major part is through imports. There has been an increasing trend in the utilization of gas as an energy resource due to environmental concerns and also due to decline in crude oil reserves .

Given the significant decline in reserves over the past few years, it is clear that unless reserves in the existing fields are significantly upgraded or new fields are discovered and brought onstream, gas production is set to decline from its current level (Table 14.9) (Deka Bhabesh et al. 1998).

The focus in this era is on unconventional energy resources, and one of the most important sources, which is highly looked upon, is "natural gas hydrates." Gas hydrates are considered to be a source of enormous energy that can possibly meet the increasing demand of energy.

One cubic meter of gas hydrate contains 164 cubic meter of methane gas at STP and a large area of deepwater has favorable conditions for formation of gas hydrates, which makes it a suitable candidate as an alternate source of clean energy. Deepwater basins in India are considered to be containing very good amount of gas hydrates.

For the conditions of the Indian offshore, a hydrate formation zone (HFZ) can be up to 600-m thick or more in the regions with water depth over 700 m. The total reserve of natural gas in the form of gas hydrates and free gas

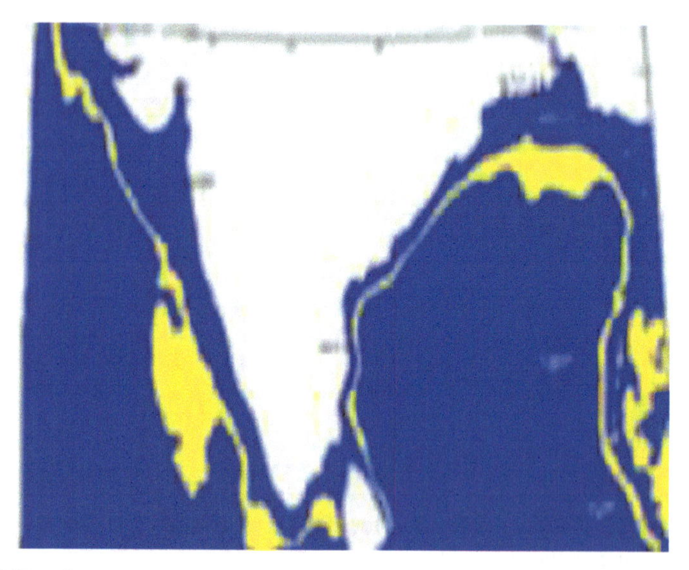

Figure 14.18 General map of the prospective regions for GHD in the Indian offshore in water depths up to 2000 m, based on the geological and thermodynamic data

below hydrate deposits in the Indian offshore for water depth up to 2000 m is estimated to be from 40–120 trillion m^3.

- India has been divided into seven hydrate zones as follows:
  1. Bombay Offshore
  2. Southern Arabian Sea Play
  3. Northern Arabian Sea Play
  4. Kerala Konkan Offshore
  5. Eastern Offshore
  6. Northern Bay of Bengal Play
  7. Southern Bay of Bengal Play
- Potential reserves of gas in the free and hydrate states can be over 6 trillion m^3 in the Andaman offshore (Figure 14.18). Other highly prospective offshore regions are located in east coast, Krishna-Godavari, and west coast of Goa at water depth over 700–800 m (Makogon Y.F. et al. 1993).
- The initiative to exploit gas hydrate resources in India have been started way back in 1996–1997 when National Gas Hydrate Program (NGHP)

Table 14.10 Energy contents of various gas resources.

Gas resource	Energy content (cu. ft. gas/cu. ft. of reservoir rock)*
Methane hydrates	50
Coalbed methane	8-16
Tight sands	5-10
Devonian shale	2-5
Conventional gas	10-20

* Assuming a reservoir with 30% porosity and depth less than 5000 ft.

was launched by Ministry of Petroleum and Natural Gas (MOPNG). Simultaneously ONGC, GAIL, and DGH started in-house studies for evaluation of gas hydrates in Indian offshore areas. As far as the amount of hydrate resources are concerned, much headway has been made by these companies in association with National Institutes like NGRI, NIO, etc. The preliminary estimates suggest that the initiatives must be taken for exploration of gas hydrates in India.

14.22.4.1 Conceptual production techniques

- A major interest to produce gas from hydrate is its high concentration of energy.
- As shown in table 14.10, a cubic foot of hydrate in a reservoir rock with 30% porosity may hold 50 cu. feet of gas, which is many times greater than that stored in other gas sources at moderate reservoir depths.
- This high energy concentration is comparable in energy content with heavy oil and tar sands and gives hope that thermal injection, pressure reduction, or other recovery methods may be applicable for hydrate production of gas from gas hydrates.
- Techniques for gas production from the hydrates reservoir are somewhat based on the same principle as those utilized for prevention of gas hydrates formation.
- Disturbing the equilibrium pressure and temperature can result in dissociation of the gas hydrate.
- Based on the two parameters (i.e., pressure and temperature), the following gas hydrates production techniques have been evolved:

    1. Thermal stimulation
    2. Depressurization
    3. Inhibition by chemical injection

The first two methods are interrelated as any change in one of these parameters changes the other.

14.22.4.2 Thermal stimulation

- It involves increasing temperature beyond hydrate formation temperature. In this technique, the heat is supplied by a source of heat to raise the temperature of the reservoir for breaking the hydrate hydrogen bond. As the temperature is increased, the conditions reach out of hydrates stability zone and the gas is released. Some of the methods for supplying heat to hydrates reservoir may be:
    - Steam injection
    - Hot water injection
    - Hot brine injection
    - Electromagnetic heating
- During heat injection (by brine, steam, or water), hydrates dissociate with temperature. Net energy balance in a closed system and in a high-quality hydrate reservoir is positive.
- The main problem with thermal stimulation techniques is that high heat losses occur in reservoirs and surrounding rock strata.
- Some investigators have calculated that heat required to dissociate hydrates is only 10% of the heating value of the produced natural gas (SPE 19810). Thus, the thermal stimulation techniques are energy efficient from the thermodynamic viewpoint. In steam injection, however, the heat losses in the wellbore and reservoir can be severe, especially for thinner hydrate zones.
- Further it has been argued by some investigators that hot water injection will yield lower heat losses than steam injection, but the injectivity of water in the hydrate reservoirs will govern the applicability of this method. Fracturing can be used to improve water injectivity, but that may result in lower heat transfer efficiency because of channeling effects.
- It has been prognosticated that injection of hot brine to dissociate hydrates is thermally more efficient than steam or hot water injection because brine also acts as a hydrate inhibitor. Brine causes reduction in dissociation temperature, which reduces reservoir heating, latent heat of dissociation, and heat losses, and increases the gas production rate.

- One of the new methods of development of gas hydrate deposits is a high-frequency electromagnetic technology. Applying the heat with a down-hole device, such as electromagnetic heating, is a much more efficient process than conventional thermal recovery techniques. In this technique, a strong high-frequency electromagnetic wave (HFEMW) is radiated from a well-bottom radiator (antenna) into the surrounding medium (productive gas hydrate layer). As the HFEMW propagates in a gas hydrate deposit, it interferes with the deposit. As a result, a heat source is generated in the productive gas hydrate layer (natural gas hydrates, Makogon Y. F.). Interaction between a strong electromagnetic field and hydrate-bearing porous rock results in its thermodynamic state changing. If the temperature reaches the value of phase transition temperature, then the hydrate starts to decompose into gas and water.

To make any progress in applying this concept of electromagnetic field, it is necessary first to study the dielectric properties of gas hydrates as a function of pressure, temperature, and frequency.

14.22.4.3 Depressurization

- It is considered that this method is feasible only when associated free gas can be produced to decrease hydrate reservoir pressure, as has been reported for the Messoyakha field, Russia (Oilfield Review, Summer 2000).
- The process of hydrates dissociation is endothermic, absorbing energy and reducing reservoir temperature.
- The process requires heat flow into the reservoir from the surrounding rock. Under normal conditions, the depressurization process alone tends to become self-limiting.
- Where hydrates overlap a free-gas zone, this technique is advantageous because hydrate dissociation can contribute significantly to gas production from a free-gas zone.

14.22.4.4 Inhibition

- Injection of inhibitors such as methanol/glycol/electrolytes shifts the pressure–temperature equilibrium so that the hydrates are no longer stable in their *in situ* conditions and methane is released.
- The main challenge associated with this technique is that inhibitor materials are expensive because large quantities of these expensive chemicals are needed to ensure sufficient gas production. Also, there may be

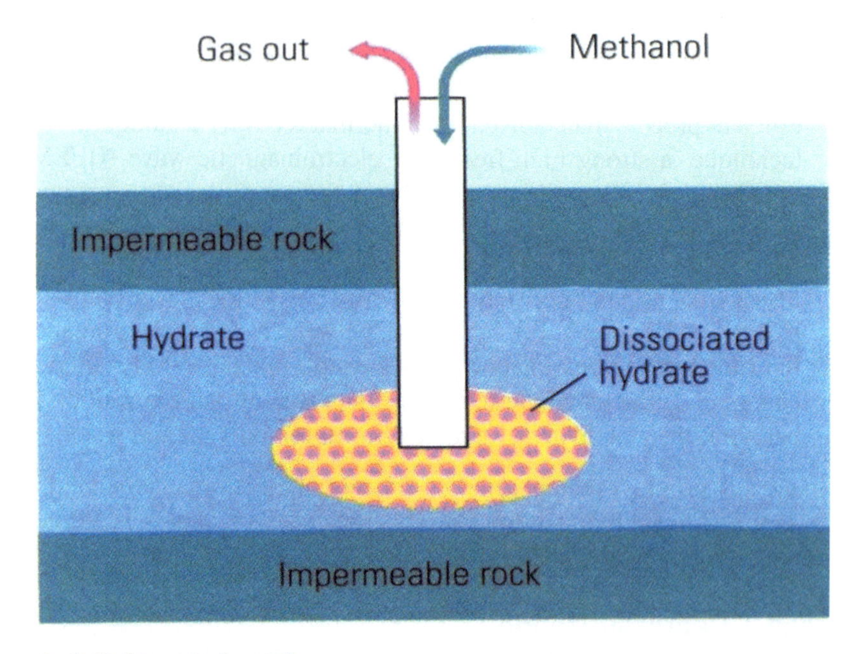

Figure 14.19 Pictorial representation of inhibitor injection into hydrate formation.

difficulty with inhibitor-hydrates surface contact. So far, no model has been developed to produce gas based on this principle of disturbing the hydrate equilibrium (Figure 14.19).

- The environmental impact of using large volumes of chemicals is also a very significant concern for developing this technique.

14.22.4.5 Mining

- The process that is thought of is to develop a mechanism at sea bottom to convert the sediments into slurry. The slurry of sediments, which contains gas hydrate crystals, is pumped up to a water depth where the hydrate starts dissociating gas. A mechanism that separates gas and sediments is installed at this level. The collected sediments are dumped back into the sea and gas is brought to a surface storage facility (Figure 14.20).
- This conceptual mechanism involves mining of seabed sediments up to a depth of 250–300 m below the seabed. The operation is to be carried

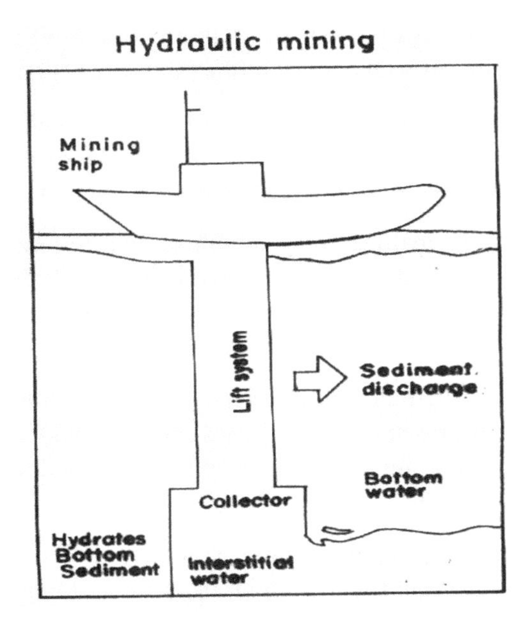

Figure 14.20 Mechanism at sea bottom to convert sediments into slurry.

out at water depths ranging from 750 to 2000 m and the area required to be mined may be a few thousand sq. km. The question is how much gas can be extracted by such a process.

- Based on the current knowledge, it is perceived that the top few hundred meters of sediments may not have sufficient gas hydrate concentration and only the lower HSZ may have appreciable gas hydrate concentrations.

14.22.5 Hydrate hazards

- Hydrates present constraints to oil and gas flow, cause drilling and subsea completion hazards, and induce risks to offshore platform stability.
- For operators drilling in deepwaters, encountering naturally occurring solid gas hydrates during drilling can pose a well control problem, if large amounts enter the borehole and depressurize.
- Furthermore, circulation of warm fluid within the wellbore can reduce the temperature in the surrounding hydrate-rich sediments, leading to hydrate melting and destabilization of the sediments holding up the well.

- Developing solid hydrates in the wellbore as a result of fluid mixing is another significant well control problem in deepwaters.
- Dissociation of hydrates can cause instability in seafloor sediments on the continental slopes. The base of the hydrate zone may represent a discontinuity in the strength of the sediment column.
- The presence of hydrates may inhibit normal sediment consolidation and compaction, and free gas strapped below the hydrate zone may be overpressured. Any technique proposed for hydrate exploitation would have to succeed without causing additional instability.
- Dissociation of hydrates may be responsible for submarine landslides. A decrease in pressure on the hydrate zone would allow hydrates at depth to dissociate and cause the unconsolidated sediments above them to slide.
- Methane increases the greenhouse effect about 20 times more aggressively than an equivalent weight of carbon dioxide. Uncontrolled release of methane into the atmosphere due to the dissociation of natural gas hydrates can lead to global warming and the destabilization of the climate.

14.23 Conclusion

- The increasing requirement for large volumes of natural gas as a relatively clean hydrocarbon fuel and the discovery of large gas hydrate accumulations in terrestrial permafrost regions of the arctic and beneath the sea along the outer continental margins of the world's oceans have heightened interest in gas hydrates as a possible energy resource of the future.
- However, significant and potentially insurmountable technical issues are required to be resolved before gas hydrates can become a techno economically viable and safe option of natural gas.

14.24 References

[1] Berecz and Balla-Achs M.: "Gas Hydrates Studies in Inorganic Chemistry." 1983, Elsevier, Amsterdam-Oxford-New York.
[2] E. Dendy Sloan Jr.: "Natural Gas Hydrates" December 1991, Journal of Petroleum Technology, 1414.

[3] Holder G.D. and Yen S. “Geological Implications of Gas Production From In-Situ Gas Hydrates.” 1980, SPE/DOE Symposium on Unconventional Gas Recovery, Pittsburgh.
[4] John L. Cox. : “Natural Gas Hydrates: Properties, Occurance and Recovery.” Butterworth Publishers, Boston, London.
[5] Kamath V.A. et al.: “Experimental Study of Brine Injection and Depressurization Methods for Dissociation of Gas Hydrates.” SPE Formation Evaluation, December 1991.
[6] Makogon Yuri F.: “Hydrates of Hydrocarbons”, 1997, Pennwell Publishing Company, Tulsa.

15

Pipeline Transportation

15.1 Pipeline Transportation in Oil Fields

There are three basic types of pipeline systems in the petroleum industry to serve:

- Gather crude oil from the wellhead.
- Transport to refineries.
- Distribute the products to process plants, retail markets, and other forms of transportation.

15.1.1 Gathering system

- The pipeline and other equipment used to transport crude oil from individual wells and other production units to a central location are called a gathering system.
- The gathering system typically consists of pipeline branches flowing into trunk pipeline stations or other locations wherefrom oil is transported to a trunkline system.
- The most common pipe sizes used in these branches are 4″–12″ diameter. The pipelines in gathering systems are short compared to trunk pipelines and range in length from a few meters to several kilometers.
- The transportation system from the well to the gathering station is referred to as well fluid transportation. The fluid in this transportation system is normally multiphase fluid. Single-phase well fluid is rarely observed in oil field.
- Increasing the length of well fluid flowlines increases wellhead back pressure on wells. However, development of multiphase pumping and

metering is a major milestone in achieving long distance well fluid transportation.

- The design of well fluid transportation and gathering system should take into consideration the following:
  - The wellhead pressure should be as low as possible to enhance self-flow and recovery of fluid.
  - Minimum pressure loss in pipeline.
  - Easy to monitor and control the wells.
  - Minimum loss of hydrocarbons.
  - Accurate metering of well fluid.
  - Flexibility for future expansion.
  - Operational safety.
  - Optimum production cost.
- The different types of gathering systems normally used in oil fields are:
  - Wellhead separation system.
  - Group gathering system.
  - Centralized gathering system.

15.1.2 Wellhead separation system

- Main gathering lines are laid in the form of a loop around the field. Individual wells have their own separation and testing facility and the oil and gas lines are connected to the main collector lines.
- Highest recovery from the field, by maintaining the lowest wellhead pressure, is possible in this system.
- This system is generally adopted in isolated small pools along with tanker-based transportation.

15.1.3 Group gathering system

- Main collector lines are laid from a central processing facility to a group gathering facility or a process platform facility.

- This type of system is generally applicable when moderate to large amounts of liquid are produced. The two-phase flow of liquid and gas causes more pressure drop in the individual flowlines and consequently high back pressure on the wells. However, this is an economical and flexible system for most of the fields and is commonly used in ONGC.

15.1.4 Centralized gathering system

- Main gathering multiphase lines that run through the field and wells are produced directly into the main line from either side. A test line is run parallel to the mainline to carry out periodic testing of wells.
- The system is generally used in oil and gas fields where getting ROU is normally difficult and the wells are densely located in the field.
- Use of clustered well location system, in both offshore and onshore, necessitated well platforms with separation and test facility in offshore or underwater manifold center (UMC) in subsea locations and group and test gathering lines in onshore locations.
- A header in a gathering of distribution system provides a means of joining several flowlines into a single gathering line. Valves are provided on each pipeline entering or leaving the header so that lines can be isolated during operation and maintenance.
- Regulations and good operational production practices require oil, gas, and water production rates to be measured for individual wells at regular intervals. In offshore, the measurement and separation facility is provided at the well platform, whereas in onshore the wells are tested in the group gathering station (GGS).

15.2 Trunkline System

- Transportation of crude oil by trunk pipeline makes it economically feasible to produce oil in areas remote from processing points and markets.
- Great advances have been and continue to be made in trunkline transportation of crude oil. These include improvements in materials and methods of pipe manufacture and better design and construction methods for both pipelines and stations.
- Major changes in operating methods have occurred with the result that many of the limitations, inherent in human control of precise,

monotonous, tiring, and otherwise burdensome operations, have been overcome by transferring the operations to devices and systems of control that are tireless, quick acting, sensitive, and capable of working in environmental conditions that human beings could not endure.

15.3 Pipe Specification

- Steel pipe used in pipeline construction is commonly called line pipe to distinguish it from steel casing and tubing installed below ground in oil and gas wells and drill pipe, used for oil and gas well drilling.
- Line pipe comes in a wide range of sizes. It is made from steels with various chemical compositions and different physical properties, using several manufacturing processes. The physical and chemical properties of steel used to make line pipe and the manufacturing processes are rigidly controlled to meet the applicable specifications. Specifications also cover pipe dimension, allowable tolerance, permissible defects, and testing.

15.3.1 API specifications

- Much of the line pipe used in the petroleum industry is manufactured according to specifications of the American Petroleum Institute (API). Key API specifications applicable to line pipe include the following:
  - API Spec. 5L covers seamless and longitudinally welded steel pipe in grades A and B.
  - API Spec. 5LX applies to high test line pipe (both seamless and longitudinally welded) in grades X42 through X70.
  - API Spec. 5LU (tentative) covers ultrahigh test, heat treated seamless, and welded pipe is grades U80 and U120.
  - API Spec. 5LS is applicable in spiral welded line pipe in grades A and B, and X42 through X70.
  - API line pipe grades are designated by their minimum yield strength in pounds per square inch (Psi). Yield strength is the tensile stress required to produce a specified total, permanent elongation in a test sample of the steel; the test sample and procedure are detailed in the specifications.

- Grade A line pipe has a minimum yield strength of 30,000 Psi; grade B a minimum yield of 35,000 Psi. In the remaining grades, X42 indicates pipes made of steel with 42,000 Psi minimum yield strength; X60 pipe has a minimum yield strength of 60,000 Psi, etc.
- Mill hydrostatic test pressure is outlined in the specifications for each grade, weight, and size of pipe. In general, the required hydrostatic test pressure increases with increasing strength (grade) and with increasing wall thickness (weight).
- API specifications also prescribe dimensions, weights, and lengths for each size and grade as well as permissible tolerances on these dimensions.

15.3.2 Schedule number and pipe sizes

- To call out a pipe, it is necessary to give both the schedule number and the weight/strength designation for both pipes and fittings. All pipes under 14 inch (35.56 cm) are designated by the nominal diameter and schedule number. Those over 14 inch are designated by the actual outside diameter and wall thickness.
- The wall thickness of pipes fluctuates according to the schedule number, whereas the outside diameter remains constant for pipes with a diameter of 14 inches or more. Modifying the thickness results in a modification of the internal diameter. The internal and external diameters will differ based on the schedule number, ranging from 1/8 inch to 12 inches in diameter. The nominal size will be equivalent to the external diameter, ranging from 14 to 42 inches.
- The American National Standard Institute, sponsored by the ASTM and ASME, published ANSI B 36.10 to standardize pipe dimensions throughout the industry. Schedule numbers from 10 through 160 were adopted for steel pipes.
- The ASA standard schedule numbers that specify wall thickness can be calculated by using the formula $1000 \times P/S$. Here P is equal to the internal pressure of the pipe (Psi) and S is the allowable fiber stress (Psi). As pressure increases, so does the pipe thickness requirement. The temperature of the line medium, besides putting thermal stress on the sphere, will also affect the pipe thickness. The S value takes into account temperature, pressure, and material.

- Stainless steel schedule numbers from schedule 5S through 80 S are published in ANSI B 36.19 for sizes up to 12 inch (30.4 cm).
- The ASTM tolerances on regular pipe products specify that wall thickness should not vary more than 12.5% under the nominal wall thickness that is specified for regular mill rolled pipe.

15.4 Design of Pipeline

- The two characteristics of a pipeline that have the most influence on the investment are diameter and wall thickness. The diameter determines the friction loss corresponding to a given throughput and hence the energy required. The thickness sets the maximum limit of operating pressure and hence the possible power distribution in terms of the spacing of pump or compressor stations along the line.
- The first task of a design engineer is to determine the pressure drop and any change in temperature along the pipeline, i.e., to determine the variation in the fluids energy and thus the amount of energy needed to maintain flow which has to be supplied by pumps or compressors along the pipeline.
- An accurate value of internal pipe diameter and the roughness of the pipe wall is important for determining accurate pressure loss.

15.4.1 Friction factor and flow types

- The energy balance required for the calculation of fluid flow behavior involves the calculation of lost work, the total irreversible energy loss to the inner pipe wall that is unavailable to move the fluid or perform any other action. The term "friction factor" is the empirical term used to assess the numerical value of these irreversible losses, usually called friction losses.
- For a Newtonian fluid, the work done in overcoming friction through a distance is proportional to the surface in contact with the fluid, approximately proportional to the square of velocity and proportional to the fluid density.

The frictional lost work, represented by the frictional resistance, proposed by Fanning is given below:

$$W_f = \frac{2fL\,V^2}{g_c d} \tag{15.1}$$

where:

f = Fanning friction factor
W_f = total friction loss (lost work)
d = diameter of pipe, ft
L = length of pipeline, ft
V = velocity of fluid in the pipeline, ft/sec
g_c = gravitational constant,
$32.17 \frac{\text{lbm.ft}}{\text{lbf.sec}^2}$

- Reynolds number, an important dimensionless number in fluid flow, may be written in many alternative forms. The standard form is "dvρ/μ."
- Pipe roughness, the distance between peaks and valleys, can be measured by modern instruments in micrometers (microinches).
- The Fanning friction factor is plotted versus Reynolds number (Re). Numerous experiments have shown that "f" is a function of Reynolds number, type of flow, and pipe roughness for a Newtonian fluid.
- It is noted that there is a gap between the laminar and turbulent flowlines. This can be defined as the critical region. In this region, the flow pattern is unstable and oscillates between laminar and turbulent characteristics. Normally, one does not design pipelines for this region.

Fully turbulent flow is desirable in commercial installations and a pipe size should be chosen to produce such a flow at specified flow rates.

Field-derived friction factor:

Simplified correlations for Fanning friction factor: The smooth tubing curve applies for small extruded tubing and the other for pipe greater than 0.20 m (8 inch) diameter.

The equations of the three lines are represented closely by the following equations.

$$\text{Re} \le 2000, \qquad f = 16/\text{Re} \tag{15.2}$$

$$\text{Re} > 4000, \qquad f = 0.042/(\text{Re}^{0.194}) \tag{15.3}$$

for smooth tubing or pipe < 20 cm (8 inches)

$$\text{Re} > 4000, \qquad f = 0.042/(\text{Re}^{0.172}) \tag{15.4}$$

for commercial pipe or pipe ≤ 20 cm (8 inches)

15.4.2 Steady-state liquid flow

For a Newtonian liquid, the basic equation is derived from thermodynamics and the friction factor calculation assumes isothermal and adiabatic flow.

The liquid flow calculation for the friction loss may take one of the three forms for a given liquid of known properties.

- Calculation of pipe diameter from known flow rate and pressure drop (ΔP_f).
- Calculation of flow rate from known diameter and pressure drop (ΔP_f).
- Calculation of pressure drop (ΔP_f) from known diameter and flow rate.

Since the friction factor and Reynolds number are a function of flow rate and diameter, solving for either of these terms becomes a trial-and-error solution unless they somehow are eliminated algebraically. The equation for pressure drop (ΔP_f) is derived from basic equations and is represented as:

$$\Delta P_f = \frac{2fLV^2\rho}{g_c d} = \frac{32f\,L\,q^2\rho}{\pi^2 g_c d^5} \tag{15.5}$$

and

$$\mathrm{Re} = \frac{dV\rho}{\mu} \tag{15.6}$$

where:

ΔP_f = frictional pressure loss, lb/fr^2
f = friction factor, dimensionless
L = length, ft
V = velocity, ft/sec
ρ = liquid density, lb/ft^3
g_c = gravitational force constant, 32.17 lbm-ft/lbf-sec^2
d = internal diameter, ft
q = volumetric flow rate, ft^3/sec
μ = liquid viscosity, lb/ft-sec

Calculation of pipe diameter:
A direct solution for diameter is possible by solving eqn (15.5) and (15.6). The final form can be written as:

$$d = 1.265q^{0.4}\left(\frac{f\,L\rho}{\Delta P_f g_c}\right)^{0.2} \tag{15.7}$$

Substituting the eqn (15.3) and (15.4) for "f" in eqn (15.7), for small pipe (commercial):

$$d = 0.6494(q^{0.379})(\rho^{0.172})(\mu^{0.036})L(0.207 / (\Delta P_f g_c)) \tag{15.8}$$

for large pipe (smooth):

$$d = 0.6474(q^{0.376})(\rho^{0.168})(\mu^{0.041})L(0.208 / (\Delta P_f g_c)) \tag{15.9}$$

Optimum pipe size is dependent on pipe cost versus pressure drop cost. For non-corrosive liquids, the optimum velocity usually is 2–3 m/s (6–10 ft/sec). If corrosive liquids like sour amine or glycol are flowing, the allowable velocity may be 0.7–1.0 m/s (2–3 ft/sec) to minimize erosion corrosion, if carbon steel pipe is used.

Calculation of line capacity:
Calculation of the capacity of a line with a given pressure drop is by trial and error when using a plot of friction factor vs. R_e. However, a direct solution is possible. Starting with the same basic equations as in the calculation of diameter, eliminating q instead of d,

$$q = \frac{0.785\, d^2}{\rho}\left(\frac{\Delta P_f g_c d\rho}{2f\, L}\right) \tag{15.10}$$

Eliminating "f" by eqn (15.3) and (15.4), for small pipe:

$$q = \frac{3.127\, d^{2.64}}{\rho^{0.453}\,\mu^{0.094}}\left(\frac{\Delta P_f g_c}{L}\right)^{0.547} \tag{15.11}$$

for large pipe:

$$q = \frac{3.180\, d^{2.66}}{\rho^{0.446}\,\mu^{0.107}}\left(\frac{\Delta P_f g_c}{L}\right)^{0.533} \tag{15.12}$$

- Approximate correlation:

A set of "quickie" charts has been developed. The values shown in this figure are based on the following assumptions:

- Curves calculated from Fanning equation
- Carbon steel pipe with a roughness (E) = 45 μm (0.00015 ft.)
- Liquid relative density = 1.0

- $R_e > 2000$ for viscous liquids
- Results are for horizontal pipe only

Maximum velocity:
Some company specification limits maximum continuous liquid velocity to

$$V_{max} = 100\, A/p^{0.5} \tag{15.13}$$

where:

A = conversion constant and is equal to 1 in English unit.

Calculation of pressure drops:
A similar derivation approach as above for the remaining variable, ΔP_f, yields the following equations for normal pipe friction, for small pipe:

Other pressure losses:
In evaluating the total pressure drop, the ΔP values for installed fittings are to be considered. Several methods may be chosen to allow for fittings loss. The total equivalent length of all fittings is then added to the actual length for the ΔP calculation.

- Complex liquid piping system
- Looping of pipelines
- Changing diameters
- Branching of pipelines (gathering system)
- Flow splitting (injection system)

For design calculations of the above cases, refer to "Petroleum Fluid Flow System" by BOYD, Campbell Petroleum Series.

Non-Newtonian liquid flow:
Oil–water emulsions containing over 10%–20% water are examples of a fluid that may exhibit sufficient non-Newtonian behavior. The viscosity of an emulsion may be many times that of oil or water alone.

The most common method of analyzing the behavior of non-Newtonian fluids is by power law equation, which is represented as

$$\text{Shear stress} = k\,(\text{shear rate})^n, \tag{15.14}$$

where k is the consistency index, a function of the liquid viscosity, and n is the flow behavior index, a measure of the degree of non-Newtonian behavior.

These two constants can be obtained by viscometer measurements. The most common viscometer is a rotating type, like the Fann V-G meter.

15.4.3 Steady-state gas flow

Steady-state gas flow is governed by the same basic energy balance principles as liquid flow. However, gas being a compressible fluid, the gas properties are a function of pressure and temperature. The effect of pressure and temperature is more critical in the case of gas than in the case of liquid.

There are two basic types of gas flow equations:

- Pipeline flow equations for basically horizontal lines.
- Vertical flow equations for wellbores, risers, and the like.

There are several equations available to interrelate capacity, diameter, and pressure drop. However, there is no single universal gas equation that is superior under all conditions, for all gases.

The factors that influence the characteristics of gas flow in gathering and transmission lines may be divided into three convenient classes:

i. Properties of the flowing gas:
   ρ = density
   μ = Viscosity
   Z = Compressibility factor

ii. Properties of the containing conduit:
   d = diameter
   L = Length
   ε = Well roughness (absolute and relative)

   (However, welds, bends, fittings, etc., are not included in ε.)

iii. Properties of the operating pipeline:
   - Operating pressure
   - Operating temperatures
   - Gas velocity
   - Elevation changes

The working equation evolving from the basic equation depends on the friction factor correlation used. These can be divided into three general classes:

- Those based on a constant friction factor.
- Those where the friction factor is some function of the diameter only.

Table 15.1 Units representation for different parameters in Metric and English Systems.

	Metric	English
Q_{sc} = gas rate at T_{sc}, P_{sc}	M^3/d	Scf/d
P = absolute pressure	kPa	Psia
P_{sc} = pressure, standard conditions	kPa	Psia
T_m = mean absolute temperature of line	kPa	°R
T_{sc} = temperature, standard conditions	K	°R
T_g = ground temperature	K	°R
D = inside diameter of pipe	m	in
L = pipe length	m	mile
μ = viscosity	Pa.s	lb/ft.s
γ = gas relative density	–	–
Z_m = mean compressibility factor	–	–
F = Fanning friction factor	–	–
E = pipeline efficiency	–	–
R_e = Reynolds number	–	–
K = Constant dependent on units used in table	–	–
F_f = drag factor from Bend index of the AGA method		

- Those where the friction factor is some function of the Reynolds number or some modified version thereof.

Table 15.1 summarizes the units used in the above stated equations.

An efficiency (E) usually is added at the end to correct for small amounts of liquid, general debris, weld resistance, valve installations, line bends, and other factors that reduce the gas flow rate below the basic equation rate. The design value of E in a clean gas line usually is estimated at 0.92. Some operators back calculated an E from line operating data and "Pig" when the value reaches a value lower than some set standard. Some companies arbitrarily use a graduated E, i.e.:

E = 1.0, new straight pipe without bends (seldom used in the design)
= 0.95, excellent conditions (with frequent pigging)
= 0.92, average to good conditions (normal design)
= 0.85, adverse, unplugged, old, dirty pipe

The Weymouth equation was devised for sizing gas lines operating at pressures from 241 to 690 KPa (35–100 Psig). By including a compressibility factor evaluated at the mean pressure and temperature of the gas in the line, the formula has been modified for the design of higher-pressure gas systems (pressure above 690 KPa (100 Psig)). Industry-wide experience indicates that:

- The friction factor used by Weymouth, in general, is too high for large-diameter lines.
- Under peak loads, predicted volumes are too low.

For gas transmission through a long pipeline, Weymouth's equation is not recommended.

Panhandle equations are useful for larger-diameter pipelines. An efficiency factor of 0.92 is used in a normal, clean, welded steel pipeline of approximately 0.609 m (24 inch) diameter. For smaller diameters, they indicated that the factor should be reduced. For larger diameter lines such as 0.914 m (36 inch) and larger, they found the efficiency may be as high as 0.910. The efficiency factor varies markedly in different sections of the line.

There is no definite proof to say which equation (Panhandle A or B) is better, although Panhandle A probably is more widely used. Neither is particularly suitable for low pressure, < 690 KPa (100 psi), small diameter, < 0.20 m (8 inch), systems operating at low Reynolds number, $R_e < 10^5$.

For the design calculations of complex gas piping systems, please refer to "Petroleum Fluid Flow Systems" by BOYD, Campbell Petroleum Series.

15.4.4 Steady-state multiphase flow

The main objective of all multiphase flow studies has been to develop a technique to relate pressure drop, flow rate, and pipe diameter. Pressure losses in gas/liquid flow are quite different from those encountered in single-phase flow.

Different investigators presented different multiphase flow correlations for different flow patterns. The details of the correlation are discussed in Chapter 1 of this manual.

15.4.5 Deepwater pipelines

Drilling and production in deepwater have now reached to 1500–2000 m water depth range. Pipeline installation in these depths benefits considerably from upcoming technology, however; practical issues confront the contractor in terms of vessel requirements, equipment layout, and the development of the special procedures needed to meet safety and reliability standards.

15.4.5.1 Design considerations

Offshore pipelines in ultra-deepwater have to meet some special design requirements. The pipeline system has to be designed and constructed to high standards as a surviving offset to the many unknowns and constant hazards that are present in the deepwater environment. The technologies associated with the design, manufacture, and installation of flowlines have to be optimized concerning cost, quality, and schedule.

The main issues that affect the design of pipelines are:

Internal diameter:
For deepwater applications, pipelines have to have a large enough internal diameter (ID) to move the volumes of crude needed to justify the field development costs. As the ID increases, the thickness of the pipe wall also has to increase to provide resistance to the external hydrostatic pressure and the internal pressures associated with high throughput volumes.

Pipe wall thickness:
This in turn significantly affects the cost and installation of the pipeline and can present challenges for welding technologies and the extra weight increases the problems for installation. The use of limit state and strain-based design methods are likely to become more widely used to allow optimization of wall thickness requirements for deepwater applications.

Buckle propagation:
A special hazard is buckle propagation, which is initiated when there is a severe combination of bending and external pressure but once started can run along the pipe, collapsing it into a dumb-bell cross-section. Buckle propagation can be driven by pressure alone, and could destroy many miles of pipeline. Buckling must be considered for both installation and operating conditions.

Flow assurance:
For deepwater applications, this is primarily concerned with the thermal properties of the flowline system and the use of passive insulation or active heating systems to ensure the continuation of product flow along the pipelines as heat losses can result in blockages caused by hydrate formation or wax deposition. The enhancement of the insulating properties of subsea pipelines can entail major expenditure, and this is especially true of deepwater development in harsh environments.

Material grade and chemistry:
This must be closely specified and controlled to ensure a cost-effective and safe design.

Internal corrosion:
The use of inhibitors as opposed to the inclusion of a corrosion allowance on the wall thickness must be considered when selecting materials.

External corrosion:
Cathodic protection systems must be carefully chosen to ensure that the required design life is met, and the effects of mechanisms such as hydrogen embrittlement, especially in deepwater applications, must be addressed.

Installation:
The stresses and strains associated with laying flowlines require careful consideration to optimize the pipe design. Welding techniques, especially for the more exotic materials (CRA, titanium) that may be used in deepwater applications, must be adequately addressed.

Bottom stability:
Designers of offshore pipelines must consider on-bottom stability problems not encountered in onshore pipelines. A pipeline resting on the bottom or in a trench must often be coated with concrete to keep it from floating when evacuated. Currents exert lift and drag forces that tend to move the pipe laterally. These forces must be resisted by making the pipe sufficiently heavy.

Weight:
The weight issue is of particular relevance to riser design as flowlines once installed rest on the seabed.

Spans:
There are many cases of underwater pipelines where significant spans occur, such as the unsupported span from a laybarge, or where the lateral loads from weight or current are carried by a combination of tension and large deflections rather than bending moments. This structural behavior is to be accounted for in piping design/analysis computer programs.

15.4.5.2 The offshore supervisory committee has been very active in sponsoring research to define safe pipeline stability design procedures

To tackle the perils of hydrostatic collapse, flowline designers resort to different types of riser armoring (which also serves to bear the tensile loading incurred during installation), but it tends to increase the overall weight of the product. Coflexip Stena Offshore has introduced the *Teta technology* (named after the cross-sectional "T" shape of the wire used in its structure). This has been designed to meet the demand for larger-diameter high-pressure pipe to withstand the severe operating conditions for deepwater.

Buckling phenomena have been the object of a great volume of research, stimulated by the need of the oil industry, both for pipelines and for well tubular. Researchers now consider combined loadings such as bending combined with external or internal pressure, bending combined with axial tension, and bending combined with torsion in their design studies. It is fair to say that the problem is well understood in terms of empirical formulas and that there is a reasonable level of confidence backed up by model tests and a small number of full-scale trials.

Some deepwater pipelines are designed so that the maximum net external pressure is lower than the propagation pressure, so that buckle propagation cannot possibly occur. This becomes an onerous requirement in deepwater, and a less conservative approach is to install buckle arresters, so that a buckle initiated by unexpectedly severed handling might propagate to the next arrester, but could not destroy a long length. The design of arresters is well understood.

One promising innovation in flowline supply is the development of chrome 13 pipe (so called because it is 13% chrome). Chrome 13 pipes have so far been used on the Asgard, Wintershall L08, and Tune developments in the North Sea. The material is about a quarter of the cost of either duplex or superduplex steel and being more resistant to corrosion. It is suitable for sour service. The downside is that it is more difficult to weld. Working on the wintershall project, Petrology claimed the first mechanized welding of Chrome 13 with 6- and 12-inch flowlines welded at Ardesier spoolbase for Stolt Offshore before being spooled into the reelship Seaway *Kestrel.*

15.4.6 Special considerations

15.4.6.1 Economic pipe diameter

- For a given flow rate of a given fluid, piping cost increases with diameter. But, pressure loss decreases, which reduces potential pumping or compressing costs. An economical balance between material costs and pumping costs is important for designing the pipelines.

- The optimum pipe size is found by calculating the smallest capitalization/operating cost, or using all the pressure drops available, or increasing velocity to the highest allowable.

- The economic diameter will be the one that makes the sum of amortized capital cost plus operating cost minimum. The total cost can be per unit of time or unit of production.

- Detailed economic study of each line in a process plant, in order to select the proper size, is difficult and usually not justified. However, in deciding between two possible line sizes particularly in the case of more expensive alloy lines or large carbon steel lines, a detailed study may be justified.

A detailed study of the factor influencing the economic sizes of pipes is given in "Optimum pipe size selection" by Nolte, Claude B., Gulf Publishing Company, Houston, TX.

15.4.6.2 Allowable pressure drops and velocities

- Some typical standards of allowable pressure drops or velocity ranges, as considered in the sizing of process lines, are shown in Tables 15.2(a), 15.2(b), and 15.2(c).

Table 15.2(a) Typical design velocities and pressure drop standards.

Service	Metric	English
Cooling and tempered water	2–5 m/s	8–15 ft/sec
General process liquid	2–3 m/s	4–10 ft/sec
Slurries	2–3 m/s	5–10 ft/sec
Steam headers	ΔP/100m = 0.03 P	ΔP/100ft = 0.01 P
Steam leads	ΔP/100m = 0.06 P	ΔP/100ft = 0.02 P
Compressor discharge lines	5–7 kPa/100m	0.2–0.3 psi/100 ft
Tower overhead vapor	1–12 kPa/100m	0.05–0.5 psi/100 ft
Nitrogen instrument and plant air headers	0.076 m	3 in
General process vapor	1–3 m/s	4–10 ft/sec
Pump suction lines	1–12 kPa/100 m for $\Delta P < 0.6$ m	0.05–0.5 psi/100 ft for $\Delta P \leq 2$ ft

Table 15.2(b) Typical design velocities and pressure drop standards.

	Reasonable velocity (tube turns)		Allowable pressure drop due to friction	
Service	Metric	English	Metric	English
Pump discharge	*d/2 + 1.2 m/s	*d/2 + 4 ft/sec	48 kPa/100 m	2 Psi/100 ft
Pump suction	One-third the above	One-third the above	–	0.5ft–lbf/ lbm/100 ft
Steam or vapor	3.23* d (5.1 m/s)	*d (1000 ft/ min)	1.6% of line pressure/100 m	0.5% of line pressure/100 ft
Gravity flow of liquids	–	–	–	0.2 ft–lbf/ lbm/100 ft
Water lines	1.5–2 m/s	5–7 ft/sec	–	–

*d = Inside diameter of pipe as 12.2 m [inches]

Table 15.2(c) Typical design velocities and pressure drop standards.

Service	Metric m/s	English ft/sec
Average liquid process	1.2–2	4–6.5
Pump suction (except boiling)	0.3–3	1–5
Pump suction, boiling	0.1–0.9	0.5–3
Boiled feed water (discharge, pressure)	1.2–2.4	4–8
Drain lines	0.5–1.2	1.5–4
Liquid to reboiler (no pump)	0.6–2.1	2–7
Vapor–liquid mixture out reboiler	4.6–9.1	15–30
Vapor to condenser	4.6–24	15–80
Gravity separator flows	0.1–0.5	0.5–1.5

Use only as a guide because pressure drop and system environment govern final selection or pipe size. For heavy, viscous fluids, velocities should be reduced to ½ values as shown. Fluids should not contain suspended solids.

- There are certain limitations in the above-mentioned standards. The velocity method does not allow for changes in fluid densities and the allowable pressure drop method does not recognize that the economical pressure drop should be greater for small pipes than for large pipes. Neither methods allow for variations in material and power costs. However, the sizing based on these standards is valid in the majority of cases.

15.4.7 Flow improvers

15.4.7.1 Treatment of crude oil to reduce pour point

- Paraffin deposition takes place as the oil cools, while moving up the well and at the surface. The deposits collect in well bore, production tubing, and flowlines and cause restricted flow, which leads to increased flowline pressure.
- If the wax content in the crude oil is low to moderate (0%–10%), the problem of deposition is periodic and is handled on "as needed" basis. With wax content above 10%, the crudes experience severe deposition and flow problems requiring constant treatment to ensure uninterrupted production.
- Various methods have been designed to treat high-wax-content crudes.
  - Dilution with low-wax-content crude.

- ○ A very specific heating/cooling/shear cycle, as was developed for improving the flow characteristics of Naharkotia (Assam) crude by OIL INDIA LTD.
- ○ Pigging of pipeline.
- ○ The addition of solutions of oil soluble surfactants (i.e., demulsifiers).
- ○ The use of wax crystal modifier.
- ○ Water emulsion.
- ○ Reduce energy losses by using drag reducer.

- The use of additives to modify the crystal structures of the precipitating waxes has become an established method of lowering the pour point of waxy crudes. This is widely used in ONGC.
- For waxy crudes, PPDs with high molecular weight are effective, while for mixed type of crudes, PPDs with lower molecular weight are effective.
- The crystal modifier, normally available in solid form, is dissolved in hot crude oil and injected into the line at controlled rate. Crude oil requiring the chemical treatment is heated in the heat exchanger to a temperature above cloud point before injection of chemical.

15.4.7.2 Drag reducers

- Drug reducers, also called flow improvers, are agents that have been proven effective in increasing the capacity of crude pipelines without adding pumping horsepower or looping the line.
- Several compounds are being considered as drag reducers, but the most applicable is high-molecular-weight hydrocarbon polymer in a hydrocarbon solvent, typically 10% active ingredient by weight in a kerosene-like solution.
- An effective drag reducer must meet several criteria. It must be effective in small concentrations in the pipeline, able to resist degradation in transit and storage, and not have a detrimental effect on refining processes.
- Though some equipment must be installed for injecting the agent into the pipeline, the investment in this equipment will be small compared with that required to install additional pump horsepower or construct a pipeline loop.

15.5 Station Design Specification

15.5.1 Station design

- When building a pipeline system, it is necessary to decide the placement of pump or compressor stations, as well as the size of the pumps or compressors within each station. The quantity and placement of stations are contingent upon the pipeline's length and the amount of energy needed to augment the fluid's movement to achieve the requisite volume at the intended delivery pressure.
- The number of booster stations varies widely in both natural gas and liquid pipeline systems. The longer the line, the more stations may be required. Most large systems with several pump or compressor stations represent a compromise between a few very large stations and a large number of small booster stations from the point of view of operation, control, and maintenance.
- Equipment in an individual station varies in both size and type, depending on the volume of fluid being handled and its properties, the size of the pipeline system, the type of monitoring and controls used, the remoteness of the station, the environment, and other factors.
- In a crude oil pumping station, the main items are the pumps and their drivers. Pump stations typically include metering equipment for measuring throughput. Major stations, where custody of the fluid is transferred from the one owner to another, contain a meter prover to calibrate the metering equipment.
- Originating stations may also have storage tanks to smoothen out variations in flow to the station so that the pumps will operate continuously at near normal capacity.
- Many stations also include scraper traps. Points along the line, such as pump stations where piping manifold are installed, make convenient points for inserting scrapers into the line and removing them.
- In addition to these key equipment items, a crude or product pump station often contains a complex array of piping and piping manifolds that permit the flow path to be directed to the pumps, to storage, or to other equipment.
- Natural gas pipeline systems that originate and booster stations contain gas compressors and their drivers. The selection of the number and size

of compressors is done based on criteria similar to that needed to choose the number and size of pumps in crude and product pump stations.

- Compressor stations also contain measuring equipment, especially where ownership of the gas changes hand. At some stations, typically originating stations, a separator is often required before the gas enters the compressor suction, to remove the liquids and sediments.
- Gas is usually compressed in stages and the heat of compression must often be removed between compressor stages. Interstage coolers are used for this purpose.
- Gas compressor stations contain piping manifolds to direct the flow of gas entering and leaving the station, and valves and valve controllers to regulate flow.
- Control equipment at pump and compressor stations also include shutdown devices. The entire station can be shut down if conditions so warrant.

15.5.2 Pump application and design

- Several types of pumps are used to handle crude and petroleum products. One basis for pump selection is the rating curve developed for each pump as a result of tests conducted by the manufacturer. Rating curves – also called efficiency curves and capacity curves – show how the pump head, efficiency, and power consumption vary with its capacity.
- Each pump has an optimum operating range in which efficiency is maximum. Ideally, a pump would be chosen that would operate within this range throughout its life. But in many cases, especially in pipeline applications, capacity and other operating conditions may change significantly.
- Two types of pumps are common in oil industry: the centrifugal pump and the positive displacement pump. The choice of pump type depends primarily on the volume to be pumped and the pressure or head that must be overcome.
- In general, centrifugal pumps are used when the volume of liquid to be pumped is relatively large and pressures are moderate; positive displacement pumps are used for pumping smaller volumes and higher pressures.

15.5.2.1 Positive displacement pumps

- Positive displacement pumps use a piston or plunger that is moved back and forth in a cylinder to increase the pressure of the liquid.
- Reciprocating pumps have been built with one to several cylinders. A three-cylinder pump (triplex) is common. Two-cylinder (duplex) pumps were used for many years on drilling rigs to pump drilling mud down the hole.
- A consideration in choosing the number of cylinders a pump should have for a specific application is the pulsation caused by the reciprocating pump. The number of cylinders has effect on the flow pulsation. The greater the number of pistons, the smoother the flow through the pump.
- Uneven flow may create the following problems:
  - Causes pulsation in metering equipment and thus affects measurement accuracy.
  - Causes vibration in piping and equipment that can result in failure.
  - Excessive vibration on offshore platforms can be transmitted to structure elements of the platform.

Methods used to reduce or eliminate vibrations caused by pulsations include bracing, use of flexible piping, and careful design of pump-connected piping.

Most reciprocating pumps have a packing element around the piston rod to seal the pump cylinder from the atmosphere. When it is not sealing properly, air leaks into the pump cylinder, reducing the pump efficiency. Other causes of reduced efficiency in reciprocating pumps include the following:

- Air or vapor in the suction line
- Air or vapor above the suction valves
- An air leak in the suction piping
- Failure of valves to close properly
- Worn valves and valve seats
- Worn cylinders or plungers
- Insufficient head (pressure) available at the pump suction

15.5.2.2 Operation of reciprocating pump

- Start-up:
  - Open the tank outlet valve and manifold valve.
  - Check oil level in crank-case.
  - Check the condition of the stuffing boxes.
  - Check the oil in the forced feed lubricator, if provided.
  - Open suction valve and delivery valve of pump.
  - Open pump bypass line valve.
  - Prime the pump with oil being pumped.
  - Start the pump.
  - See that the pump picks up full speed and then close the bypass valve. Check suction pressure and temperature ensuring that vapor formation does not occur to avoid fluid knock hammer.

- Running:
  - Check the discharge pressure.
  - For operation at constant speed, pump may be operated at the optimum load of 90% of rated head.
  - Pumps operating at a variable (controlled) speed may only be operated within the range indicated in the pump operating diagram provided by the manufacturer.
  - Check the vibration.
  - Check the stuffing box; it should not get over-heated.

- Shut-off:

  Switch off the driver and watch the pump run down smoothly to a stand-still.

- Trouble shootings of plunger pump:
  - Low discharge pressure may be due to worn or fluid cut valve assembly, valve propped open, improper filling, fluid slippage, and erroneous gauge reading.

- Low suction pressure may be due to low head, suction strainer blocked and retarded fluid flow, and erroneous gauge reading.
- Fluid knock hammering may be due to air entering suction line, air entering suction stabilizer, and air/gas in pumped fluid. It may be due to probable causes of low discharge pressure and low suction pressure.
- Discharge line vibration may be due to discharge pulsation, dampener unsupported to low discharge pressure and low suction pressure.
  - i) Repair, replace, or charge dampener.
  - ii) Provide supports or hanger.
  - iii) Remedies are the same as in the case of low discharge pressure and low suction pressure.
- Short valve life may be due to abrasive fluid valve not seating, pump not filling, and pulsation damper malfunction.
- Short piston and liner life may be due to abrasives in fluid, and short liner life due to deteriorated piston.
- Short plunger packing life may be due to abrasives in fluid, friction wear due to over tightening of packing, misalignment, longitudinal wear, or scoring.
- Diaphragm leakage may be due to worn or damaged extension rod, corroded extension rod, worn wiper rings, wiper rings improperly sealing, worn lantern ring, oil baffle misplaced, or pressure in crank case.
- Oil seal leakage may be due to worn sealing lip, damaged sealing lip O.D. not seated, foreign material at seal point, or pressure in crank case.
- A knock in the power end may be due to incorrect pump rotation, a loose piston/rod plunger, a loose extension rod, a loose connecting rod cap, and a loose bearing housings/covers, a worn crosshead pin bushing, or a worn crank pin bearing.

- Trouble shootings of piston pumps:
  - Pump pressure not maintained due to inlet line clogged, leak in inlet line, air in pump head, leaking pump spring guide "O" rings, leaking

unloader or regulator, leakage in pump packing, worn bypass valve seal, worn pump valve seat or valve disc(s), faulty pressure gauge.

- The pump gives abnormal sound due to partially or fully closed valve in inlet line, clogged inlet line, air leak in inlet line, air in pump head, insufficient water supply, cavitation, oil level too low, worn connecting rod bearing shell or bushing, water in crankcase, worn piston shaft bearings, pump valve seat loose.
- The pump overheats due to overload, oil level too high or too low, incorrect oil type, water in crankcase, worn bearings or gears.
- The pump does not prime due to suction vale closed, suction line clogged, insufficient water supply, leak in inlet line, air in pump head, foreign debris caught in pump valves.
- Oil leak at crosshead shaft seals may be due to oil level too high, worn out crosshead shaft seals, crosshead shaft surface damaged.
- The pressure gauge reads excessive pressure due to air in pump head, unloader (or regulator) piston jammed or worn out, unloader check valve jammed or worn out, foreign matter caught in pump valves, loose pump valve seat, worn out unloader (or regulator) regulating rod, worn out unloader check valve seat, worn out unloader packing or "O" rings, unloader adjusting screw excessively tightened, small leak in discharge lines or shut-off devices, faulty pressure gauge.
- Excessive leak at stuffing box may be due to worn out packing, worn out stuffing box "O" rings, loose plunger.

15.5.2.3 Centrifugal pump

- Rather than operating with a reciprocating motion, the centrifugal pump rotates. It consists of an impeller and a casting. The impeller is turned by the pump's driver through a shaft and throws the liquid into the pump casing, increasing the energy of the liquid by centrifugal force.

  The flow of liquid is continuous under the following conditions:

  - Liquid must flow into the impeller at the same rate it is being discharged from the pump.
  - Pressure in the suction pump must exceed the pressure at the impeller inlet by an amount great enough to overcome suction line resistance and the difference in elevation or lift, from the sump to the impeller.

  - Pressure inside the impeller should not fall below the vapor pressure of the liquid.
  - Total head or energy developed by the impeller must be great enough to overcome the resistance of the system downstream of the pump.
- Impeller designs vary according to the manufacturer, the type of service, and operating conditions. Different casing designs also are available and casings can be either one piece or two pieces (split).
- In addition to capacity, centrifugal pumps are classified according to their specific speed, which relates flow rate, head, and operation speed of the pump.
- The main reasons of reduced centrifugal pump efficiency are air leaks and air or vapor pockets in the suction line or in the pump. Another important consideration in centrifugal pump operation is alignment of the pump with its driver. Misalignment causes vibrations, overheating, and undue stresses on the shaft and other pump components.
- In many pumping operations, cavitation is an important concern. Cavitation occurs when pressure is reduced below vapor pressure of the liquid in a localized area of the pump. It causes excessive vibrations, reduced pump efficiency, and in some cases failure of pump components.

15.5.2.4 Operation of centrifugal pump

- Start-up:
  - Open the tank outlet valve and manifold valve.
  - Check the oil level in pump bearings.
  - Check the condition of the stuffing boxes.
  - In case of a mechanical seal with internal circulation, open flow controller fully.
  - Turn on cooling liquid supply and check whether it flows freely.
  - Open suction valve fully.
  - Leave isolation valve in discharge line closed for the time being.
  - The pump must be completely primed with the oil being pumped.

- Check suction pressure and temperature, ensuring that vapor formation does not occur to avoid vapor locking.
- For initial start-up or start-up after long duration, check the rotating direction of the pump by running it for a very short duration.

- Running:
  - Check the discharge pressure.
  - For operation at a constant speed, the pump may be operated at the optimum load of 90% of its rated head.
  - For pumps operating at a variable (controlled) speed, the pump may only be operated within the range indicated in the pump operating diagram provided by the manufacturer.
  - While operating the pump, the discharge valve should be partially closed to ensure that the pressure does not go below the minimum discharge pressure corresponding to the particular speed or capacity at which the pump is operated at the time.
- Shut-off:
  - Close the isolation valve in the discharge line.
  - Switch off the driver and watch the pump run down smoothly to a stand-still.
  - Turn off the sealing, circulation, or flushing liquid (if applicable).
  - Throttle cooling liquid, supply partially, and turn it off fully only after temperature measurement at the pump nozzle had dropped below 80 °C.
- Problems and their remedies:
  - No discharge and no head generation may be due to suction valve closed, delivery valve closed, reverse direction of rotation, foreign matter in the impeller passage, air pocket in suction line, suction strainer blocked, or suction and delivery pipe clogging/chocking. Rectify one by one.
  - Reduced and/or irregular discharge may be due to air leakage in the suction line, delivery head higher than rated, air pocket in suction line, running speed being much lower than the motor synchronous speed, clogging of impeller.

  - Overloading of motor (excessive power consumption) may be due to gland packing being too tightly compressed, mechanical defect, i.e., bending of shaft rotor rubbing against casing, seizure of any bearing.
  - Heavy leakage at gland may be due to gland nut being loose, wearing of gland packing, incorrect fitting.
  - Heavy leakage at mechanical seal may be due to worn out "O" rings, worn out carbon ring.
  - Noise and vibration may be due to misalignment, insecure or improper foundation/loose foundation bolts, worn out rubber disc between coupling valves, or foreign particle inside impeller passage. Rectify one by one.

- Net positive suction head:
  - The net positive suction head (NPSH) is a crucial parameter in the design of a liquid pump. It represents the pressure differential between the source of liquid (tank or pipe) and the pump suction.
  - Each pump has its own requirement of NPSH, usually expressed in feet of head. The NPSH required is normally shown on the manufacturer rating curve for each pump.
  - Net positive suction head available to a pump in a specific application must be equal to or greater than the required NPSH specified by the manufacturer. If enough suction head is not available at the desired flow rate, vapor lock, cavitation, and pump damage may result.
  - NPSH is determined from the pressure in the tank or pipe, the atmospheric pressure, the vapor pressure of the liquid, the specific gravity of the liquid, the friction losses in piping and valves, and the difference in elevation between the fluid in the tank and the pump. For reciprocating pumps, the motion of the pump piston may be also considered for the calculation of NPSH.

- Suction piping design:
  - Factors to be considered in the design of suction piping for a reciprocating pump include the following:
    - The pump should be as close to the fluid supply as possible.
    - Use full opening gate valves and avoid valves that constrict flow.

  - The ideal piping arrangement is short and direct, using no ells. If ells are required, use a 45° radius instead of 90° ells.
  - If a reducer is required in the suction line between the main line and the pump, an eccentric reducer rather than a concentric reducer is used. The straight side of the eccentric reducer should be on top.
  - Slope the suction line downward uniformly from the fluid supply to the pump to avoid air pockets.
  - If a bypass is installed, it should be liquid to the source vessel and the suction line.
  - Suction lines should be firmly attached or buried to avoid strain on them and to help prevent vibration from acting on the pump.
  - When two or more pumps are connected to a common suction header, the size of that common suction header is important.
  - Install a gate valve on the suction to allow the pump to be isolated for maintenance.
- It is necessary to place a gate valve in the discharge line. To achieve isolation, just close the gate valve on the suction line. It is advisable to install a check valve on the discharge line in order to prevent vibrations that may arise from fluid flow originating from other units connected to the same piping system.
- Ideally, the slope of the discharge line should be straight from the pump until a change in direction occurs. It is important to securely anchor the line.
- The reciprocating pump pipework mentioned above can also be used for centrifugal pumps. Pulsations and vibrations are important factors to consider when installing centrifugal pumps, and the use of dampeners is often necessary to mitigate their effects.

- Pressure and volume capability:
  - The primary factors to consider when choosing a pulse application are the amount of fluid being pumped and the pressure difference that needs to be overcome. The pressure in the pipeline is determined by the required hydraulic delivery pressure and the pressure loss between the pump outlet and the point of interest.

- The flow and heat reciprocating pumps are determined by several physical characteristics, such as the size of the piston and cylinder. A larger piston and cylinder result in a greater flow. The capacity of centrifugal pumps is influenced by the dimensions and geometry of the impeller and casing.
- The combination of volume and pressure differential determines the energy that must be supplied to the liquid by the pump. The hydraulic horsepower that must be supplied by the pump is

  where:

  HHP = hydraulic horsepower.
  H = differential head (ft).
  Q = liquid flow (gal/min).
  Sp.gr. = specific gravity of the liquid.

  To determine the horse power that must be supplied by the pump driver, hydraulic horsepower must be corrected to account for pump efficiency.
- The following information is needed to properly select and size a pipeline pump:
  - Characteristics of the fluid, including specific gravity at pumping temperature, pumping temperature, vapor pressure at pumping temperature, and the presence of any corrosion material.
  - Desired pumping rate and expected future changes in volume requirements.
  - Pressure conditions include suction and discharge pressure, availability of NPSH, expected future pressure conditions, and whether the pump will operate in series or parallel with other pumps.
  - Preferred type of the pump and type of the shaft seal for centrifugal pumps.
  - Special metallurgy is required to handle high temperatures, corrosive fluids, or other severe conditions.
  - Type of the pump driver to be used and any space limitations.
- Station monitoring and control is critical for the reliable operation of the equipment. Individual pumps may be designed to shut down automatically, when high temperature, low flow, excessive pressure, high lube oil temperature, or other such conditions occur.

15.5.3 Compressor design and operation

- Compressors, like pumps, add energy to the flowing fluid to cause it to move through the pipeline. Like pumps, compressors can be divided generally into reciprocating and centrifugal units. Reciprocating compressors generally operate at slower speeds than centrifugal units and are used in applications where relatively high pressures are required.
- As is the case with positive displacement pumps, reciprocating compressors also produce a pulsating flow. A reciprocating compressor installation must be designed to avoid equipment and piping damage resulting from pulsations and vibrations.
- Reciprocating compressors:

  Many reciprocating compressor units used for natural gas pipeline service are integral, i.e., the compressor driver and the compressor are contained in a single unit. In a large multicylinder compressor, several compressor cylinders and the engine cylinders are connected to the same crankshaft.

  There are also reciprocating compressors that do not have integral drivers. These compressors are generally smaller than integral machines and are often used for auxiliary services.

  Reciprocating compressor cylinders contain suction and discharge valves to permit the flow of gas into and out of the cylinder. Gas flowing into the cylinder through the suction valve at the suction temperature and pressure is compressed in the cylinder and discharged at a higher pressure through the discharge valve.

  The volume that the unit can compress under the given pressure conditions depends on the size of the cylinder, the length of the piston stroke (cylinder size and stroke length determine the piston displacement), and the clearance volume within the cylinders.

  Reciprocating compressors and their drivers also require cooling and lubrication systems to prevent excessive buildup of heat and damage to piston and cylinders.
- Trouble shooting of reciprocating gas compressor:

  Compressor will not start due to power supply failure, switchgear or starting panel malfunction, low oil pressure shut-down switch, control panel problems.

  Motor will not synchronize on applicable units due to low voltage, excessive starting torque, incorrect power factor, excitation voltage failure.

Lower oil pressure may be due to oil pump failure, oil foaming from counter-weights striking oil surface, cold oil, dirty oil filter, interior frame oil leaks, excessive leakage at bearing, improper low oil pressure switch setting, malfunctioning oil relief valve, defective pressure gauge, plugged oil sump strainer.

Frame knocks may be due to loose/worn main crankpin or crosshead bushing, low oil pressure, cold oil, incorrect oil.

Crankshaft oil seal leaks may be due to faulty seals, clogged drain hole.

Noise in the cylinder may be due to loose piston, piston hitting outer head or frame end of cylinder, loose crosshead lock nut, broken or leaking valve(s), worn or broken piston rings or expanders, valve improperly seated/damaged seat gasket, free air unloader plunger chattering.

Excessive packing leakage may be due to worn packing rings, improper lube oil and/or insufficient lube rate (blue rings), dirt in packing, excessive rate of pressure increase, packing rings assembled incorrectly, improper ring side or end gap clearance, plugged packing vent system, scored piston rod, excessive piston rod run-out.

Packing overheating may be due to lubrication failure, improper lube oil and/or insufficient lube rate, insufficient cooling (water-cooled packing cases).

Excessive carbon on valves may be due to excessive lube oil, improper lube oil (too light, high carbon residue), oil carryover from inlet system or previous stage, broken or leaking valves causing high temperature, excessive temperature due to high pressure ratio across cylinders.

High discharge temperature may be due to excessive pressure ratio on the cylinder due to leaking inlet valves or piston rings on the next higher stage, fouled intercooler/gas piping, leaking discharge valves or piston rings, high inlet temperature, fouled water jackets, improper lube oil and/or lube rate.

Piston rod oil wipers leak may be due to worn wiper rings, rings incorrectly assembled/installed, worn/scored rod, improper fit of rings or rod/side clearance.

Safety valve popping may be due to faulty safety valve, leaking inlet valves or piston rings on the next higher stage, obstruction (foreign material, rags), blind or valve closed in discharge line.

- Centrifugal compressors:

  A centrifugal compressor discharges gas at high velocity into a diffuser, where gas velocity is reduced and its kinetic energy is converted to pressure.

  Centrifugal compressors consist of a housing, an impeller mounted on a rotating shaft, bearing and seal to prevent gas from escaping along the shaft. The shape and size of the diffuser and impeller vary, depending on operating conditions and on the manufacturer's design.

  The output from a centrifugal compressor is smooth as compared with the pulsating flow of reciprocating compressor. Because of this feature, they are often considered for installation on offshore platforms where vibrations must be minimized.

  Centrifugal compressors are not capable of as high a compression ratio as reciprocating machines, but they can be arranged in series so that each is only required to develop a portion of the total differential pressure required. Their continuous flow characteristic makes this series arrangement practical.

- Compression ratio:

  A key consideration in designing any compressor installation is the compression ratio. If the overall compression ratio is high, several compressor stages may be required. Compression ratio must be limited to avoid excessive temperature that would result from too high a ratio.

  Recommended operating temperatures for compressor cylinders are used as a guide in determining maximum compression ratio per stage.

  Compression ratio required, per stage, is calculated using the overall ratio of the installation and pressure losses in suction and interstage cooling piping. A close estimate of the compression ratio per stage for a two-stage compressor can be obtained by calculating the square root of the station discharge pressure divided by the station suction pressure. If a three-stage compressor is to be used, the cube root of the ratio of station discharge pressure to suction pressure gives the approximate compression ratio per stage.

  For final design, suction losses and losses between stages must be considered. Maximum allowable compression per stage is usually based on recommended limits on compression ratio or operating temperature.

The temperature increase that will result from compression from a given suction pressure to a given discharge pressure can be calculated using the temperature of the gas at suction conditions, the suction and discharge pressure, and the heat capacity of the gas.

To limit the gas temperature to recommended values, it is often necessary to cool the gas between compression stages. Interstage cooling between compression stages can be done by air cooling, by cooling with water in a heat exchanger, or by exchanging heat with the inlet gas in a gas-to-gas heat exchanger.

Depending on the type of gas, separations may also be needed between compression stages to remove any liquid condensed by interstage cooling.

- Capacity and horsepower:
  - A common method is to express capacity in cubic feet per unit of time at suction temperature and pressure. Because gases are highly compressible, their volume changes directly with temperature and pressure.
  - A compressibility factor must be included in volume calculations to account for the deviations of a gas from the ideal gas flow.
  - The volume that a given reciprocating compressor can handle depends on the piston displacement and the volumetric efficiency of the cylinder. Piston displacement depends on whether or not the piston is single acting (compresses only at one end of the cylinder) or double acting.
  - The capacity of a centrifugal compressor depends on the size and speed of its impeller and the pressure against which the compressor is discharging. Centrifugal compressor capacity varies directly with speed. Manufacturer's charts depict how volume, head, and compressor speed are related.

To select a suitable compressor, it is necessary to determine how much compressor horsepower will be required to handle the required volume. Factors that affect the brake horsepower requirement include the volume to be compressed, suction and discharge pressure (compression ratio), the heat capacity of gas, and the efficiency of the compressor.

For both reciprocating and centrifugal compressors, charts have been developed that show horsepower required to compress 1 mmcfd of a gas with a given ratio of specific heats at various compression ratios. These charts are used for a preliminary selection of compressor sizes.

Then calculations are made to confirm that the compressor is capable of handling the required volume.

The best compressor for each installation is the one that represents the most desirable combination of capital cost; annual operating and maintenance cost; fuel efficiency; expected increase in operating, maintenance, and fuel costs; and the specific advantages and limitations of each alternative.

Considerable time and expertise must be devoted to provide auxiliary services, including lubrication, cooling, monitoring and control instrumentation, interstage cooling, and liquid removal.

The design of interstage cooling involves sizing heat exchangers by calculating heat flow and the sizing of vessels if liquid removal is required. Additional design work may involve pulsation dampeners, the selection of valves to regulate and distribute flow, and station piping.

15.5.4 Prime mover design

The purpose of a pipeline prime mover is to supply the shaft horsepower required by a pump or compressor to move fluid through the pipeline. Types of prime movers include electric motor, gas turbine, and diesel internal combustion engines.

Basic selection parameters are horsepower output and efficiency. The designer should consider the availability of energy to power the prime mover and cost of energy. A good estimate of fuel and other operating costs is necessary to compare prime mover alternatives accurately.

The type of reliable energy source available to power the installation is the key consideration in choosing among driver types. The projected cost of fuel or electricity over the life of the project is also to be considered.

The type of supervisory control system to be used at the pump or compressor station and the pipeline overall monitoring and control configuration are also considered when selecting prime movers.

Recommended operating speed of the pump or compressor to be driven also influences the type of the driver. The prime mover speed must be compatible with the speed of the machine to be driven.

Other considerations in selecting the proper type of prime mover include efficiency, availability, and the expected time between major inspections and overhauls.

A prime mover must be capable of supplying the shaft horse power required by the pump or compressor and be selected on the basis of initial cost, fuel cost, maintenance and operating cost, control adaptability, flexibility of operation, and other factors.

- To calculate the break horsepower, the following equation is used.

$$\text{bhp} = \frac{H\,Q(\text{sp.gr})}{3960\,E} \tag{15.15}$$

where:

bhp = brake horsepower
H = head, ft
sp.gr.= specific gravity of the fluid
E = pump efficiency
Q = liquid flow, gal/min.

Calculating the amount of horsepower required to drive a compressor involves determining the theoretical horsepower required to increase gas pressure from suction pressure to discharge pressure and then accounting for the losses in the compressor.

Assuming adiabatic compression (compression in which no cooling occurs and gas temperature rises steadily), theoretical horse power for a reciprocating compressor can be calculated from this formula:

$$\text{Theoretical HP} = \frac{P_1 V_1 K\left[(P_2 \,/\, P_1)^{(K-1)/K-1}\right]}{229(K-1)} \tag{15.16}$$

where:

K = ratio of specific heats of the gas (C_p/C_v)
P_1 = suction pressure, psia
P_2 = discharge pressure, psia
V_1 = suction volume, cu ft/min.

The actual horse power required is then obtained by multiplying theoretical horsepower by a factor that accounts for losses due to pressure drop through valves and piping and the friction of piston rings and rod packing.

15.6 Installation and Testing of Pipeline

15.6.1 Installation

The installation of a pipeline includes the following major steps:

- Clearing the right of way as needed.

- Ditching.
- Stringing pipe joints along the right of way.
- Welding the pipe joints together.
- Applying coating and wrapping to the exterior of the pipe (whole length, except a portion of the pipe at each end, is sometimes coated before being delivered to the job site).
- Lowering the pipeline into the ditch.
  - Backfilling the ditch.
  - Testing the line for leaks.
  - Cleaning and drying the pipeline after testing to prepare it for operation.

15.6.2 Testing overview

Hydrostatic:

All completed pipelines must be tested before being put into operation. On long pipelines, the line will normally be tested in sections; on short lines, the entire pipeline may be tested as a unit.

A common approach is hydrostatic testing – filling a closed pipeline section with water, then pressurizing the line to a specified pressure to check for leaks. Temporary connections for filling and draining the pipeline are used, and a pump is used to "pressure up" the line. The pressure is maintained on the line for a specified time. If pressure declines, a leak is indicated.

The pressure to be reached and the time period, which must be monitored, are specified in the construction plans. The pipeline test pressure to be used is based on the pipeline location, its function, its design operating pressure, and other factors. Hydrostatic test pressure may be specified as 125% of the maximum design operation pressure of the line.

After the pipeline has been tested, it is important that the moisture and foreign materials be removed from the pipeline before it is put into operation. Such materials could damage pumping, compression, and other equipment if swept into them when the pipeline is put into service. Water, sand, dirt, welding slag, and even some stronger materials have been removed from newly completely pipelines.

It is often necessary to dry natural gas pipeline after hydrostatic testing to prevent the formation of hydrates when the pipeline is put into service. Gas hydrates are complex chemical compounds, formed when free water is available in the presence of hydrocarbon gases and, if allowed to build up on the walls of natural gas pipelines, would reduce flow efficiency by increasing friction and reducing the effective diameter of the pipe.

- **Non-destructive:**

Pipeline welds must be inspected visually to ensure that they are performed in accordance with the welding procedures and that the welds are acceptable under the appropriate specifications.

Non-destructive testing is also required on welds made on a pipeline that will be operated at a pressure that results in a hoop stress of 20% or of the specified minimum yield strength (SMYS).

- At least 10% of the welds made by each welder during each welding day in a liquid pipeline must be non-destructively tested over the entire circumference of the weld. In many locations like any offshore area, within railroad or public road rights of way, at overhead road crossings and in tunnels, at pipeline tie-ins, within any incorporated subdivision of a state government, etc., 100% of the welds must be tested non-destructively.
- Testing of offshore pipeline is similar to that of onshore pipeline. Welds are X-rayed at a station on the lay barge and the completed pipeline is hydrostatically tested to check for leaks.

15.7 Operation and Control

- Each pipeline system has unique characteristics that direct the type of control system that is most suitable. The primary goal of a pipeline control system is to obtain the highest throughput at the lowest cost without exceeding pressure limits in the system and to deliver the required product volumes to the customer on schedule.
- Primary control systems can protect pipeline and equipment by monitoring and adjusting pressure and other operating variables, providing alarms when limits on operating conditions are exceeded, scheduling the shipment and delivery of different products, monitoring machinery performance and wear, controlling pressure surges in the pipeline, providing leak detection, and performing other functions.

15.7.1 Supervisory control

- Pipeline supervisory control systems regulate pressure and flow, start and stop pumps or compressors at stations along the line, and monitor the status of pumps, compressors, and valves. In a large pipeline system, many of the supervisory functions can be performed from a central location.

- In general, modern computer-based pipeline control systems consist of the following elements:
  - The computer complex includes computers, computer peripherals, and interfacing equipment for the man/machine systems and remote stations.
  - The man/machine system includes devices necessary for the operator to communicate with the computer, such as the video display units, keyboards, and loggers.
  - Remote stations are connected to the computer complex via a communication channel microwave, telephone, radio, or other means.
  - Field service pumps and motor operated valves are controlled and monitored by the remote stations. Field instrumentation includes pressure and temperature transmitters, tank gauges, and similar components.
  - The effects of component failure must be considered during the design phase. This is important for ensuring system's reliability.

Personnel must be thoroughly trained to operate and maintain the system.

15.7.2 Scheduling

- One of the most important functions of the pipeline operator, especially in the case of liquid pipelines, is to schedule the volume of each product transported by the pipeline to ensure delivery to the customer at the desired time.
- Manual scheduling can involve complex, repetitive calculations by pipeline operations personnel. But the same can be automated by the use of computers. The assignment of shipment and delivery time is done in computer, which also calculates the hydraulic rates at which the product moves through the pipeline.
- Hydraulic rates are a function of pump configuration, product mix, and its characteristics. The scheduling system should examine all possible pump configurations and choose the one that moves the liquids in the desired time while minimizing the cost of power.

- In computers, calculating a schedule consists of six steps:
  1. Create or modify batch codes and volumes.
  2. Verify batch codes and volumes.
  3. Calculate starting conditions.
  4. Generate rate profile.
  5. Calculate shipment times.
  6. Calculate delivery times.
- To calculate hydraulic rates for a given segment of the system, the computer searches for the maximum suction pressure and maximum discharge pressure. Pressure loss caused by friction in the pipeline and changes in elevation are calculated and suction and discharge pressures are determined for the specific product at the given rate.

15.7.3 Metering

- Gas measurement:
  - Natural gas may be measured with orifice, positive displacement, turbine, and other types or meters. These devices measure the volume of gas flowing in the line.
  - Measurement of natural gas volume requires that the condition at which the volume is determined must be stated. Measurements are normally adjusted to a base temperature 60 °F and pressure 14.7 psi. The exact base conditions are spelled out in each gas purchase contract and vary only slightly from area to area and from contract to contract.
  - To calculate the gas volume passing through a meter, the data required includes flowing and base temperatures and pressures, gas-specific graving, constants that have been determined for the specific meter, and super compressibility of the gas at flowing conditions.
- **Liquid measurement:**
  - Liquid stream may also be measured with orifice meters by applying appropriate correction factors. The factors differ from those used when measuring natural gas.

  - Positive displacement meters and turbine meters can be used to measure liquid rates. In these meters, fluid passes through the meter in successive isolated quantities by filling spaces of fixed volume. A counter registers the total quantity of fluid passing through the meter.
  - In a turbine meter, the force of the flowing fluid turns a bladed rotor that is parallel to the direction of flow and the rotor's speed of rotation is proportional to the flow rate. Using proper gearing, the revolutions of the rotor are related to volume to provide the flow rate.
  - In addition to the orifice meter, positive displacement meters, and turbine meters, other devices like flow nozzle, venturi meter, and vortex meter are also used for fluid flow measurements.
- Metering of oil and gas has been described in detail in Chapter 7.

15.7.4 SCADA system

- Pipeline automation systems are becoming more sophisticated and integrated. Traditional pipeline SCADA systems are being replaced by new concepts and capabilities.
- A summary of the major points regarding these new concepts and capabilities are described below:
  - Satellite communications, which are cost effective and highly reliable, are replacing traditional methods of remote communication such as microwave and leased telephone lines.
  - High technology standard hardware such as programmable logic controllers (PLCs) at remote sites and work stations as man–machine interfaces are replacing the traditional central processor systems.
  - Control system applications are involving the integration of SCADA, optimization and business applications, utilizing a common data base with data security.
  - The basic architecture of SCADA system is changing to one based on local and wide area communication, similar to the architecture of factory automation systems.

15.8 Maintenance and Repair

15.8.1 Cathodic protection

- Pipeline corrosion can result in damage to the pipeline that requires repair or replacement of pipe, loss of product through leaks, damage to property along the pipeline and downtime.
- Underground corrosion of steel pipelines can result from the flow of electrical current between areas of different electric potential. The current flows from an area of higher potential through an electrolyte to an area of lower potential. The area of higher potential (the anode) will be corroded.
- In a cathodic protection system, anodes are installed and an electric current is made to flow between the pipe and the anodes through the soil. The pipeline becomes the cathode of the system and the anode is corroded.
- The design of a cathodic protection system includes a current requirement survey, the selection and sizing of current drainage points, and the detailed design of the ground (anode) beds.
- The magnitude of the corrosion currents for a given potential difference between two electrodes (cathode and anode) depends on several factors like soil resistivity, chemical constituents of the soil, separation between the anode and cathode, anode and cathode polarization, and relative surface areas of the cathode and anode.
- Cathodic protection also has been discussed in Chapter 9.

15.8.2 Leak detection

- Traditionally, pipelines were inspected visually, traversing the route on the ground or patrolling the pipeline route in a light aircraft/helicopter.
- Comparatively larger leaks can be detected by direct observation of pressure drop and volume loss, based on comparing flow into a segment of pipeline and flow out of the segment.
- In addition to monitoring inflow and outflow in a segment of the system, other leak detection systems for liquid pipelines include acoustic commission inspection systems, instrumented pigs, and ultrasonic methods.

- Natural gas pipelines can be inspected for leaks with surface sampling instruments using the flame ionization principle.

15.8.3 Line repair

- Onshore pipelines are often plugged temporarily, in case of a leak, on either side of a problem area and flow is redirected through a bypass so that the work can be done on the isolated area.
- In a typical plug and bypass operation, the line is uncovered and weld fittings are installed on the pipe. Temporary valves are bolted onto the fittings and a hot tap is made through each, penetrating the wall of the pipe.
- In addition to mechanical plugging methods, another technique has been used in which a plug is frozen into place in the pipeline to isolate a section for repair or maintenance.
- Leaks can also be located using the ice plug technique by isolating successive sections of the line until the leak is pinpointed. Then, two freeze plugs can be used to seal off the damaged section.
- There are a variety of methods available for repairing submarine pipelines; they generally fall into three categories:
  - Surface repair
  - Underwater hyperbaric welding
  - Mechanical connectors
- Surface repair involves lifting the pipeline to the surface of the water and welding in a new piece to replace the damaged section, or welding flanges or fittings onto each end of the pipe after removing the damaged section.
- Underwater welding techniques can be used to repair a damaged pipeline without lifting it to the surface. Welding can be done by a welder/diver completely enclosed in a dry habitat, or the diver may work in the wet with only the work area being enclosed in a controlled environment.
- Choice of an offshore pipeline repair method depends on location; water depth; pipeline size, age, and amount of burial; design and operating pressures; traffic in the area; special hazards such as unusual currents or mud slides; and weather conditions.

15.9 Pigging Operation

- Operations accomplished by pipeline pigging are as follows :
  - Periodic removal of dirt and water accumulation from the operating pipelines.
  - Product separation to reduce the amount of interface in the transition zone between different types of flowing crude oils or refined products.
  - Control of liquids inside a pipeline, including two-phase pipelines, when filling pipelines for hydrostatic testing, dewatering pipelines following hydrostatic testing and during drying and purging operations.
  - Inspection of pipelines for detection of dents, buckles, or excessive corrosion, using gauging pigs and/or intelligent pigs.
- Types of pigs:
  - Non-intelligent pigs:

    Sphere
    - To dewater/remove liquids
    - To remove debris
    - Chemical swabbing
    - Batching

    Cup-type pigs
    - To clean
    - To remove debris (and wax)
    - Chemical swabbing
    - Batching

    Foam pigs
    - To clean
    - To dewater/remove liquid
    - To remove debris (and wax)

Gel pigs

- To remove debris
- Batching
- Dewatering/dehydration
- Pigging multiple diameter line

- Intelligent pigs:

  Cup types

  - Magnetic flux principle
  - Ultrasonic sound principle

- To move a pig through a pipeline, a pressure differential is required across the pig. The force to move the pig is dependent on several factors such as travel uphill or downhill, friction coefficient and force between the pig and pipe walls, and the lubrication available such as dry gas or crude oil service.
- Non-intelligent pigging operations on gas transmission pipeline systems are performed primarily to maintain efficiency by cleaning and swabbing the pipeline for emergency situations. Pipeline section downstream of compressor stations will require periodic pigging to remove lubricating oil from compressor units.
- Two-phase flow pipeline systems use mechanical pigs to keep the liquid drop-out at a level that will maintain the design efficiency of the system. Cleaning pigs are also run through these pipelines to remove foreign materials that may accumulate.
- Material cleaned by the pig will follow the flow at the receiving trap. Most of the material will go through the side valve rather than into the trap barrel. Pump stations and compressor stations on pipeline systems utilizing pigging should be protected from damage due to slugs or foreign materials, as a result of pigging.
- Intelligent pigging:

  These pigs are used to gather information regarding the condition, configuration, or performance characteristics of a pipeline and are of following types:

- Magnescan:

  The magnescan tool is a sophisticated application of magnetic flux leakage technology. Powerful permanent magnets, coupled to the pipe wall by high density brushes, induce a magnetic field in the pipeline steel with a resultant leakage field on the inside pipe wall. Where metal loss has occurred, on the inside or outside surface of the pipe, a change takes place in the leakage field.

- Ultrascan:

  The ultrascan tool applies the precision of ultrasonic measurement technology to pipeline inspection. The ultrascan tool consists of a number of units linked together by robust universal joints. The tool is free swimming and is driven by the leading unit. The other units are supported by rollers or cups with bypass holes. In a typical tool, the leading unit carries the battery that provides the operating power, the second unit carries data processing and recording equipment, the third unit carries electronic sensor, and the final unit carries the ultrasonic transducers. All the data is processed and stored in digital form.

15.10 Underwater Inspection, Maintenance, and Repair

- The inspection, maintenance, and repair (IMR) group is primarily responsible for underwater installations, i.e., jackets, risers, pipelines, SBMs, PLEMS, subsea templates, wellheads, and Christmas tree. IMR also provides safety coverage, as per statutory and mandatory requirements, for offshore operations.
- The underwater installations are subjected to:
  - Environmental forces, viz. sediment transport and scourings, sand waves, waves and currents, submarine landslides, and corrosion.
  - Human activities causing damages, for example, impact caused by dragging/falling of anchors on pipelines.
- Objectives:
  - At the time of fabrication of offshore installations, life period is fixed for each installation. For the safe working on these platforms, for the earmarked life, a constant monitoring of the integrity of the platform is required to be undertaken, which is being done by the IMR group.

- The inspection and repair jobs involving diving operation (both air range and saturation range) for dynamically positioned vessels is very costly. To reduce the costs of inspection, maintenance, and repair, the monitoring of marine growth, frequency of maintenance, effective deployment and utilization, all play major roles.
- The integrity of offshore structures and pipelines is being monitored by visual and instrumental techniques. The data thus acquired is correlated and the frequency of regular checks is then decided.
- The experiences of other companies with respect to their environmental conditions suggest a yardstick for the procedures and frequency of inspection based on the experience in North-sea in the Norwegian set-up.

- Major tasks being performed by IMR section:
  i) Inspection of offshore jackets, risers, and pipelines.
  ii) Maintenance of offshore jackets, risers, and pipelines.
  iii) Repairs of offshore jackets, risers, and pipelines.
  iv) Fire-fighting and safety.
  v) Pollution control.
  vi) Production assistance.
  vii) Drilling assistance.

15.10.1 Inspection

- The inspection work, generally carried out by surface vessels and divers, is to monitor the following aspects:
  - As-built status verification.
  - Cathodic protection potential level survey.
  - Weld inspection to detect cracks.
  - Magnetic particle inspection to detect invisible cracks.
  - Riser and riser clamp survey.
  - Boat fender, boat landing attachment inspection.

- Marine growth inspection.
- Scour and debris survey.
- Inspection of crossings, PLEMS.
- Pipeline inspection.
- Inspection and maintenance of all single-point mooring systems.

15.10.2 Maintenance

Marine growth removal and underwater cleaning must be conducted for:

- Stress relief of structures by removing all marine deposits. Large amounts of deposits increase the weight of the structure and the drag caused by marine currents (fatigue cracks in weld seams are a result of such overloads and have to be repaired at extremely high cost).
- In hard bio-fouling like Barnacles Oyster, muscles are removed from structures to keep the growth within design limit parameters for the structures to remain safe for working.
- Anode installation: It is carried out to augment the cathodic potential level.
- Supporting suspended sections of pipelines: Free spans are generally caused by natural erosion of seabed. As remedial measures, suitable supports are to be provided by sand bags or grouting to prevent sagging and to stabilize pipelines.
- Riser thickness monitoring in splash zone: As the splash zone is more prone to corrosion, regular thickness, monitoring, and visual survey are very essential to avoid any sudden leakage (which would lead to stoppage of production) and remedial measures are to be planed, as required.
- Single-point mooring (SPM) system maintenance and repair:
  This mooring point is for the tankers to moor for receiving crude. The following aspects are taken care of to maintain the system:

1. Buoy body inspection.
2. Buoy bearing maintenance.
3. Cathodic protection survey of buoy.
4. J-tube and expansion joint inspection.

5. Subsea hose survey and replacement.
6. Mooring rope replacement.
7. Chain angle measurement.
8. Overhauling of buoy.

15.10.3 Repairs

- Repair on structures:
  Repair on structure is generally called for, due to the following reasons: Cracks, dents, bent bracings, wire rope scars, holes in braces/legs/piles.
  - Cracks:
    In case of major cracks on nodes or members, structural analysis of the jacket is carried out assuming that the concerned member is not in existence. On the basis of the analysis, the member is completely replaced, the joint is strengthened, or the damaged area is left as it is, depending upon the redundancy provided in the system.
  - Dents:
    In case of major dents, similar structural analysis is carried out as in the case of cracks and the member is replaced, if called for. Minor dents are however monitored during inspection of structure.
  - Bent bracings:
    - Degree of bent is first thoroughly evaluated with respect to its original shape and dimensions. Connecting joints with other structural members are also thoroughly checked up for any possible cracks.
    - In case a major bent is detected or crack is found in the connecting joints, the whole truss system has to be subjected to structural analysis. Depending upon the result of the analysis, the whole member is to be replaced or the truss system is strengthened with the help of another member placed in the vicinity of the affected zone.
    - Minor bents are monitored closely for any further deformation or metal loss.
  - Wire scar abrasion:
    Degree of abrasion, i.e., the extent of metal loss affecting reduction in thickness of members, decides the mode of repair to be undertaken. In case of excess metal loss, it may lead to hole in the member. Minor wire scar is a very common phenomenon. Left over debris, in

general, is responsible for such damages. Such situations are closely monitored.

- Holes in braces/legs/piles:
  - Presence of large holes on structural members may lead to total damage of whole member, thus requiring complete replacement. However, structural analysis is carried out prior to replacement.
  - Minor holes, however, do not pose any significant threat to the integrity of the structure.
  - In case of minor holes, their uniformity of size and any possible presence of cracks in the vicinity are checked up.
  - In case of uniformly sized minor holes, they are left as it is and monitored during jacket inspection. In case of uneven or cracked holes, a large-sized uniform hole is drilled to arrest any possible crack propagation from the affected area.

- Riser repair:
  Damages commonly observed are as follows:
  - Leakage
  - Heavy corrosion
  - Riser pulled outwards
  - Bare metal
  - Leakage:
    Minor leakages are repaired with the help of mechanical clamps. Major leakage calls for replacement of the section by mechanical connectors or wet welding. Since wet welding is a costly exercise, often the other alternative of installing mechanical connectors is implemented.
  - Heavy corrosion:
    - Though the most affected area, i.e., the splash zone, is coated with monel metal sheathing, heavy corrosion is observed in most of the risers within one year's time and the remedial measures may be required.
    - Heavy corrosion leads to loss of thickness in the riser. Such sections are replaced with a new section.

  - In case of minor external corrosion, the depth of pittings is measured and the remaining effective thickness is compared with as-built thickness.
  - Internal corrosion or internal laminations are detected with the help of USM II scanner and the effective thickness is measured against the as-built specifications.

- Riser pulled outwards:
  - When the localized deformation due to unwanted pull by anchor or similar device is major in nature, the section is replaced with a fresh one.
  - In the event of minor pulling, the riser is straightened to a degree possible by tirfors or left as it is, if it does not hinder the flow of hydrocarbons considerably. A clamp is placed in the area of deformation to ensure proper integrity with the structure.
- Bare metal:
  It is generally left as it is for the purpose of monitoring cathodic protection potential as well as measuring thickness.
- Clamp misalignment:
  A new clamp is generally installed in place of the old one with proper electrical connections with the main structure.

• Repairs on pipelines:

- Repairs on pipelines depend on the particular type of damage suffered by the pipeline.
- Types of damages, in general, are anchor pull, pipeline leakage, free spans, coating damage, anode loss, etc.
  - Anchor pull:

  Anchor pull causes sharp bend in the line, abruptly reducing the pipe diameter necessitating complete replacement of that particular section. Replacement is done with a similar piece of pipe with the help of gripper connectors.

  - Leakage:

  Leakages of minor nature are repaired with the help of mechanical clamps placed over the area of the leakage. However, in the case of major leakage, the affected section is replaced by a fresh one with the help of mechanical connectors.

- Coating damage and anode loss:

Both the phenomena reduce the level of cathodic protection of the pipe line, which is made up by installing sacrificial anodes.

- Emergency pipeline repair (EPR):
  - For crucial pipelines, which require that their repair time be crushed down to minimum, special repairing methods are adopted. These are called HAS (habitat alignment system) and CHAS (combined habitat alignment system). Special types of fixtures, suitable particularly for pipeline vendor repair, are constructed and lowered into the damaged site.
  - A special housing is also lowered at the site for carrying out hot welding on the pipeline. So far, these facilities are not available with ONGC; however, these are being acquired as an insurance against major rupture of any of the oil/gas pipelines.

16

Oil Storage, Treatment, and Transportation

16.1 Introduction

The fluid produced from a well is usually a mixture of oil, gas, water, and sediment in varying amounts. The oil alone is a complex mixture of many hydrocarbon compounds, including compounds that enter the gas phase during the production process. Gas may also be produced from the reservoir in varying amounts and travel up the tubing as bubbles entrained in the oil. Alternatively, oil droplets may be entrained as a mist in the gas. Formation water may be carried in the gas as water vapor, emulsified as droplets within the oil, or produced as free water containing dissolved gas, dissolved salts, and entrained oil. Sand and silt from the formation and rust or scale from the tubing or casing may also be included in the produced fluid stream, along with CO_2, H_2S, and other non-hydrocarbon contaminants. These elements may be present in a wide range of relative proportions.

Surface production facilities must be designed to separate and turn this mixture into separate streams of clean, dehydrated oil and gas, and safely disposable water and solids. The oil and gas must be metered and sold to a pipeline or delivered to a plant or refinery for further processing.

Production and separation of well fluids:

Figure 16.1 is a typical block diagram that outlines the major components of a typical production facility.

The system begins at the wellhead, which for flowing wells will usually include a choke to control the flow rate. The pressure drop across the choke is determined by the flowing tubing pressure (upstream) and the initial process vessel operating pressure (downstream). Varying the choke size varies the flow rate.

When two or more wells are commingled at a central processing facility, it is necessary to use a *production manifold* to combine the flow streams or to divert a single stream for testing or special treatment.

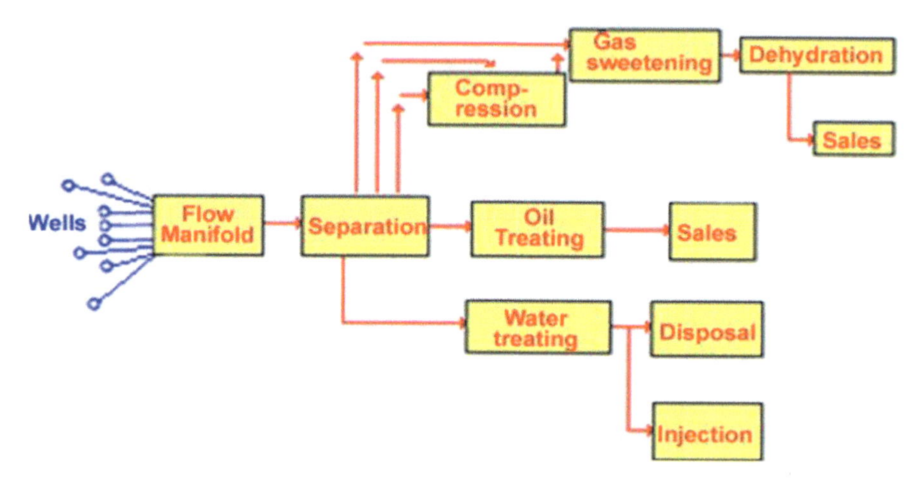

Figure 16.1 Block diagram that outlines the major components of a typical production facility.

The first component of the processing facility the produced fluid encounters is typically some type of *separator*. Separators manipulate the stream of produced fluid to take advantage of the density differences that exist among gas, oil, and water and that cause these phases to separate. A two-phase separator separates gas and liquids; a three-phase separator goes one step further and separates oil and water as well. Separators come in a variety of configurations and operating pressures depending on the degree of gas–oil separation desired and the producing pressure of the production stream. Stage separation, which typically involves high, intermediate, and low pressure separators in series, can maximize oil volumes and allow several wells with a variety of flowing tubing pressures to be serviced by the same facility. Because a separation vessel is normally the initial processing vessel in any facility, improper design of this component can cause a bottleneck and reduce the capacity of the entire system.

16.2 Oil Treating

Separated crude oil must often be treated to remove impurities, particularly water, which exists as an *emulsion* within the oil.

Figure 16.2 is a photomicrograph of a water-in-oil emulsion, which shows the tiny droplets of water tightly bound in a continuous body of oil. These droplets are slow to coalesce and separate under the influence of gravity alone. The three things necessary to cause an emulsion, two immiscible phases, agitation, and emulsifying agents, are commonly present in producing

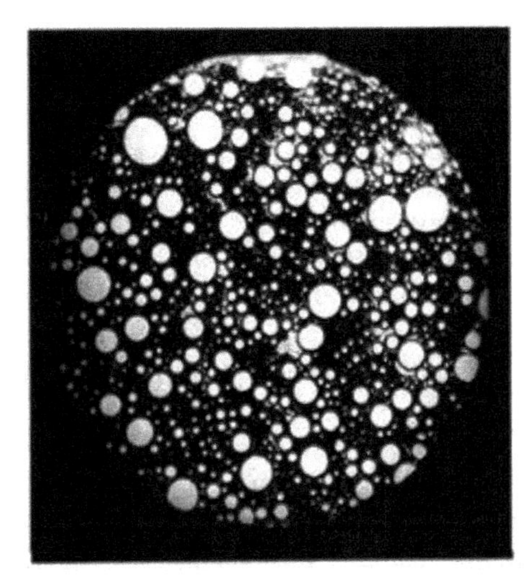

Figure 16.2 Photomicrograph of a water-in-oil emulsion.

systems. Fine solids, carbonates, and sulfate compounds can react with certain crude oils to form surface films around the water droplets, which become very stable and difficult to break.

In order to break the emulsion, an agent must be added to the system to speed up the water droplet coalescence and separation. Generally this energy is added by using *heat*, *chemicals*, or *electricity*, or a combination of these agents. In general oil field practice, heat is the primary treating method, while chemicals and electrostatic fields are used to supplement the process. In the case of a very minor emulsion, simple chemical injection upstream of the separator may solve the problem.

16.2.1 Oil-treating equipment

In selecting a treating system, a number of factors should be considered to determine the most desirable methods of treating the crude oil to contract requirements. Some of these are:

- the "tightness," or stability, of the emulsion;
- the specific gravity of the oil and produced water;
- the corrosiveness of the crude oil, produced water, and gas;
- the scaling tendencies (hardness) of the produced water;
- the quantity of fluid to be treated and percent water in the fluid;

- the paraffin forming tendencies of the crude oil;
- the desirable operating pressures for the equipment;
- the availability of a sales outlet and the value of the gas produced.

Heat assists in the emulsion breaking process by increasing the temperature of the immiscible liquids, reducing viscosity, deactivating the emulsifying agents, and allowing the dispersed droplets of water to collide. As the droplets collide, they grow in size and begin to settle. If the vessel is designed properly, the water will settle to the bottom of the treater due to the difference in specific gravity.

The process of coalescence requires that the water droplets have adequate time to contact each other. It also assumes that the buoyant forces on the coalesced droplets are such that these droplets will still be able to settle to the bottom of the treating vessel. Consequently, design considerations should take into account temperature, time, viscous properties of oil that inhibit settling, and the physical dimensions of the vessel that determine the speed at which settling must occur.

Laboratory analysis of an emulsion sample, in conjunction with field experience, should be the basis for specifying the configuration of treating vessels.

16.2.2 Vertical treaters

The most commonly used single-well lease treater is the vertical treater (Figure 16.3).

Flow enters the top of the treater into a gas separation section. The liquids flow through a downcomer to the base of the treater, which serves as a free-water knockout section. If the treater is located downstream of a free-water knockout (a large three-phase separator), the bottom section can be very small. If the total well stream is to be treated, this section should be sized for three to five minutes' retention time for both the oil and the water, to allow the free water to settle out. This will minimize the amount of fuel gas needed to heat the liquid stream rising through the heating section. The end of the downcomer should be slightly below the oil–water interface to "water wash" the oil being treated. This will facilitate the coalescence of water droplets in the oil.

Treated oil flows out the oil outlet. Any gas flashed from the oil due to the increase in temperature flows through the equalizing line to the gas space above. Pneumatic or lever-operated dump valves maintain the oil level. The

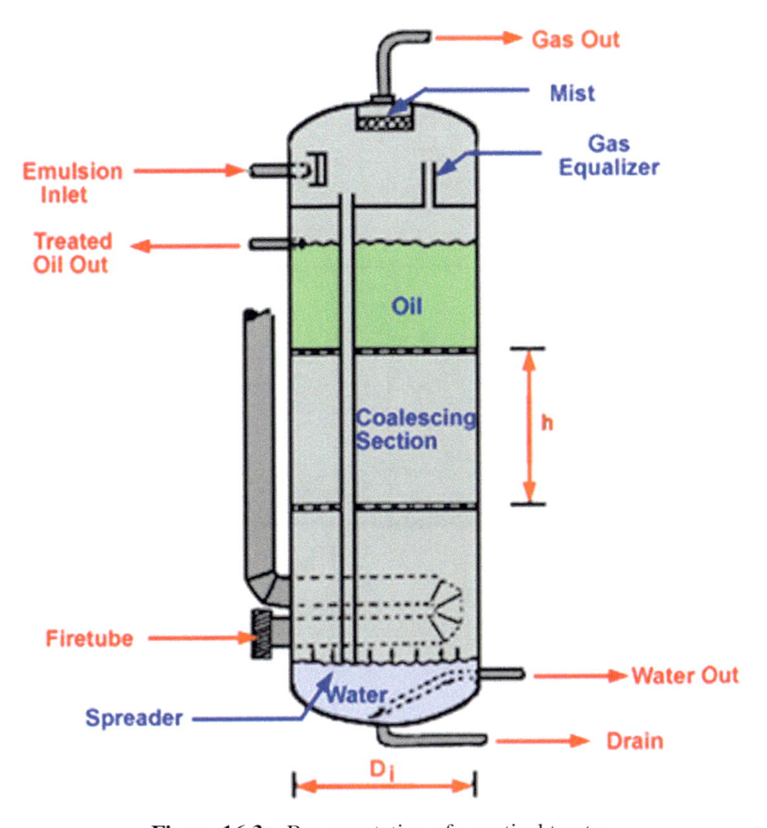

Figure 16.3 Representation of a vertical treater.

oil–water interface is controlled by an interface controller, or an adjustable external water leg.

16.2.3 Horizontal treaters

Horizontal treaters are normally required for most multiwell situations. Figure 16.4 shows a typical design of a horizontal treater.

Flow enters the front section of the treater, where gas is evolved. The liquid is deflected around the outside of the fire tube to the vicinity of the oil–water interface, where the liquid is water-washed and free water is separated. The oil and emulsion is heated as it rises past the fire tubes and is skimmed into the oil surge chamber. The oil–water interface in the inlet section of the vessel is controlled by an interface level controller, which operates a dump valve for the free water. From the oil surge chamber, the heated oil and emulsion is forced to flow through a spreader into the rear,

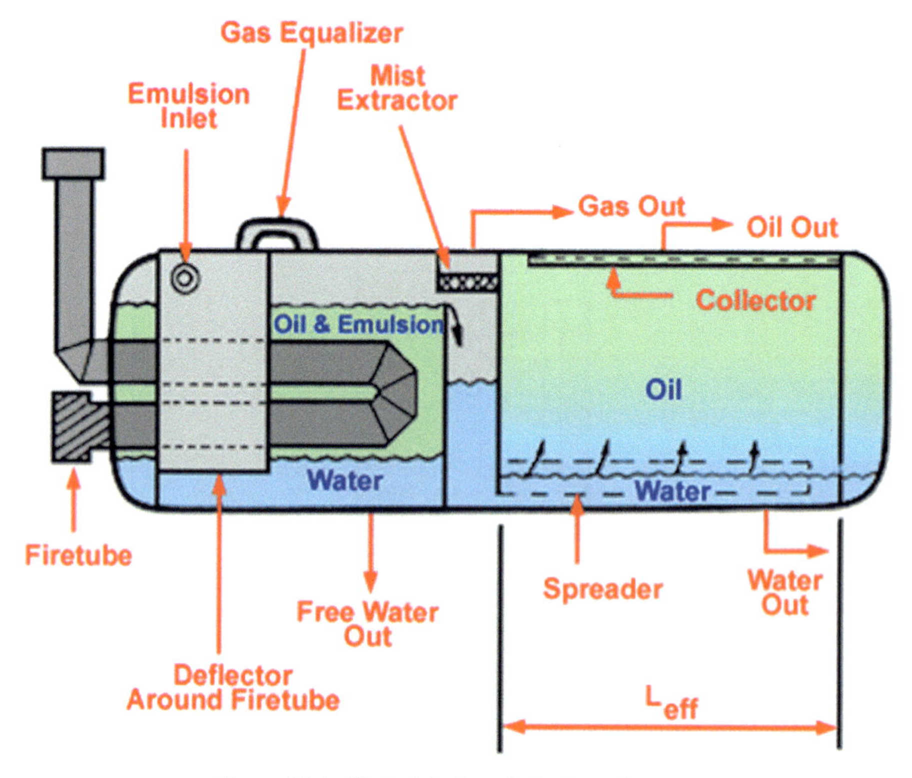

Figure 16.4 Typical design of a horizontal treater.

or coalescing, section of the vessel, which is kept full of fluid. The spreader distributes the flow evenly throughout the length of this section. Treated oil is collected at the top through a collection device sized to maintain uniform vertical flow of the oil and avoid re-emulsification. The coalescing water droplets fall countercurrent to the rising oil continuous phase in this section. The oil–water interface is maintained by a level controller and dump valve. Figure 16.5 is a schematic of a typical horizontal electrostatic treater that uses an AC/DC electrostatic field to promote coalescence of the water droplets by charging them and allowing opposite charges to attract the droplets together.

The flow path in an electrostatic treater is generally the same as that for a horizontal treater. Field experience tends to indicate that electrostatic treaters are efficient at reducing water content in the crude to below 0.5% BS&W. This makes them particularly attractive for desalting applications,

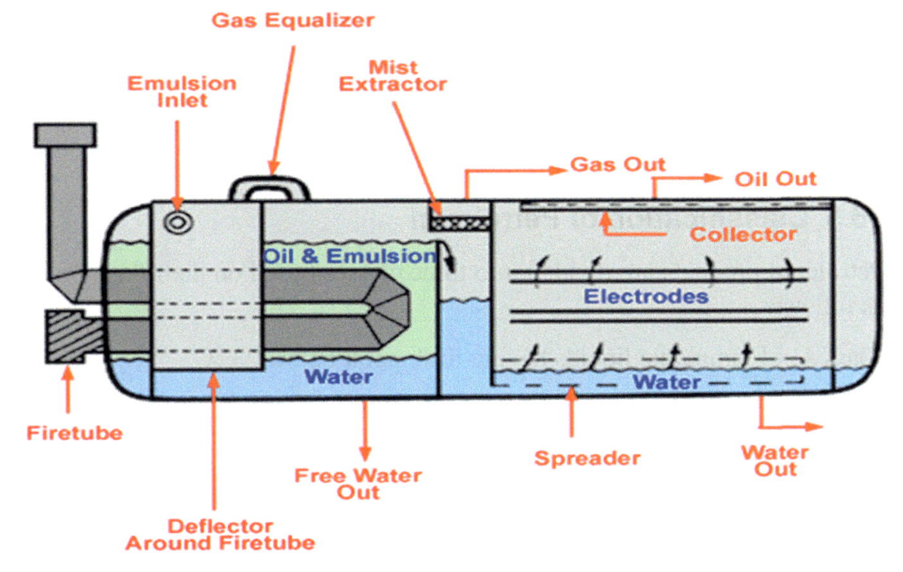

Figure 16.5 Schematic of a typical horizontal electrostatic treater.

and consequently they are often found in refinery and field desalting operations.

16.3 Chemical Treating

Chemical emulsion breakers may be added at any time during processing, but are usually most effective when added as far upstream of the treating equipment as possible. Unfortunately, the wide variety of hydrocarbon compositions, formation water compositions, and emulsion characteristics prevent the easy selection of an appropriate treating chemical. The practical approach is to perform a bottle test, in which samples of emulsion to be treated are mixed with a variety of commercial chemicals that attack the surface films holding the water droplets apart.

16.4 Oil Storage and Handling

Crude oil, the feed to all petroleum refineries, is received and stored in tanks to build up enough inventory prior to processing. This takes care of contingencies like delays in crude receipt and avoids interruptions in crude oil processing. Tanks are also provided to store intermediate products/finished products prior to transfer to terminals for further distribution. Ultimately, distribution of petroleum products is done by wagons/trucks/pipeline/tankers/barges, etc.

16.5 Classification of Petroleum

Petroleum products are divided into three classes based on their flash points as follows:

Class A – Flammable liquids having flash point below 23 °C.

Class B – Flammable liquids having flash point of 23 °C and above but below 65 °C.

Class C – Flammable liquids having flash point of 65 °C and above but below 93 °C.

Excluded – Liquids having flash point of 93 °C petroleum and above.

Types of storage tanks:
Tanks are classified based on their roof design. Normally, atmospheric tanks are of fixed roof or cone roof or floating roof or fixed cum floating roof (with or without nitrogen blanketing) type tanks and low-pressure nitrogen blanketed tanks.

Floating roof:
Floating roof may be single deck pontoon roof, double deck, or pan roof. Pan roof shall not be used as these are unsafe. For designing these tanks, API 650 guidelines may be followed.

Floating-roof tank:
The floating-roof tank is the most widely used storage method to safely control product evaporative emissions from volatile petroleum products. Floating-roof tanks remain the safest, most cost-effective method of storing petroleum and many petrochemical products.

Fixed roof:
Fixed roof may be of cone type or dome shaped. The tank may be pressurized (to a few inches of water) type with breather valves. Alternatively, tanks may be provided with fuel gas or inert gas blanketing to prevent oxygen/moisture ingress. Fixed-roof tanks for light products (e.g., Motor Spirit) breathing into a neoprene balloon is not acceptable. For designing atmospheric/low pressure tanks, API 650 or API 620 may be followed based on the type of the tank.

Fixed cum floating roof:
These tanks have a fixed roof over a floating roof. They are used for products having very stringent water content specifications like Aviation Turbine Fuel and products sensitive to oxygen like light intermediate feed tanks. Where oxygen ingress is to be avoided, it is preferable to provide nitrogen blanketing.

Selection of roof:
The selection of type of roof generally depends on ambient conditions and the product handled.

Product handled:
The following guidelines should be used for specific cases:

a. Tanks used to store finished Aviation Gasoline/Turbine Fuel shall be floating cum fixed roof to avoid entry of water into product.

b. Where product degradation due to air/moisture ingress is a problem and fixed-roof tanks are used, such tanks should be provided with inert gas blanketing.

c. Nitrogen blanketing for internal floating-roof tanks/fixed-roof tanks should be considered for storing hazardous petroleum products like benzene, etc.

Location:
Tanks may be above ground, on elevated ground, or underground. In case of underground storage tanks, protection to the external surfaces of fixed tanks/pipes shall be provided by a glass or synthetic fiber-reinforced hot applied bitumen and by surrounding the heated tank with a backfill of selected sand. This is preferred over solid skin of fine concrete through contraction or subsidence. Cathodic protection should be provided where very high standard of protection is required due to soil condition/geographic location.

Corrosion allowance:
Corrosion allowance should be taken depending on the nature of petroleum products to be stored, its impurities level, atmospheric conditions, etc.

Tank appurtenances:

1. **Ladders and handrails:**
   Individual tanks are provided with access to the roof. A platform with railing provided from the top of the stairway to gauge well and roof ladder. On floating-roof tanks, non-sparking self-leveling tread-type rolling ladders with suitable earthing connection are provided.

2. **Stairs:**
   Stairs should be made of grating. All staircases shall have resting/landing platform preferably for every 5-m height.

3. **Manholes:**
   Number of manholes shall depend on the diameter of the tank (refer API 650 for details). Minimum of one flush type clean-out manhole should be provided for tanks under dirty services.

4. **Drains:**
   i) **Bottom drains**
   Drains should be provided in all tanks for draining water and also for emptying out the tank for cleaning. Besides, these are also useful for draining water after a hydrotest or initial flushing during a start-up operation. Refer API 650 for number and details of such drains. Apex down tank bottom shall have one drain connection located at the lowest point near the center of the tank in addition to normal circumferential drains.

   ii) **Floating-roof drains**
   Maximum rainfall rate on hourly basis for the past 15 years should be considered for designing the number and size of drains for open floating-roof tanks. Also drains shall pass the design rainfall when roof is resting at the lowest position. The primary roof drain system shall be closed type using pipe and swing joints and shall include a suitable outlet valve. The inlet for these drains shall have a swing-type check valve to prevent the product from flowing into roof if pipe drain leaks/fails.

   iii) **Emergency roof drain**
   Emergency drain for floating-roof tank shall be provided on the roof to take care of drainage problem and drainage of total water in case of plugging of normal roof drain.

5. **Dip hatch:**
   Dip hatch or gauge hatch is used for gauging the height of the liquid in a tank as well as to take out samples for testing. Gauge hatch shall be non-sparking (or lined with non-sparking material) and self-closing type. Storage tank having pressure while in normal operation may pose problem in sampling or taking manual dip. For such tanks, it is suggested to resort to slot dipping devices. This accessory permits sampling/dipping in tanks having pressure up to 300 mm WG. For

operating pressures beyond this, it may be necessary to provide appropriate instrumentation with redundancy.

Gauge well pipe (with slots) should be provided for all types of tanks. This should have continuous contact using strips with bottom plate of the tank. Continuous contact makes the tank safer concerning static charge accumulation and acts as a support for the gauge well pipe.

6. **Walkway on the roof:**
   Walkway with a handrail on the roof of the tank should be provided to facilitate inspection/checking of vents/flame arrestor, etc., so that movement of personnel on the roof is safer.

7. **Instrumentation:**
   i) **Level**
      Tanks should be provided with level instruments along with high/ low level alarms with independent primary sensing device being recommended.

   ii) **Temperature**
      When rundown temperatures are likely to be higher than 100 °C, a remote temperature indicator with alarm should be provided in addition to local indicators.

8. **Tank protection:**
   Every storage tank, including its roof and all metal connections, should be electrically continuous and be effectively earthed. In case of floating-roof tanks, stainless steel shunts may be provided across the peripheral seals to ensure earthing of the floating roof. Alternatively, the pontoon, ladder, and shell of the floating-roof tank shall be continuously bonded (electrically continuous) with copper cable and the shell shall be independently earthed. Refer OISD-RP-110 on Recommended Practices on Static Electricity.

16.6 Design Consideration for Tank Farms/Manifolds

1. **General:**
   The dyke wall may be of earth, masonary, or stone. The purpose of a tank dyke is to contain the petroleum product, in the event of the tank rupture. For details of tank farms, refer OISD-STD- 118 on "Layout for Oil and Gas Installations."

2. **Tank farm drains:**
   Tank farm drainage/spillages/rain water shall be routed either to oily water sewerage or storm water channel. Provision should exist for diversion valves located outside the dyke. In the case of clear rain water, the same shall be diverted to open channel. Should a tank rupture, the contents shall remain within the bundwall and gradually be diverted to oily water sewer. In the case of high-wax-content products or high pour crude, the tank oil drains could be separated and pumped to crude/slop tanks. Depending on capacity, a group of tanks can be considered. The separator shall have steam heating arrangement and auto start/stop for pump can be provided for. In this regard, refer OISD Standard 109 on "Blow down & Sewer System."

3. **Fire protection:**
   The details of the fire protection are covered under OISD Standard-116 on "Fire Protection Facilities for Petroleum Refineries & Oil/Gas Processing Plants" and OISD Standard-117 on "Fire Protection Facilities for Petroleum Depots and Terminals." Where large tank farms are involved (especially in refineries/crude terminals or marketing installations in thickly populated areas), hydrocarbon detectors may be located in selected tank farms with remote alarms in control stations.

4. **Manifolds:**
   For safety considerations, it is desirable to keep the number of inlet/outlet connections to the tank shell to minimum. This reduces the number of flanges/valves close to the tank. In case of more number of lines, it is desirable to take a single header and form as manifold away from the tank.

   Tank manifolds shall be located outside the dyke area. The floor underneath should be paved, have curb walls, and be connected to drainage system. In crude and other tanks, where water contamination can lead to unit upsets, additional suction at two elevations may be considered so that the top outlet can be lined up initially. Alternatively, floating suction shall be installed. After tank settlement, a depression is normally formed on tank pad along the circumference. The same should be effectively made up with proper slope to avoid rain water accumulation and subsequent corrosion. Where large settlement is anticipated, it is desirable to use flexible joints/spring supports for piping to nozzles.

16.7 Tank Heaters/Mixers

Tank heating can be accomplished either by steam heating or electric tracing or hot oil circulation.

1. Steam heating
2. Electric heating
3. Mixers

Heating fuels using fired burners are not recommended as these are not safe.

16.7.1 Tank operation

Entry on floating roof is permitted only if all the following conditions are fulfilled:

1. The roof is at least half way to the top.
2. Gas test shows no presence of H2S and gas concentration is below 10% of the lower explosive limit.
3. One man is standby at the top of platform with a cannister mask/breathing apparatus readily available.
4. A lifeline with safety belt is used for the man going on the roof. The other end of the line is held by the standby at the top platform.
5. The tank is not under receipt or delivery.
6. No gauging or sampling of tanks should be undertaken during thunder or hail storms.
7. Flow velocity at tank inlet should not exceed 1 m/s until the inlet is completely submerged.
8. Conductive footwear, e.g., leather soles or electrically conducting rubber soles, should be worn while gauging, sampling, or taking temperatures. Nylon rope shall not be used for lowering sample bottles in the tank.
9. One of the most common sources of leaks and spills is mobile storage tanks, such as diesel fuel tanks used for construction machinery. It is desirable to dig a small pit or construct temporary dyke around the tank.

10. In the case of large tank farms, effective communication is essential. Telephone with loud hooters may be provided on roadside at various locations.

16.7.2 Loading/unloading facilities

i) Pumps shall be located in an exclusive paved area with drainage facilities. To avoid wide variation in pressure, leading to a "kick" or "hammering" in header and hoses, it is necessary to choose pumps with flat characteristic curves.

ii) It is desirable to have separate pumps for truck loading and not combined with wagon loading as the latter are normally of much higher capacity.

iii) Receiving lines as well as discharge lines shall be provided with thermal safety relief valves to relieve pressure due to ambient temperature rise. Whenever isolation valves are used to isolate TSV, isolation valve with lock open provision should be considered.

iv) Safety relief valves may vent into a tank or may be piped to a collector drum having level indicator/alarm or to OWS located in safe area. When connected to tanks, it should be provided with isolation valve on either side and break flange/union on tank side.

16.8 Oil Transportation

There are three basic types of pipeline system in petroleum industry to serve:

- Gather crude oil from wellhead
- Transport to refineries
- Distribute the products to process plants, retail markets and to other forms of transportation

16.8.1 Gathering system

- The most common pipe sizes used in these branches are 4″ to 12″ in diameter. The pipelines in gathering systems are short compared to trunk pipelines and range in length from a few meters to several kilometers.

- The transportation system from the well to the gathering station is referred to as well fluid transportation. The fluid in this transportation system is normally multiphase fluid. Single-phase well fluid is rarely observed in oil fields.
- The design of well fluid transportation and gathering system should take into consideration the following:
  - The wellhead pressure should be as low as possible to enhance self-flow and recovery of fluid.
  - Minimum pressure loss in pipeline.
  - Easy to monitor and control the wells.

16.8.2 Trunkline system

- Transportation of crude oil by trunk pipeline makes it economically feasible to produce oil in areas remote from processing points and markets.
- Great advances have been and continue to be made in trunk line transportation of crude oil. These include improvements in materials and methods of pipe manufacture and better design and construction methods for both pipelines and stations.

17

Effluent Management

17.1 Introduction

Petroleum reservoirs always contain some water along with hydrocarbons. This water is produced with oil and gas and contains some impurities such as.

- Oil, grease, and other floating materials
- Suspended solids, inorganic and organic colloidal suspensions
- Biodegradable organics
- Dissolved inorganic solids such as salts of sodium, calcium, etc.
- Toxic organic compounds and heavy metals

These impurities hurt the environment if effluent is discharged into potable water sources or irrigation waters. These are suspended solids and colloids, which show turbidity and dissolved organic matter and give coloring.

Generally, the effluent has the following parameters (Table 17.1)

Table 17.1 List of parameters to be considered for effluent water treatment.

S. no.	Parameters
1	pH
2	Specific gravity (at room temperature)
3	Salinity as NaCl (PPM)
4	Calcium (PPM)
5	Potassium (PPM)
6	Oil content (PPM)
7	Viscosity (CPS)
8	Total dissolved solids (PPM)
9	Suspended solid (PPM)
10	Turbidity (NTU)
11	BOD: 5 days

Different methods for the disposal of produced water include:

- Injection into permeable underground formations containing saline water, through disposal wells.
- Reuse for supplementary recovery operations like water injection.
- Disposal by evaporation.
- Surface disposal in river, lake, sea, etc.

Many of the impurities in produced water are harmful for the above disposal systems:

- The impurities such as oil sludge and higher salinity (super saturated brines) can plug back permeable underground reservoirs. This makes underground water injection into permeable layers, or water disposal, problematic.
- Disposal by evaporation is not practical for large quantities of produced water. Nowadays, pollution control board also do not allow for evaporation of effluents.
- Environmental regulations require maintenance of specified quality of water before disposal as follows.

17.2 Minimal National Standards

Minimal National Standards (MINAS) for liquid effluent are as follows (as per Central Pollution Control Board, 1995).

17.2.1 For onshore facilities

i) For marine disposal, onshore discharge of effluents, proper marine outfall has to be provided to achieve the individual pollutant concentration levels below their tolerable limits as given below, within a distance of 50 meters from the discharge point. For protection of the marine aquatic life (Table 17.2)

ii) Oil and gas drilling and processing facilities, situated on land and away from saline water sink, may opt either for disposal of treated water by onshore disposal or by re-injection in abandoned well, which is allowed only below a depth of 1000 m from the ground level. In case of re-injection in abandoned well, the effluent has to comply only with respect to suspended solids and oil and grease at 100 and 10 mg/l, respectively.

Table 17.2 List of permissible limits of various parameters in discharged effluents.

Parameters	Tolerable limit,
pH	5.5–9.0
Oil and grease	10.0 mg/l
Suspended solids	100.0
BOD at 20 °C	30.0
Chromium, as Cr	0.1
Copper, as Cu	0.05
Cyanide, as Cn	0.005
Fluoride, as F	1.5
Lead, as Pb	0.05
Mercury, as Hg	0.01
Nickel, as Ni	0.1
Zinc, as Zn	0.1

Table 17.3 List of permissible onshore discharge standards for various parameters.

S. no.	Parameters	Onshore discharge standards
1	pH	5.5–9
2	Temperature	40 °C
3	Suspended solids	100 mg/l
4	Zinc	2 mg/l
5	BOD	30 mg/l
6	COD	100 mg/l
7	Chlorides	600 mg/l
8	Sulfates	1000 mg/l
9	TDS	2100 mg/l
10	% Sodium	60 mg/l
11	Oil and grease	10 mg/l
12	Phenolics	1.2 mg/l
13	Cyanides	0.2 mg/l
14	Fluorides	1.5 mg/l
15	Sulfides	2.0 mg/l
16	Chromium (Cr + 6)	0.1 mg/l
17	Chromium (Total)	1.0 mg/l
18	Copper	0.2 mg/l
19	Lead	0.1 mg/l
20	Mercury	0.01 mg/l
21	Nickel	3.0 mg/l

iii) For onshore disposal, the permissible limits are given below (Table 17.3).

17.2.2 For offshore facilities

For offshore discharge of effluents, the oil content of the treated effluent without dilution shall not exceed 40 mg/l for 95% of the observation and shall never exceed 100 mg/l. Three 8-hourly grab samples are required to be collected daily and the average value of oil and grease content of the three samples should comply with these standards.

17.3 General Specification for Water Injection

Table 17.4 List of operating limits and optimum values for various parameters during water injection process.

Sl. no.	Parameters	Optimum value	Operating limit
1	a) Suspended solids (mg/l)	NIL – 2.5	< 3.5 mg/l
	b) Turbidity (NTU)		
	c) Suspended solids		
2	a) Dissolved oxygen (mech. deaeration)	0.2 mg/l	< 0.5
	b) Dissolved oxygen (chemicals)	0.015 mg/l	–
3	Bacterial count/ml		
	a) Sul. reducing	NIL	1.0–10.0
	a) Gen. aerobic	100–10,000	< 10,000
4	Iron content	0.1 ppm	< 0.2 ppm
5	Corrosion rate (non-pitting)	1 mpy	< 2.0 mpy
		1 mpy	< 2.0 mpy
6	Salinity (mg/l)	At least half of formation water salinity	–
7	Oil content	10 ppm	10–25 ppm

17.4 Effluent Treatment Methods

In order to achieve the tolerance limits, the effluent is treated in suitable treatment plants. The selection, design, and operation of the treatment and disposal facilities depend upon the nature of the effluent (Table 17.4), i.e., on the nature of the contaminants present in it.

The important contaminants to be removed from the effluent are (Table 17.5) :

- Oil, grease, and other floating material
- Settle-able suspended solids
- Non-settle-able (filterable) suspended solids
- Biodegradable organics
- Dissolved inorganic solids

- Toxic organic compounds and heavy metals

17.4.1 Contaminants and methods of effluent treatment

Table 17.5 The variety of contaminants that are removed by different operations, process and treatment system.

Contaminant	Unit operation, unit process of treatment system
1. Free oil, grease, settle-able suspended solids, and other floating material	Primary treatment: • Gravity/sedimentation, floatation • filtration
2. Non-settle-able (filterable) suspended solids, oily colloids, and other colloidal material	Secondary/physical chemical treatment: • Chemical polymer addition • Flocculation/coagulation followed by sedimentation • Filtration
3. Biodegradable organics	Biological treatment: • Activated sludge treatment • Trickling filters treatment • Aerated lagoons treatment • Sand filtration treatment • Physical–chemical treatment
4. Heavy metals	• Chemical precipitation/ion exchange
5. Dissolved organic and inorganic solids	• Ion exchange/reverse osmosis/electro dialysis

17.4.2 Contaminants, methods of effluent treatment, and equipment

- **Contaminants:** Free oil, grease, and settle-able suspended solids and other floating materials.
- **Treatment (primary treatment):**
  - Gravity/sedimentation
  - Floatation
  - Filtration
- **Equipment:**
  - Hold-up tank
  - Equalization tank

  - Gravity separation/sedimentation (tilted plate interceptor (TPI))
  - Floatation
  - Filtration

- **Contaminants:** Non-settle-able (filterable) suspended solids, oily colloids, and other inorganic and organic colloidal suspensions.
- **Treatment (secondary treatment/physical chemical treatment):**
  - Chemical polymer addition
  - Flocculation/coagulation followed by sedimentation
  - Filtration
- **Equipment:**
  - Hold-up tank
  - Equalization tank
  - Gravity separation/sedimentation
  - (tilted plate interceptor (TPI))
  - Dosing tank, dosing pumps
  - Flash mixer
  - Flocculation tank
  - Floatation (DAF)
  - Clarifier tank
  - Filtration
- **Contaminants:** Biodegradable organics
- **Treatment (Biological Treatment):**
  - Activated sludge treatment
  - Trickling filters treatment
  - Aerated lagoons treatment
  - Sand filtration treatment
  - Physical–chemical treatment

- **Equipment:**
  - Hold-up tank
  - Equalization tank
  - Gravity separation/sedimentation
  - (tilted plate interceptor (TPI))
  - Dosing tank, dosing pumps
  - Flash mixer
  - Flocculation tank
  - Floatation (DAF)
  - Bio tower
  - Aerated lagoons
  - Clarifier tank
  - Filtration
- **Contaminants:** Heavy metals
- **Treatment:** Chemical precipitation/ion exchange
- **Equipment:**
  - Electrostatic forces
  - Hydrogen cation exchange
  - Anion exchanger
- **Contaminants:** Dissolved organic and inorganic solids
- **Treatment:** Ion exchange/reverse osmosis/electro dialysis
- **Equipment: treatment units (RO system):**
  - Effluent storage tank
  - Feed pump
  - Pressure sand filter
  - Acid dosing system
  - SMBS dosing system
  - Anti-scalant dosing system

- Micron cartridge filter
- RO (membranes)
- Membranes cleaning in place system

17.4.3 Description of some of the equipment and techniques

- **Hold-up tank:** This tank gives the settling to the oil and other floating materials from the effluent.
- **Equalization tank:** For removing oil, floating solids and to minimize flow surges to further treatment.
- **Gravity separation/sedimentation techniques:** Free and floating oily materials are separated from the effluent by gravity and float to the surface to be skimmed off and removed. In addition to hold-up and equalizing tanks, TPI is used. The design of oil separator, tilted plate interceptor (TPI) as developed by the American Petroleum Institute, is based on the removal of free oil globules larger than 0.015 cm in diameter (150 microns). If oil particles are very small, then it is more advantageous to add some coagulants and coagulate the oily materials with heavier compounds so that the oil and the coagulants settle to the bottom of the tank. Air flotation technique is also used to lower the gravity of oily compounds by attachment of air bubbles to the oily particles (Figure 17.1).
- **Gravity systems/sedimentation techniques have the following features:**
  - Simplest and economical available system.
  - Oil separates from water because of its lower density and then is skimmed at the water surface.
  - The material with high density will settle down and will be removed from the bottom.
  - An enlarged separating surface is achieved by placing various smaller plates on the top of each other in tank. These plates are fitted in an inclined position.
  - Oil/solids are separated between the plates in the counter-current flow.
  - Efficiency is limited by the size of oil particles (>150 microns).

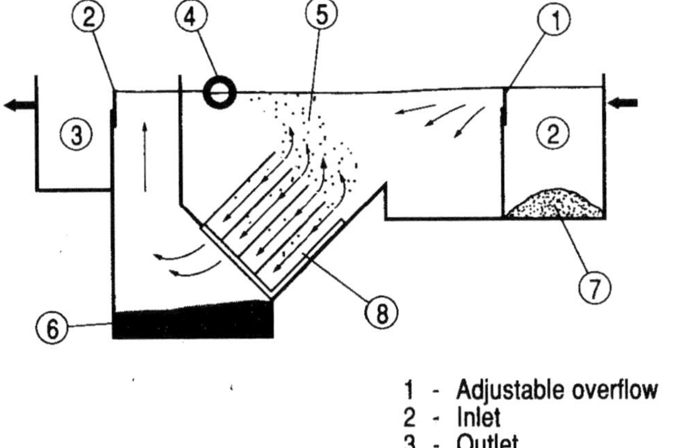

Figure 17.1 Tilted plate separator.

- Gravity systems can accept any amount of oil in water, even sludge of pure oil.
- It is a typical primary treatment, which can remove large quantities of oil but frequently does not reach the required standards.

- **Gas floatation techniques:**
  - An accelerated gravity separation technique, in which floatation of oil (and suspended solids) is accomplished by numerous microscopic gas bubbles. Froth layer of oil is removed by a skimming device.
  - Particles having both higher and lower densities (oil suspension in water) than the liquid can be made to rise.
  - Air bubbles are formed or added by two main methods:
    - i. Dissolved gas floatation (Figure 17.2)
    - ii. Dispersed gas floatation.

- **Chemical treatment technique (coagulation–flocculation–sedimentation technique):**
  For removing suspended solids, phosphate, color, turbidity, BOD, and COD, some coagulants such as lime, alum, or ferric salts are added to the inorganic and organic colloidal suspensions in the effluent that

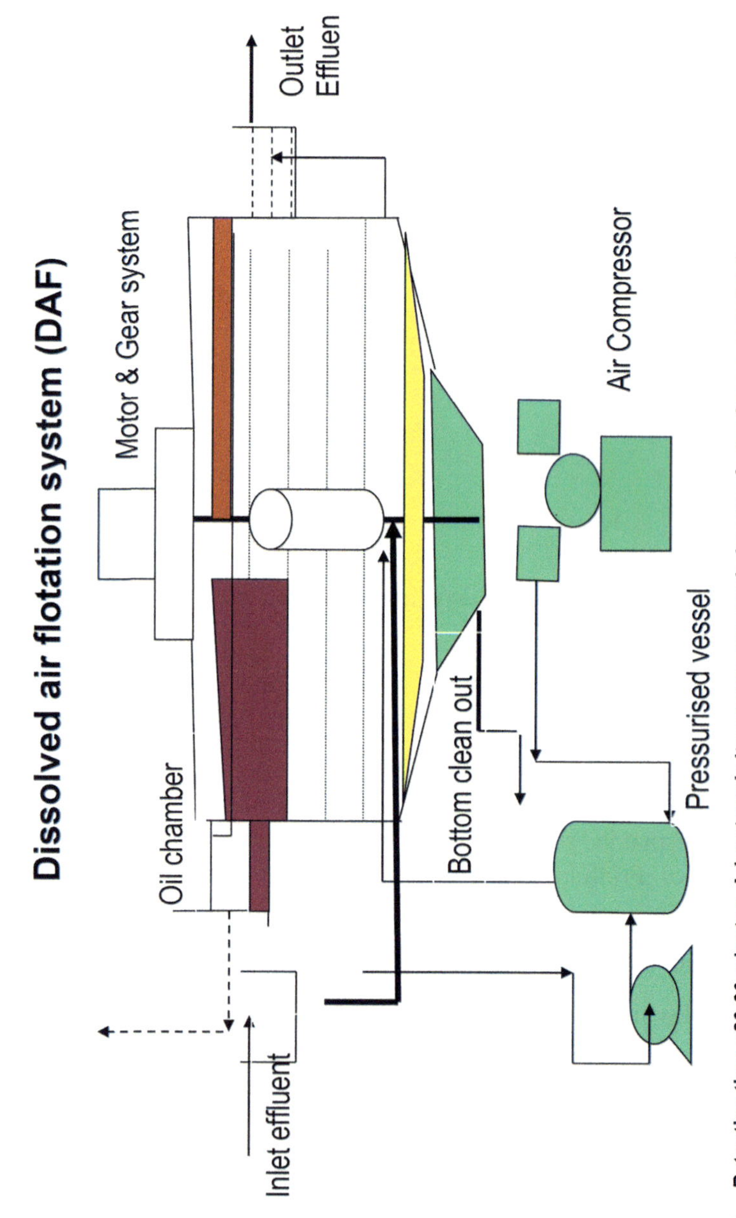

- **Retention time : 20-30 minutes. Advantage is it removes suspended matter faster &more completely**
- **Chemicals are used to break emulsion**
- **Oil removal efficiency : up to 95%**
- **Air pressure maintained : approx. 3.5 kg/cm²**

Figure 17.2 Flow diagram representing dissolved air flotation system (AFS).

results in particle destabilization by the reduction of surface charges and formation of complex hydrous oxides that form flocculation suspensions and get removed from the effluent(Figure 17.3). Chemicals are used in conjunction with separation, coalescing, floatation, or filtration. Some of the chemicals that enhance the efficiency are:

- ○ Lime is used for the removal of phosphorous from waste water and will also remove some BOD and suspended solids as well as hardness present in waste water; they will also remove metals.
- ○ Alum is a coagulant and when added to water reacts with the available alkalinity (carbonate, bicarbonate, and hydroxide) and phosphate to form insoluble aluminum salts.
- ○ Ferric chloride or $FeCl_3$ is also a chemical coagulant, which when added to waste water reacts with the alkalinity and phosphates forming insoluble iron salts. Polymer addition or polyelectrolytes are high molecular weight compounds and are used as coagulants, coagulant aids, filter aids, or sludge conditioners. They carry charges either positive, negative, or neutral and act as bridges reducing charge repulsion between the colloidal particles and dispersed flock particles and increasing settling velocities.
- ○ Chemicals are used according to the technique and type of impurities to be removed from the effluent water.
- ○ Selection of the right chemical is often the most important requirement and is done through lab and field trial studies.

- **Filtration techniques:**
  - ○ The main objective is to remove the residual suspended solids present in the effluent.
  - ○ Main methods employed are layer, membrane, and fibrous media filtration.
  - ○ Filtration through sand, crushed anthracite coal, or activated carbon is most common in the oil industry. A backwash system is always provided, requiring hot water/surfactants or with fresh water and air scouring. This type of filtration is also known as "granular media filtration."

- **Ion exchange**: It is employed for the removal or exchange of dissolved inorganic salts such as hardness (Ca and Mg) or metal ions (Cr and Zn),

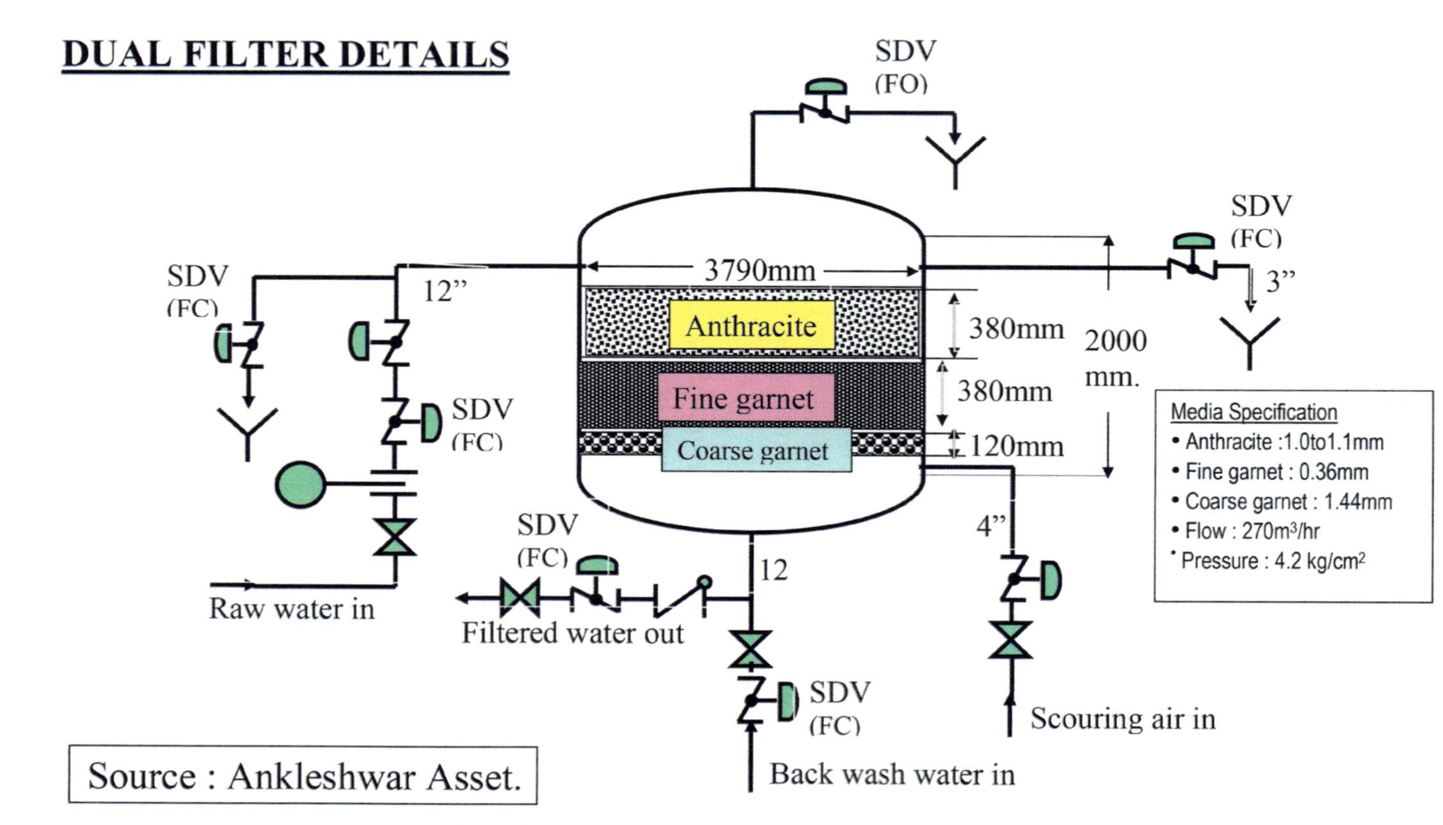

Figure 17.3 Process mechanism involved in flocculation dual filter

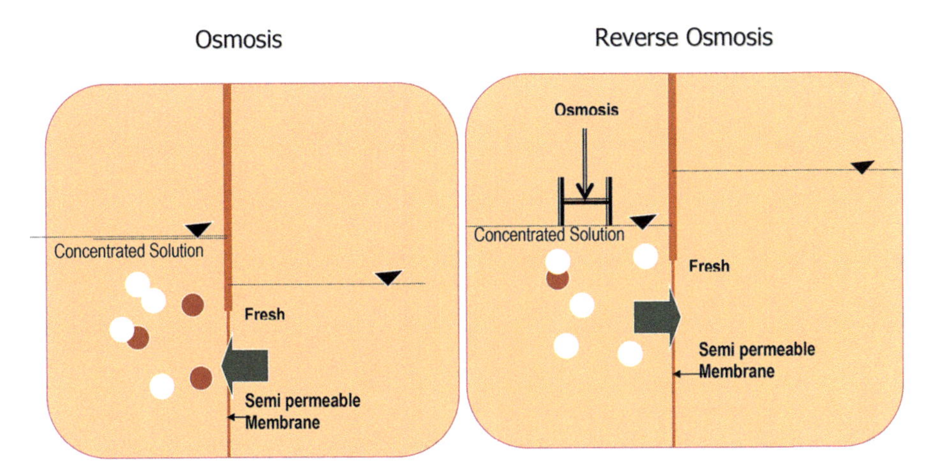

Figure 17.4 Process mechanism involved in Osmosis and Reverse Osmosis.

etc., in the effluent water. In this process, electrostatic forces hold ions on the surface of a solid and exchanged for ions of a different species in solution. Ion exchange can be considered if the total dissolved solids are less than 1000 mg/l in the effluent.

- **Reverse Osmosis?**
  - In the osmosis process, a salt solution (solution of high concentration) is separated from the pure solvent (solution of less concentration) by a semi-permeable membrane (Figure 17.4).
  - Certain pressure is required to bring thermodynamic equilibrium conditions to stop the flow from low concentration to high concentration. This pressure is called the osmosis pressure of the solution.

When an excess of osmosis pressure of solution is applied to the more concentrated solution chambers, pure solvent (solution of less concentration) flows out from the solution of high concentration chamber. This phenomenon is based on the reverse osmosis process (Figure 17.5).

A schematic representation of an effluent treatment plant installed at the Cauvery project is represented in Figure 17.6 .

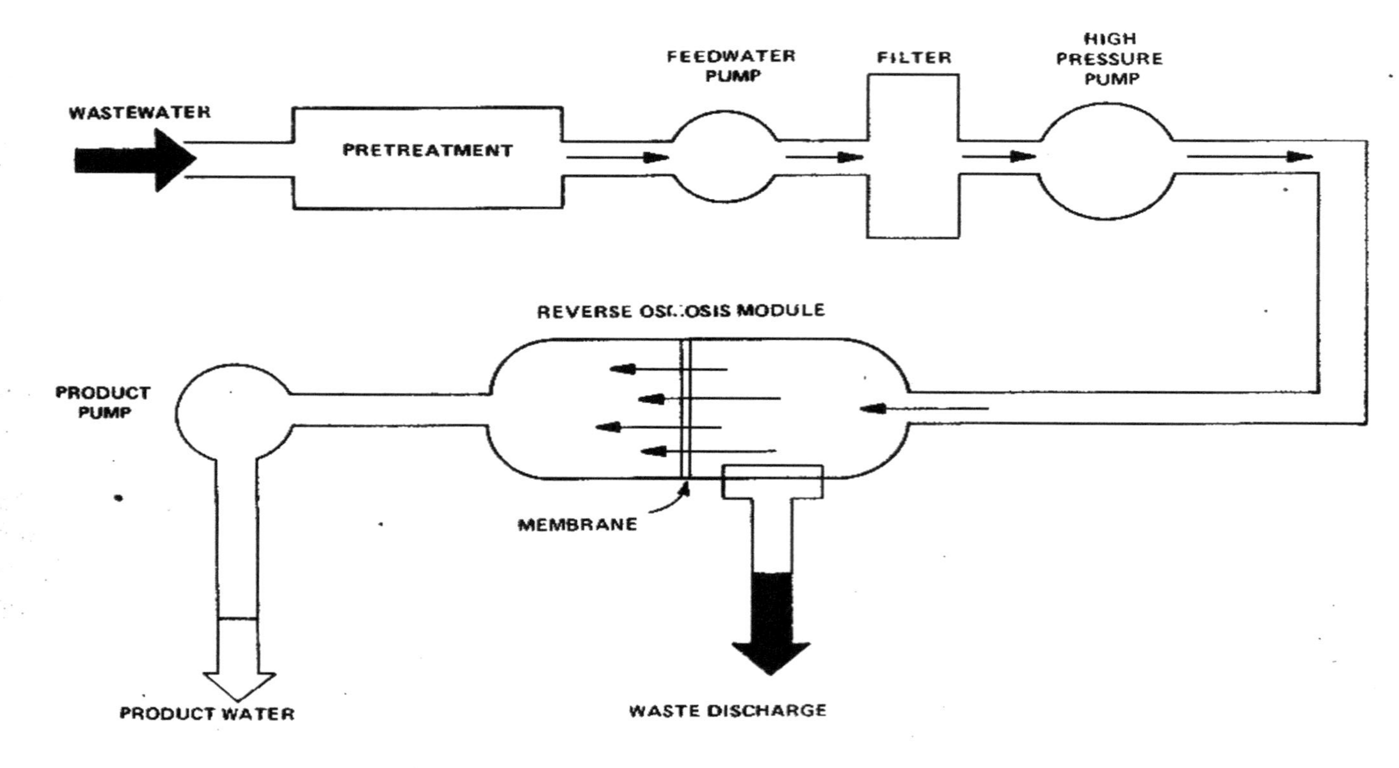

Figure 17.5 Schematic of basic reverse osmosis process.

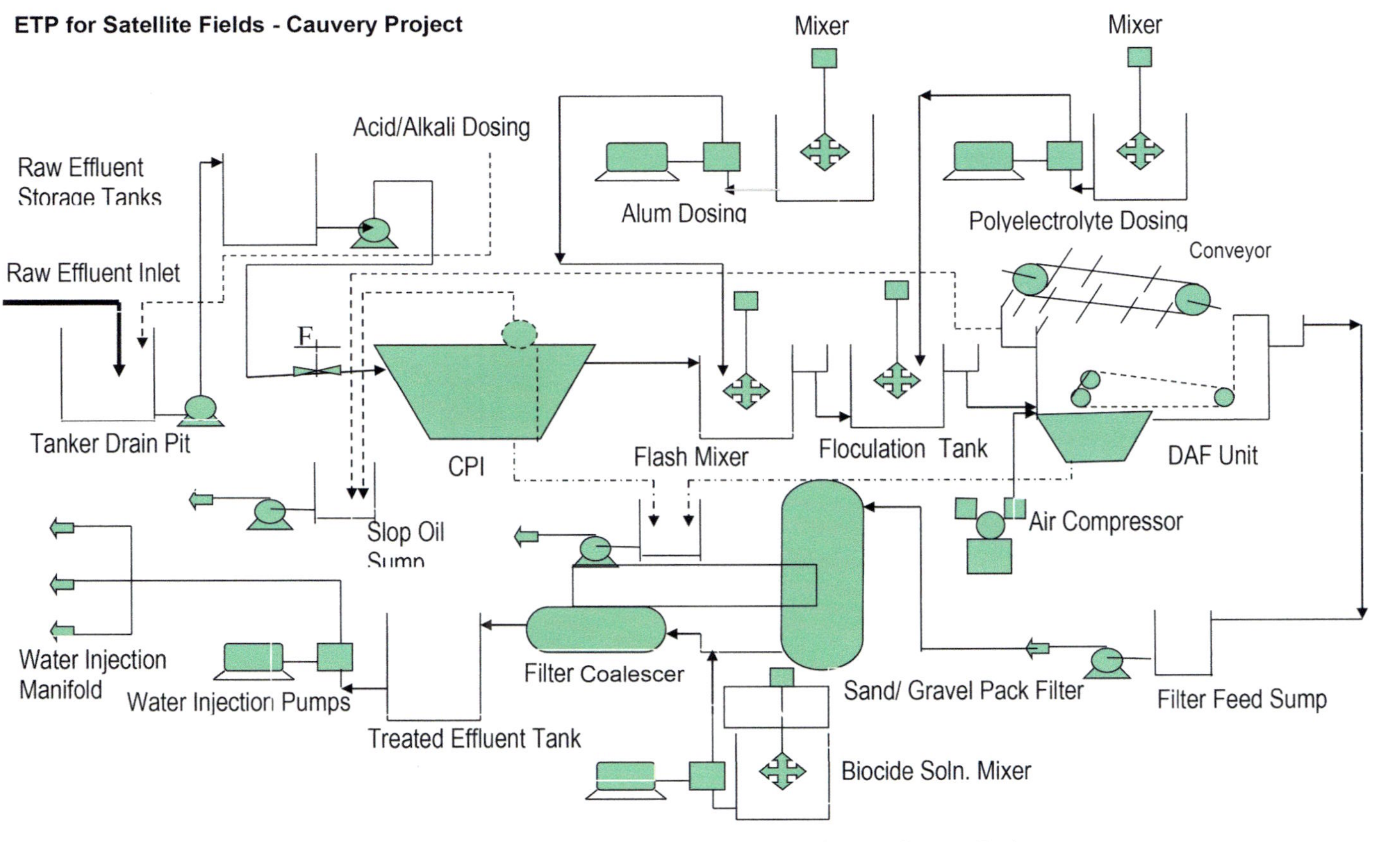

Figure 17.6 Effluent treatment plants for satellite fields – Cauvery Project.

18

Injection Water Treatment

18.1 Need for Pressure Maintenance

Producing a field with its natural energy may yield a maximum of 15%–30% of original oil in place (OOIP). These reservoir natural energies are referred to as reservoir drive mechanisms, which push fluid from reservoir to wellbore.

There are three types of primary drive mechanisms:

- Water drive.
- Solution gas or depletion drive.
- Gas cap expansion.

The total production by these natural drive mechanisms is referred to as primary recovery of the field.

Reservoir pressure declines due to withdrawal of fluid and creation of void space. Reservoir pressure decline adversely affects oil production. Therefore, it is required to maintain reservoir pressure.

1. **Artificial means of pressure maintenance:**
   There are broadly three artificial means by which reservoir pressure is maintained and they are:

   a) Water injection.

   b) Gas injection (natural gas).

   c) Any other miscible fluid.

2. **Pressure maintenance by water injection:**
   This is the most commonly used method of pressure maintenance to increase recovery of oil from the established fields having normally less viscous oils. The pressure maintenance in oil reservoirs is carried out by injecting water into the formation.

Water is selected as displacing and pressure maintenance fluid mainly because:

- It is the cheapest fluid.
- It is the most efficient displacing agent.
- It is available in abundance.
- Less treatment is required to make it compatible with formations.
- It co-exists with crude oil in reservoir and is, therefore, largely compatible.
- Since it co-exists with crude oil, contamination of the produced crude oil and its treatment is not a problem.
- Fewer problems in disposal.

3. **Water source for injection:**
   Raw water to be used as an artificial means of pressure maintenance may be obtained from the following sources:

   - River or pond water.
   - Alluvium water.
   - Formation water.
   - Seawater.

4. **Quality of injection water:**
   The raw water is processed by a processing unit and chemically treated to make it suitable for injection. Injection water should have the following qualities before its injection into the wells.

   a) Compatibility:
   It should be compatible with the formation water, i.e., it should not contain dissolved solid, which forms insoluble precipitates when it comes in contact with the formation water. It should not result in formation damage due to clay swelling.

   b) Suspended solids or oil:
   It should not contain suspended solids or oil above certain specified limits. This can plug the conductive pores and impair the permeability of wellbore/formation.

c) **Corrosion/dissolved gases:**
It should not contain dissolved oxygen, CO_2, or H_2S, which cause corrosion of metallic pipes and equipment fitted in the injection system.

d) **Scale formation:**
Water should not contain barium, strontium, iron, and calcium in solution with sulfate, carbonate, and oxide radicals. The presence of these ions can result in precipitation of compounds (scales), which can plug the formation face.

e) **Micro-organism:**
Water should be free from algae, iron, bacteria, and capsulated bacteria. All these organisms tend to collect as large masses of slime that can plug the central water injection system and water injection wellbore. Water should not contain sulfate reducing bacteria.

18.2 Process and Treatment of Injection Water in Offshore

Water injection systems in offshore draw raw water from the sea, and it is processed and chemically treated before injection into the wells. The processing of injection water at offshore is divided into the following steps. A typical case of SHW platform is described below.

- Seawater lifting
- Coarse filtration
- Fine filtration
- Deoxygenation
- Dosing of chemicals
- Pumping of treated water to injection well

18.3 Seawater Lifting

The major components are seawater lift pumps and chlorinators.

1. **Seawater lift pumps (SWLP)**
The raw seawater is lifted by three seawater lift pumps (SWLPs) from the depth of 30 ft of MSL. These are multistage centrifugal pumps. Out of the three pumps, one remains standby.

2. **Chlorinators:**
   The onsite electrolytic NaOCl generators generate nascent chlorine for treatment at the intake of each SWLP to control microbial growth. Free chlorine acts as a primary biocide. Continuous monitoring of chlorine is done at the inlet of the coarse filter and at the outlet of the fine filters.

18.4 Coarse Filtration (CF)

There are three coarse filters. One remains standby. The coarse filter is composed of sieves of 80 microns. Seawater lift pumps feed water at the bottom of coarse filter. It removes all suspended particles of sizes greater than 80 microns.

18.5 Fine Filtration (FF)

There are seven numbers of fine filters; five in operation, one remains standby, and one is kept in backwash. These are vertical pressure type, down-flow filters with multi-filtering media of pea grave at the bottom followed by coarse garnet, fine garnet, and anthracite at the top. The thicknesses of each layer are 330, 125, 400, and 400 mm, respectively. Fine filters are designed to remove 98% of all particles of more than 2 microns.

To aid in the filtration process, coagulant and polyelectrolytes are added before the water enters the fine filter.

18.6 Deoxygenation Tower (DOT)

There are two deoxygenating towers, each equipped with one vacuum pump. One always remains standby. The outlet of fine filters, which is devoid of suspended solids and bacteria but still containing dissolved gases (O_2, N_2, CO_2, and Cl_2), now goes to DO towers, after dosing defoamer, separately through Train-A and Train-B. The first stage vacuum brings down pressure to 48 mm and the second stage to 19.8 mm of Hg.

The water enters the tower through a distributor above the top section. In the top section, most of the dissolved gases (O_2, N_2, CO_2, and Cl_2) and water vapor are taken out and their concentration is reduced from 7.7 to 0.1 ppm. Oxygen scavenger is added through spray nozzles to reduce the oxygen content to less than 0.02 ppm at the outlet of the DO tower.

For the safety of the handling system, i.e., injection water pipeline, the outlet of the DO tower is now treated with (a) bactericide (amine and aldehyde type), (b) corrosion inhibitor, and (c) scale inhibitors (type-I and type-II).

18.7 Booster and Main Injection Pumps (MIP)

The water from the DO tower is sent to the booster pumps to provide net positive suction head for main injection pumps. There are three numbers of booster pumps. One remains standby.

To supply high-pressure injection water to various injection wells, there are five main injection pumps with discharge of 9.3 M^3/min. and discharge pressure of 140 Kg/cm^2.

Injection points and function of different water injection chemicals are given in Table 18.1 and suggested doses in Table 18.2. The specification, operation limit, and testing schedule of different quality parameters are presented in Table 18.3.

18.8 Injection Water Quality Measurement

The parameters, which determine the quality of injection water, are briefly described below.

18.8.1 Residual chlorine

Take 2–3 drops of standard o-toluidine solution in a glass sampling tube and fill it with approximately 10 ml of water to be tested. Shake it, and water turns to yellowish color. In another sample tube, take chlorine free water, i.e., reference water (FF water kept in open beaker for more than 24 hours will act as reference water). Place both the sample tubes in a color comparator and match the color. The observed reading will give the concentration of residual chlorine in ppm directly.

In case of high concentration of Cl_2 residue, reduce the volume of water to be tested by half, one-third, or one-fourth and measure the concentration as above by making it up with reference water. Multiply the result by 2, 3, or 4 to get the actual value of concentration.

Excessive concentration of Cl_2 results in higher corrosion, so its concentration at FF outlet is maintained between 0.2 and 0.6 ppm and it should be nil in train-A and train-B and thereafter.

It has been observed that in the absence of chlorine, even 90% removal efficiency of particles less than 2 microns is difficult to achieve.

18.9 Measurement of Particle Count

Measurement of particle count is one method of measurement of amount of contamination by suspended solids in water sample.

Principle:
The sample solution is drawn through a small aperture by means of a vacuum source. A current passing through the aperture between two electrodes enables the particles to be sensed by the momentary changes in the electrical impedance as they pass through the aperture. Since each particle displaces its own volume of electrolyte solution, the change in the impedance pulse is essentially proportional to the volume of the particle, which produced it. These pulses are then amplified, counted, and allocated to the correct size.

The instrument can be set either in time mode wherein a sample is drawn through the aperture for a fixed interval of time or in volume mode wherein a fixed volume (0.1, 0.2, 0.5, or 1.0 ml) of sample is sucked through the aperture. Various aperture tubes having different aperture diameters are also available.

For analysis of injection water, an aperture diameter of 70 and an aperture length of 32.5 microns are required. This aperture gives a detection range from 1.4 to 42 micron particles, which is adequate for our needs.

Measurement:
Once the Coulter counter has been set up with the proper aperture tube, the determination of particle count in the injected water sample is simple. The sample is placed in the sample stand such that the aperture is at least 1 cm below the surface of water. The measurement knob is then turned on as per procedure and the number of particles counted in a specified volume is recorded and then calculated for 1 ml. Fine filter outlet water shall be used for filling reservoir after filtering through a 0.22-micron filter paper.

A minimum of three analyses should be done for each sample and its average be taken as the counts for the sample. The total number of particles above 2-micron size should be recorded. Normally, it should not be more than 300 per ml after the fine filter.

18.10 Turbidity

Turbidity also gives a measure of amount of contamination by suspended particles in injection water sample.

Principle:
It is based on the principle of comparison of the intensity of light scattered by particles in a sample under test condition with the intensity of light scattered by a standard reference suspension of formazine.

Apparatus:
Nephalo Turbidity Meter and sample tube of clear colorless glass.

Standardization with calibration standard:
Switch on the instrument. Allow stabilizing for 30 minutes. Use turbidity-free distilled water (prepared by filtering distilled water through a 0.45-micron membrane filter) to set zero of scale in all ranges with zero set knob. Using a range selection switch, standardize the instrument using standard solutions (prepared or readymade), say 10 or 1 NTU. Turbidity of seawater used for injection is less than 1 NTU. So only the setting in the range of 0–0.02 and 0–1 NTU are required for all practical purposes.

Measurement:
Fill the sample tube with sample water. Select the correct range for accurate reading. Note the turbidity. Results are expressed in nephalometric turbidity unit (NTU).

Precautions:
Dirty glass, the presence of air bubbles, the effect of vibration that disturbs the surface visibility, and colored solution give erroneous results. Sample tube filled with water shall not be left in the instrument after the measurement.

18.11 Millipore Filtration

Measurement of millipore filtration rate along with calculation of wt % of suspended solids is another and third way of measurement of amount of contamination by suspended solids in injection water.

Apparatus:
(1) A filter paper holder with hole at the bottom through which filtered water can pass, (2) a top cap with screw and vent valve on top, (3) a fine mesh stainless steel screen, (4) a rubber O-ring, (5) a filter membrane of type: HA, pore size: 0.45 micron, filter thickness: 180 micron, diameter: 47 mm, porosity: 35%, (6) a pressure gauge to maintain 20 psi water pressure in line connected to top of millipore apparatus, and (7) a clean graduated vessel to collect filtered water.

Measurement:
Purge the sample line and tubing thoroughly to remove any accumulated dirt or rust. If the sample line has remained closed for a long period, flow water through it for 2–3 hours before the test. Weigh the filter paper nearest to 0.0001 g. Errors in weight can arise from variation in moisture content of the membrane filter with the relative humidity of the atmosphere. Place a fine-mesh stainless steel screen in the holder and then put the membrane filter paper and cover it with rubber O-ring. Moisten the membrane with filtered (0.45 microns) distilled water to ensure that there is no air bubble on the membrane; otherwise, it will lead to erroneous results of millipore and TSS. Connect the apparatus to the top cap fitted in sample line and allow some water to flow from open vent to remove any trapped air bubble. Close the vent valve. Apply 20-psi pressure in a sample line connected to millipore apparatus. Collect the filtered water in graduated vessel. Continue the experiment for 30 minutes.

Result:
Report the quantity of water collected as millipore, L/30 minutes.

18.12 Total Suspended Solids (TSS)

Stop the water flow into the millipore test assembly, disconnect and open the holder, and carefully remove the membrane with tweezers. Keep this membrane on another holder connected to a vacuum pump. Start the vacuum pump and wash the membrane several times with 0.45 μ filtered distilled water to remove the water-soluble salts present. Care should be taken while washing so that the trapped suspended solids on the membrane surface are not washed away.

After washing, remove the membrane carefully, put in a Petri dish and dry in an oven at 85 °C for 30 minutes. Cool the membrane in a desiccator and determine its weight to the nearest 0.1 mg.

Calculation:

Weight of the membrane before test: a gram
Weight of the membrane after test: b gram
Correction factor for the membrane: c gram
Millipore: d liter/30 minutes

$$\text{TSS (mg/l)} = \frac{b - a + c}{d} \times 1000.$$

The correction factor for the membrane having weight ranges as follows (Table 18.1):

Table 18.1 Correction factors under various membrane weight ranges.

Membrane weight (g)	*C* g
0.082 – 0.0850	0.0011
0.0851 – 0.0900	0.0012
0.910 – 0.0950	0.0013
0951 – 0.1000	0.0014

18.13 Dissolved Oxygen

Oxygen is monitored upstream and downstream of the DO tower to find out the removal efficiency. The presence of residual sulfite ensures the removal of oxygen positively. The residual sulfite content should be 0.5 ppm (minutes) at the MIP discharge and wellhead.

Dissolved oxygen is a very important parameter for water injection as it enhances corrosion of all pipes, and hence it must be maintained at prescribed limits (<0.02 mg/l). It is measured by two methods.

1. **Online meter:**
   Continuous online monitoring of dissolved oxygen at the DO tower and MIP outlets is necessary and online oxygen meters like those supplied by M/s Orbisphere Laboratories or equivalent are recommended. Installation, calibration, and maintenance instructions for these online meters are given in the instruction manuals supplied with these meters.

2. **Dissolved oxygen determination kit:**
   These kits comprise sealed ampoules containing oxygen reacting chemicals together with a color standard or comparator to determine the DO content. The procedure is simple and is convenient.

Procedure:
The sampling funnel supplied with the kit is purged free of air bubbles with flowing water. With the sampling funnel held in flowing water, the ampoule is inserted inside and its tip is broken. Water flows into the ampoule, reacts with the chemical, and develops color. The sample is removed, with its tip quickly covered with thumb taken upside down a number of times. The developed color is then matched in the comparator or with color standards. The result is reported as DO, mg/l.

18.14 Corrosion Monitoring

This is monitored using three methods by measuring the amount of iron in water.

1. **Measurement of total iron:**
   It is measured by the spectrophotometric method and gives the total iron concentration in the water.

Principle:
The method is intended for total iron (up to 1500 mg/l) that includes Fe(II) and Fe(III). Fe(III) is reduced to Fe(II) with hydroxylamine hydrochloride. Iron when treated with TPTZ (2,4,6-tripyridyl-s-triazine) in the pH range of 3.4–5.8 is converted into a violet complex, which is colorimetrically measured at 590 nm. Cobalt, nickel, and copper interfere.

Calibration/multiplication factor (MF):
Transfer O (reagent blank), 1, 2, 4, 6, 8, 10, 15, 20, and 25 ml of the working standard of known concentration of iron into 50-ml volumetric flasks, add 2 ml of NH_2OH. HCl + 2 ml of TPTZ + 2 ml of sodium acetate and dilute to make up point. Measure absorbance of iron complex vs. blank at 590 nm. Plot a standard curve of absorbance against concentration of Fe (mg/l). A straight line is obtained.

Multiplication factor = Concentration divided by absorbance

Sampling and measurement:
Take 15 ml of water sample in a flat bottom flask, add 0.2 ml of 4.5 M solution of H_2SO_4, and keep the sample for aging for half an hour at 60 °C. Then allow it to cool at room temperature. Add 2 ml of hydroxylamine hydrochloride, 2 ml of TPTZ, 2 ml of sodium acetate solution and dilute it to 25 ml. Record the absorbance at 590 nm.

Calculation and result:

Iron conc. = Absorbance × MF × Dilution Factor
(mg/l) = Absorbance × MF × (25/15)

Calculation of MPY (mills per year)

$$\text{MPY} = 0.003895 \times \text{iron conc. (ppm)} \times Q/(D \times L)$$

where:

1 MPY = 1/1000 inch
Q = flow rate of oil, bbls per day
L = length of oil line, km
D = diameter of oil line
Specification of MPY: <2

2. **Weight loss coupon –** By weighing the coupons installed online and in the wellhead platforms.

3. **Corrator probe –** Online probes based on polarization admittance instantaneous rate (pair) technique provide corrosion rate in MPY.

18.15 Scaling

Coupons are installed at the process and well platforms. These are removed periodically and visually inspected and in case of scale deposits, these are analyzed chemically.

18.16 Residual Sulfite

Principle:
The water sample is treated with fuschin-formaldehyde reagent. Sulfite gets oxidized with formaldehyde and sulfur dioxide is released, which reacts with fuschin to produce a violet red color. The intensity of color is measured colurimetrically at 560 nm. The aerial oxidation of sulfite, which is extremely rapid in the presence of catalyst, such as metal ion, is overcome by introducing the sample directly into acid fuschin-formaldehyde reagent. Under acidic conditions, the reaction rate of sulfite with oxygen is negligible.

Apparatus:
Spectrophotometer, narrow bore stainless steel sample line, and plastic to be avoided to prevent the ingress of O_2.

Reagent:
Fuschin-formaldehyde solution

Calibration/multiplication factor:
The intensity of the developed color is depressed by chloride ion. It is therefore necessary to carry out the calibration in the presence of that volume of seawater, which will be used for testing the sample. In 25-ml volumetric flasks, pipette 7 ml of fuschin formaldehyde reagent. Add using 100-μl syringe 0 (blank), 6, 12, 24, 36, 48, and 60 μl of sulfite standard solution injecting below the surface of the reagent to avoid atmospheric oxidation. Dilute to 25-ml mark with untreated filtered seawater (free from Cl_2 and SO_3). Mix by inverting and allow it to stand for 30 minutes at 20–25 °C. Transfer to a 1- or 2-ml cell and measure the absorbance at 560 nm. Plot absorbance against the concentration of sulfite.

Multiplication factor = Concentration divided by absorbance

Sampling and measurement:
Allow the stainless steel sample line to flush thoroughly. Add 7 ml of fuschin-formaldehyde reagent into a 25-ml volumetric flask. Momentarily stop the flow and quickly place the 25-ml flask in position so that the outlet tube is beneath the surface of the reagent. Start the sample flow and gradually withdraw the tube. Stop the flow at 25-ml mark. Mix immediately by swirling. Allow to stand for 30 minutes at 20–25 °C. Measure absorbance at 560 nm against freshly prepared blank using 7-ml reagent and 18-ml sulfite free seawater.

Result:

i) Sulfite conc. (ppm) = Absorbance × Multiplication factor

Or

ii) Directly read from the graph plotted between concentration vs. absorbance.

18.17 Microbiological Analysis

Both aerobic and anaerobic bacterial content of processed water is determined by API-RP-30 serial dilution method.

Introduction:
Typical oil field brine/injection water may contain different kinds of bacteria. These can be separated into two groups: (a) aerobic bacteria, which can survive only in the presence of oxygen, and (b) anaerobic bacteria, which survive where no oxygen is present.

If the water contains oxygen, the major problems will be molds, algae, fungi, and other slimy growth that plug formulations and foul equipment.

Sulfate reducing bacteria, which are found in many oil field brines, are anaerobic and can cause corrosion. These bacteria digest sulfate in water and convert the sulfates to corrosive hydrogen sulfide, and the iron sulfide formed is a very troublesome plugging agent particularly in combination with small amounts of oil in water injection and disposal systems.

A knowledge of the microbial population in injection water/produced water is necessary to determine whether a microbial problem exists and to evaluate the effectiveness of the chemical treatment.

1. **General aerobic bacteria:**

   Principle:

   Suitable dilution of the sample of interest is prepared in culture media. The bacteria are incubated in an incubator under aerobic conditions and within +5 °C of recorded temperature of water when sampled. Culture media are observed for growth after seven days of incubation. If culture media become turbid, it means bacteria growth is positive, and if not then negative.

   No single growth medium or set of physical and chemical conditions can satisfy the physiological requirement of all bacteria in water sample. Consequently, the number of colonies may be substantially low.

   Preparation of culture medium:

   Beef extract : 3 g
   Tryptone : 5 g
   Dextrose : 1 g
   Agar : 15 g
   Distilled water : 1000 ml
   pH : 7+/−0.1 (to be adjusted) with NaOH before sterilization

The ingredients for the GAB culture medium are dissolved by gentle heating with constant stirring. After dissolution has occurred, the pH of the medium is adjusted to 7.0. Cleaned serum bottles of 10 ml nominal capacity are filled with 9 ml of the nutrient broth. The bottles are stoppered, using butyl-type rubber stoppers. A disposable metallic cap is used to protect and seal the rubber stopper in place. The filled and sealed bottles are then sterilized at 15-psi steam pressure at 121 °C for 15 minutes.

Sampling procedure:

1. Sample bottles should be thoroughly cleaned with detergent and finally rinsed with distilled water and then sterilized.
2. The tap from which the sample is to be drawn should be allowed to flow for at least 5–10 minutes before sampling.
3. The sample should be taken in such a manner as to avoid contamination from external sources.
4. The sampling bottle should be kept unopened until the moment it is to be filled. During sampling, do not touch the inside of the stopper and

the neck of the sampling bottle. Hold the bottle near the base, fill it without rinsing, and replace the stopper immediately.

5. The sample should be filled to the top of the sample bottle without any air gap inside. The presence of air can be detrimental.

6. The time, date, location, temperature, and other details should be recorded at the time of sampling.

7. The bacteriological examination of a water sample should be started promptly after collection. If the sample cannot be processed within 1 hour after collection, the sample should be preserved under refrigeration. The maximum time between sampling and examination should not exceed 24 hours.

Sample dilution and inoculation:
The first vial is inoculated with 1 ml of the water sample, using a sterile disposable syringe; the syringe is then discarded. The inoculated bottle is turned upside down four times and using another sterile syringe, 1 ml of the inoculated broth is withdrawn. This 1 ml is injected to the second dilution bottle and the procedure is repeated. Subsequent serial dilutions of each water sample are made in the same manner until the dilution factor is 10 E-4. All work should be done in duplicate.

Incubation and bacteriological examination:
Incubation may be done at a temperature +5 °C of water sampled. Bottles, which become turbid within seven days, will be considered positive indication of bacterial growth. If no turbidity, then it is taken as negative. Data are reported as the highest dilution indicating growth, as compared to the lowest dilution showing no growth.

2. **Sulfate reducing bacteria:**
Principle:
The principle for SRB is the same as for GAB except that the bacteria are incubated in an incubator under anaerobic conditions and within 5 °C of the recorded temperature of water when sampled. Culture media are observed for growth after 28 days of incubation. If culture media becomes black, it means bacteria growth is (+) and if not then (–).

No single growth medium or set of physical and chemical conditions can satisfy the physiological requirement of all bacteria in water sample. Consequently, the number of colonies may be substantially low.

Preparation of Culture Medium

Sodium lactate, USP	: 4.0 ml
Yeast extract	: 1.0 g
Ascorbic acid	: 0.1 g
$MgSO_4$ $7H_2O$	: 0.2 g
K_2HPO4 (anhydrous)	: 0.01 g
Sodium chloride	: 10 g
Fe (NH_4)2 $(SO_4)_2$	: 0.2g/100 ml*
Agar	: 15 g
Distilled water	: 1000 ml

* Should not be autoclaved, but filtered through 0.22 membrane filter.

Resazurin 0.001 g/l may be added as an indicator for the presence of oxygen.

Preparation method is the same as for GAB except that serum bottles are stoppered and caped in an inert gas atmosphere to exclude oxygen contamination.

Sampling procedure:
It is the same as mentioned for GAB.

Sample dilution and inoculation:
It is the same as mentioned for GAB.

Incubation and bacteriological examination:
Incubation may be within +5 °C of the water being sampled. Bottles, which turn black, will be used to define the sulfate-reducing bacterial population. Bottles, that turn black within 2 hours, are not to be considered positive since this probably will be due to the presence of sulfide ion in the sample. Subculture of these false positive samples may be made after one week. Cultures are to be held a minimum of 28 days and the data reported as the highest dilution indicating growth as compared to the lowest dilution showing no growth. The data are reported as a range of numbers, i.e., 100–1000 SRB bacteria per ml (if 10 E^{-2} is positive and 10 E^{-3} is negative).

18.18 Sulfide Analysis

Take 25 ml of water to be tested in a beaker and add 3 drops of activating solution. Insert the ampoule meant for sulfide analysis in the beaker and break it at the bottom and allow it to develop its color. After 5 minutes, match the developed color in ampoule. The presence of sulfide is a positive indication of the presence of SRB.

Table 18.2 Function and injection point of water injection chemicals.

Sr. no.	Chemicals	Function	Injection point
1.	Chlorine	Biological Control	SWLP intake
2.	Coagulant ($FeCl_3$)	Coagulate suspended solids	FF intake
3.	Polyelectrolyte	Coagulate and stabilize suspended solids	FF intake
4.	Defoamer	Prevent foaming in DOT	DOT inlet
5.	O_2 scavenger	Remove dissolved O_2	DOT outlet
6.	Scale inhibitor-I	Prevent $CaCO_3$ scale	DOT outlet
7.	Scale inhibitor-II	Prevent barium and strontium scale	DOT outlet
8.	Corrosion inhibitor	Prevent corrosion	DOT outlet
9.	Bactericide (amine)	Control bacterial growth	DOT outlet
10.	Bactericide (aldehyde)	Control bacterial growth	DOT outlet

Table 18.3 Suggested dose of water injection chemicals.

Sr. no.	Chemicals	Dose
1.	Chlorine	0.5 ppm
2.	Coagulant ($FeCl_3$)	0.2–0.3 ppm (used as 5.7% diluted solution)
3.	Polyelectrolyte	0.5–1.0 ppm
4.	Defoamer	0.5–1.0 ppm
5.	O_2 scavenger (NH_4 HSO_3)	2–3 ppm
6.	Scale inhibitor-I	3–5 ppm
7.	Scale inhibitor-II	3–5 ppm
8.	Corrosion inhibitor	3–5 ppm
9.	Bactericide (amine)	150–200 ppm (4 hours in a week)
10.	Bactericide (aldehyde)	150–200 ppm (4 hours in a week)

Various chemicals added to water perform different functions and the point of injection also has main role on water treatment (Table 18.2).

In addition to the type of chemical and injection point added to the water treatment process, the optimum concentration depending on the type of chemical will also play an important in effective water treatment plants (Table 18.3).

Table 18.4 Specification and testing schedule of quality parameters.

Sr. no.	Parameters	Location	Spec limit	Operation limit	Testing schedule
1	Chlorine (ppm, minutes)	CF outlet	1.2	–	Every 2 hours
		FF outlet	0.6	0.2	Every 2 hours
		TR outlet	Nil	Nil	Every 2 hours
		MIP outlet	Nil	Nil	Every 2 hours
2	Particle count (No/ml, above 2 micron, max)	FF outlet	300	600	Every 2 hours
		MIP outlet	300	600	Every 2 hours
3	Turbidity (NTU)	FF outlet	0.15	0.15	Every 2 hours
		MIP outlet	0.2	0.2	Every 2 hours
		IP	0.2	0.2	Monthly
4	Millipore filtration (L/30 minutes, 20psi, minutes)	FF outlet	10	7	Daily
		MIP outlet	–	7	Daily
		IP	–	7	Monthly
5	TSS (mg/l, max)	FF outlet	0.05	0.2	Daily
		MIP outlet	0.2	0.2	Daily
		IP	0.2	0.2	Monthly
6	Dissolved O_2 (ppb, max)	Tr outlet	20	30	Every 2 hours
		MIP outlet	20	30	Every 2 hours
		IP	20	30	Monthly
7	Residual sulfite (ppm, minutes)	Tr outlet	0.5	0.1	Daily
		MIP outlet	0.5	0.1	Daily
8	Total iron (ppm, max)	FF outlet	0.01	0.02	Daily
		MIP outlet	0.02	0.02	Daily
		IP	0.02	0.02	Monthly
9	Sulfide (ppm, max)	FF outlet	Nil	0.02	Daily
		MIP outlet	0.02	0.02	Daily
		IP	0.02	0.02	Monthly
10	Bacteriological test (GAB, SRB)	FF outlet	Nil	Nil	Monthly
		MIP outlet	Nil	Nil	Monthly

The effectiveness of added chemicals for water treatment is analyzed under different time periods and at locations (Table 18.4).

19

Flow Assurance for Organic Solid Deposition

19.1 What is Flow Assurance?

The term "flow assurance" is believed to have been coined by Petrobras in the early 1990s as "Garantia de Fluxo," literally translated as "Guarantee the Flow," or flow assurance.

Flow assurance can be defined as the ability to produce fluids economically from the reservoir to process facility (and beyond to sales point), over the life of a field in any environment. It is a rigorous engineering process that works with an objective to maximize production by ensuring unrestricted production flow path throughout the field life with lifecycle costs at minimum.

19.2 Need for Flow Assurance

The crude oil is produced from the reservoir through sub-surface and surface facilities. As the oil moves upwards, reduction in temperature and pressure starts. This can cause precipitation and deposition of organic solids, jeopardizing flow, system integrity, and ongoing operations.

These solids have the potential to deposit anywhere from the near wellbore and perforation to the wellbore, topside surface facilities, and pipelines. For cost-effective and successful development of oil fields, development of reliable guidelines for design and operation of systems and mitigation of these flow hazards is crucial through extensive analysis and flow assurance modeling. To assure that the entire system can be designed, built, and operated successfully and economically, system designers must consider and/or control reservoir characteristics and production profiles, produced fluids properties and behavior, the design of major system, operating strategies, and other system variables from the point of view of flow assurance.

Wax, asphaltenes, and gas hydrates are the primary organic solids, which cause flow-related problems.

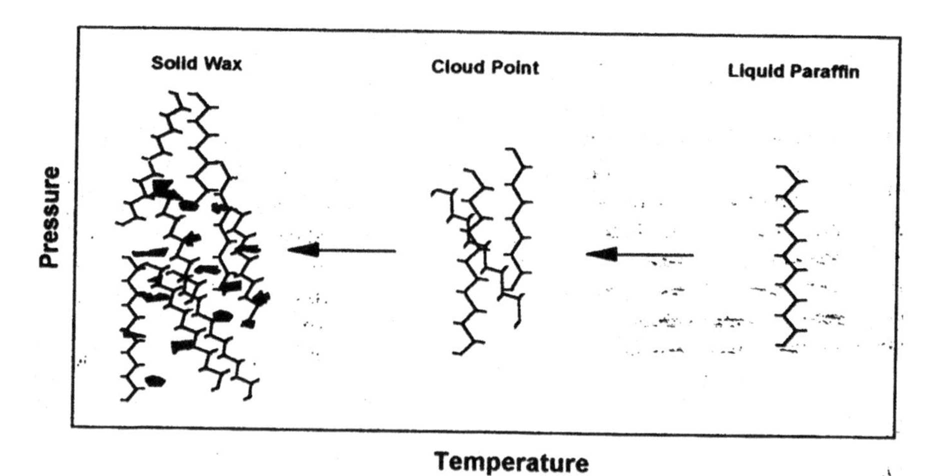

Figure 19.1 Mechanism of wax deposition.

19.3 Organic Solid Deposition – Mechanism

Paraffins, asphaltenes, and hydrates are the primary culprits that cause problems to offshore and onshore production and process facilities. For a solid deposition problem to occur during transport or shut-in conditions, the produced fluid must enter thermodynamic conditions inside their deposition envelopes. However, the production facilities may be engineered to operate outside these envelopes.

19.3.1 Wax deposition

Waxes are essentially mixtures of long-chain hydrocarbons (n-paraffins) with carbon numbers between 16 and 75+. They are crystalline in nature and tend to crystallize/precipitate from crude oil as the temperature is lowered below their melting points. Crystallization is the process where ordered solid structure is produced from a disordered phase, such as a melt or dilute solution (e.g., crude oil). It usually involves two distinct stages, namely nucleation and growth.

As the temperature of liquid solution or melt is lowered to the freezing point, the energy of molecular motion becomes increasingly hindered, and the randomly tangled molecules in the melt tend to move closer together and form clusters of adjacently aligned chains. The paraffin molecules continue to attach and detach to and from these ordered sites until the clusters reach a critical size and become stable (Figure 19.1). This process is termed *nucleation* and the clusters are called *nuclei*. These nuclei are only stable below the melting temperature of the paraffin as they are disrupted by thermal motion above this temperature. Once the nuclei are formed and the temperature is

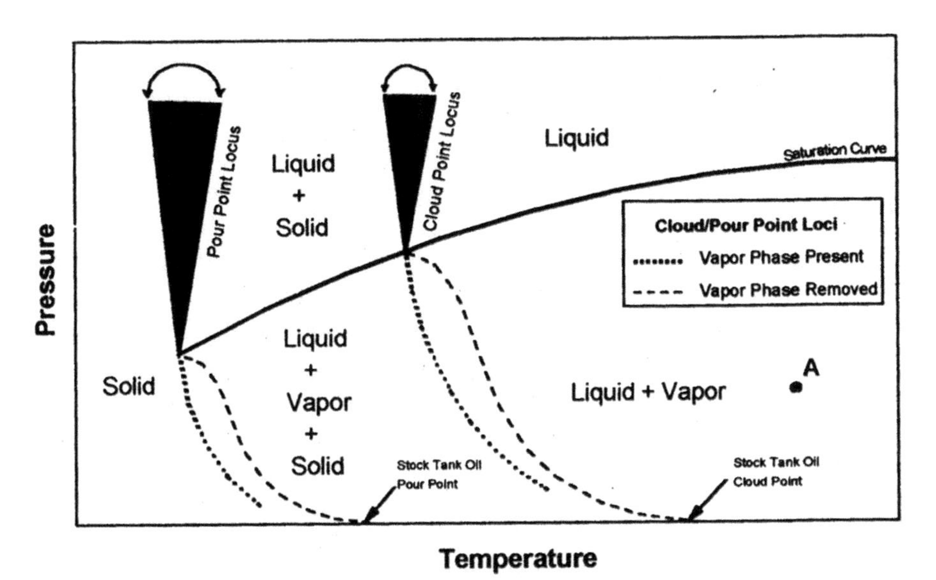

Figure 19.2 Wax deposition.

kept at or below the freezing point, additional molecules are laid down successively on the nucleation sites and become part of the growing lamellar structure. This mechanism is called the *growth process*.

Nucleation can be either homogeneous, meaning that the sample is pure or the nucleation sites are time-dependent, or heterogeneous, which implies that all nucleation sites are activated instantaneously. The latter type is the most common in crude oils where impurities such as asphaltenes, formation fines, clay, and corrosion products act as nucleating materials for wax crystals. These nucleating agents tend to lower the energy barrier for forming the critical nucleus and induce the growth process.

19.3.2 Typical P-T wax diagram

A schematic P-T phase diagram for a typical waxy crude oil is presented as Figure 19.2. In this figure, the solid line corresponds to the bubble point pressure curve for the oil under consideration; whereas the dashed lines represent the onset of wax precipitation (i.e., cloud point temperature locus) and/or the pour point temperature boundary.

From the single-phase liquid region, it can be seen that an isobaric decrease in temperature will result in the first formation of a "wax" solid phase at the cloud point boundary. This first formation of wax will correspond to the heavier molecular weight wax components in the crude oil and will not be sufficient to crystallize and congeal the fluid. Upon additional cooling of the oil, wax precipitates increasingly until the pour point phase boundary is reached. At this condition, the entire hydrocarbon phase congeals into a waxy solid (Figure 19.2).

The wax appearance temperature (WAT) represents a thermodynamic boundary at which crystalline paraffins form stable nuclei and precipitate from the bulk hydrocarbon fluid.

From a rheological perspective, the WAT represents a transition point between Newtonian and non-Newtonian flow regimes. In the absence of wax precipitates, crude oils normally exhibit shear-independent viscosity. Conversely, below the WAT, waxy crude oils cease to be homogeneous and become instead a polymeric-like suspension with an apparent viscosity that is higher than the viscosity of the continuous (liquid) phase.

The nucleation of wax crystals is followed by crystal growth, which is a time-dependent process of accretion. If the temperature is lowered below the WAT, under quiescent conditions, the suspended crystals may network to form a stable gel, i.e., an interlocking matrix structure containing occluded residual liquid. The temperature at which incipient gelation occurs is referred to as the "pour point," the precise definition of which is often operational.

Gelation is associated with the acquisition of Hookean mechanical properties; that is, a gelled crude cannot be induced to flow unless a force is applied that exceeds a certain minimum shear stress. This minimum yield stress required to restore steady shear flow is referred to as the "gel strength." Once the fluid yields, suspension flow results. The yield stress is a material property that determines restart requirements for a gelled flowline.

19.3.3 Factors influencing wax deposition

Although the cloud point/pour point are specific thermodynamic properties for a given crude oil, their relative positions/boundaries within a P-T diagram as well as the rate (and amount) of wax deposition and accumulation between them are dependent upon several factors, which include:

- Composition
- Temperature/cooling rate
- Pressure
- Paraffin concentration
- Molecular mass of paraffin molecules
- Occurrence of nucleating materials such as asphaltenes, formation fines, and corrosion products

- Water–oil ratio
- Shear environment

Recognizing that the onset of wax deposition is sensitive primarily to the waxy crude composition, it is important to know the properties as well as the relative amounts of the four major oil components, namely the paraffins, the napthenes, the aromatics, and the polars. For instance, both n-paraffins as well as iso-paraffins are flexible hydrocarbon molecules and, hence, tend to cluster together and precipitate from crude oil as wax solids. Being branched molecules however, the iso-paraffins tend to delay the formation of wax nuclei and usually form unstable wax solidifies. Aromatics, on the other hand, are known to be good solvents for paraffinic waxes. Naphthenes, also known as cyclo-paraffins, are stiff and bulky in nature; they tend to disturb and/or disrupt the wax nucleation and/or growth processes. Unlike in the case of asphaltenes precipitation, polars do not have a direct effect on wax deposition. Finally, the presence of impurities such as clays and/or other amorphous solids (like asphaltenes) in the oil usually induces wax nucleation process (but not necessarily the subsequent growth) as they tend to lower the energy barrier for forming wax nuclei.

The primary effect of water–oil ratio is related to changes in the rate of fluid production with changes in water-cut. Increased fluid production involves higher wellhead temperature and less wax deposition; whereas decreased fluid production would have the opposite effect. Water tends to increase the water wettability of metal surfaces, thereby reducing the likelihood of wax and crude contact with the metal surface. However, water-wet metal surfaces signal the onset of corrosion and, hence, the benefit of preventing wax deposition may be more than offset by corrosion products (which may act as nucleating agents) and related costs.

Pressure has different effects on the wax formation in a single-phase system and multiphase system. In a single-phase oil system, since the wax phase is denser than the oil, an increase in pressure slightly increases the wax deposition tendency. In a multiphase system, an increase in pressure drives the light ends of the mixture into the liquid phase and tends to decrease the cloud point, therefore tending to reduce the amount of wax formed at a particular temperature.

Paraffin precipitation is normally associated with changes in the physical environment surrounding the crude oil. Due to the normal subsurface temperature gradient, and the hydrostatic pressure variation in crude oil, when the oil is produced up the tubing, a pressure and temperature change occurs. This allows the lighter hydrocarbons to break out of solution and become a gas phase. These lighter hydrocarbons help keep the

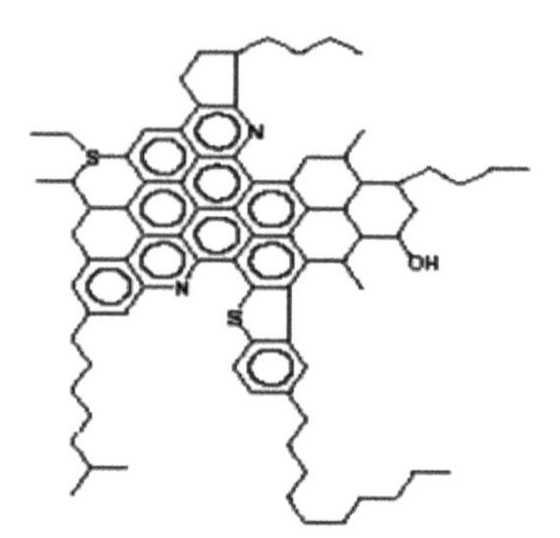

Figure 19.3 Chemical structure of asphaltenes.

heavy end paraffins in solution. Sudden pressure drops may also promote precipitation.

The shear environment can either delay or induce wax deposition depending on the regime (i.e., turbulent regime), and the ability of wax constituents to congeal and reach a pour point or to plug a tube is diminished as the nucleation process is disrupted; whereas, in a low shear environment (i.e., laminar regime and/or closer to equilibrium/static condition), the flexible paraffinic molecules tend to be aligned adjacent to one another in the direction of flow, thereby inducing/speeding the nucleation/clustering process. Hence, the severity of flow fine plugging is usually observed to increase for low shear environments.

A wax precipitate from a crude oil will re-dissolve into the single-phase liquid or the V-L region upon an isobaric increase in temperature beyond the cloud point.

19.4 Asphaltene Deposition

Asphaltenes are defined as high molecular weight aromatic organic substances that are soluble in toluene but are precipitated by alkanes (n-heptane/n-pentane). They are complex polar macrocyclic molecules that contain carbon, hydrogen, oxygen, and sulfur. Asphaltenes are aromatic and occur in a partly dissolved and a partly dispersed colloidal suspension stabilized by non-polar resins and maltene fraction of the crude (Figure 19.3).

Generally, asphaltenes tend to remain in solution or in colloidal suspension under reservoir temperature and pressure conditions. They may start to precipitate once the stability of the colloidal suspension is destabilized, which is caused by the changes in temperature and/or pressure during primary depletion. On the other hand, asphaltenes have been reported to become unstable as a result of fluid blending (co-mingling) of fluid streams as well as by gas injection during improved oil recovery (IOR) operations.

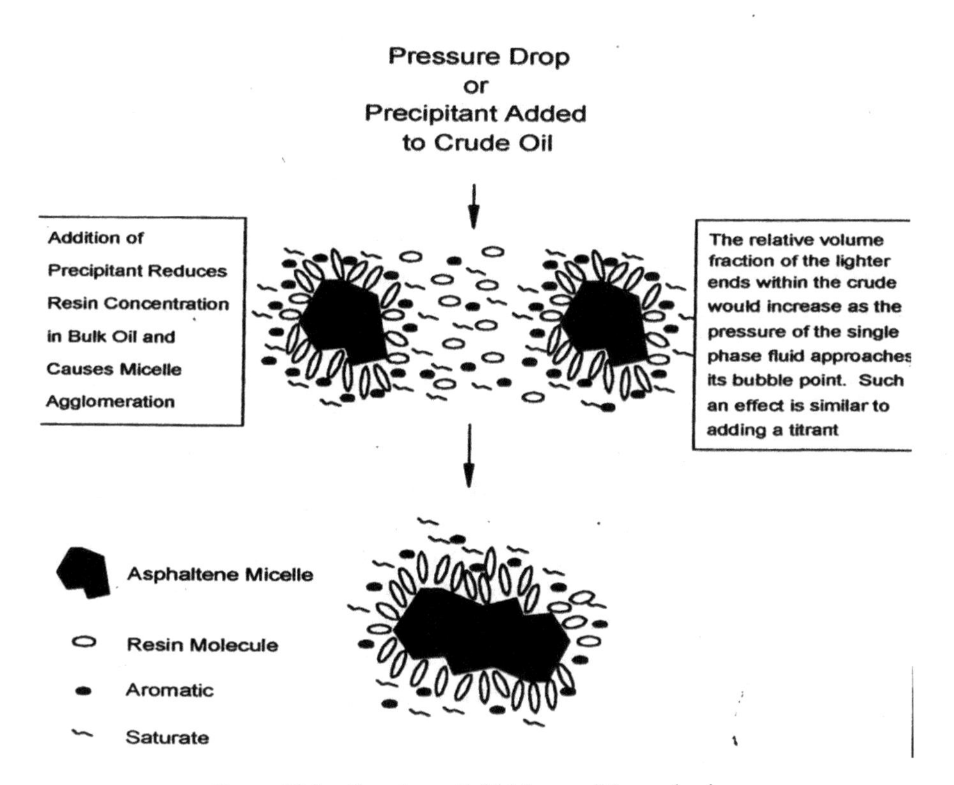

Figure 19.4 Complex colloidal irreversible mechanism.

19.4.1 Mechanism: colloidal and/or thermodynamic

The available laboratory and field data indicates that asphaltenes separated from crude oils consist of various particles with molecular weight ranging from 1000 to several thousands. Such an extensive range of asphaltene size distribution suggests that asphaltenes may be partly dissolved and partly suspended/peptized in the crude oil. While the first scenario is a relatively well understood reversible thermodynamic process, the latter is a more complex colloidal irreversible mechanism (Figure 19.4).

Asphaltene molecules (micelles) are believed to be surrounded by resins that act as peptizing agents; that is, the resins maintain the asphaltenes in a colloidal dispersion (as opposed to a solution) within the crude oil. The resins are typically composed of a highly polar end group, which often contains hetero-atoms such as oxygen, sulfur, and nitrogen, as well as long, non-polar paraffinic groups. The resins are attracted to the asphaltene micelles through their end group. This attraction is a result of both hydrogen bonding through the hetero-atoms and dipole–dipole interactions arising from the high polarities of the resin and asphaltene. The paraffinic component of the resin molecule acts as a tail making the transition to the relatively non-polar bulk of the oil where individual molecules also exist in true solution.

Because asphaltenes are stabilized as colloidal particles by resins, alterations of a chemical, electrical, or mechanical nature may de-peptize the asphaltene micelles, thereby inducing their flocculation and/or precipitation. Field experience and experimental investigations clearly indicate that asphaltene stability and precipitation are sensitive to changes in temperature, pressure, and chemical composition of the crude combined with streaming potential effects due to flow in tubing and porous media.

The addition of compounds with molecules that differ greatly from resins in terms of size and structure, and therefore, solubility parameter, shifts the equilibrium that exists in the non-asphaltene portion of the crude oil. For example, normal alkane liquids (such as pentane or hexane) are often added to crude oils in an attempt to reduce heavy oil viscosities. The result of this introduction is an alteration in the overall characteristics of the crude oil making it lighter. In response, resin molecules desorb from the surface of the asphaltenes in an attempt to re-establish the thermodynamic equilibrium that existed in the oil. The desorption of peptizing resins forces the asphaltene micelles to agglomerate in order to reduce their overall surface free energy. If sufficient quantities of the particular solvent are added to the oil, the asphaltene molecules aggregate to such an extent that the particles overcome the Brownian forces of suspension and begin to precipitate. As such, the quantity and type of solvent added to the crude oil is crucial to the amount and characteristics of the asphaltenes deposited.

19.4.2 Factors influencing asphaltene deposition

Asphaltene precipitation during petroleum production and processing, e.g., in well tubing, flowlines, and/or process equipment normally proceeds under elevated temperature and pressure conditions. It has been established that the effect of composition and, in turn, of pressure on asphaltene deposition is stronger than the effect of temperature. However, there still exists some disagreement in the literature regarding the effect of temperature on asphaltene precipitation (Figure 19.4). In general, the parameters and properties that are reported to affect asphaltene deposition and/or flocculation include:

- Composition
- Pressure

- Asphaltene and resin concentrations in the reservoir fluid
- Electro-kinetic effects induced by streaming potential generation during reservoir fluid flow
- Temperature
- Water-cut
- pH concentration

With respect to pressure, there is considerable agreement in the literature that the greatest asphaltene precipitation occurs in the proximity of the bubble point curve of the asphaltenic crude oil. Some studies of asphaltene deposition in oil-well tubing indicated that deposition occurs below the depth, at which the pressure is above the bubble point pressure, i.e., the oil was in single phase. This phenomenon was largely ascribed to the different compressibilities of the lighter ends and the heavier components of live crude oil. Actually, the relative volume fraction of the lighter ends within the crude would increase as the pressure of the single-phase reservoir fluid approaches its bubble point. Such an effect is similar to adding a light hydrocarbon (precipitant) to a crude causing asphaltene de-peptization. Below the bubble point, the low molecular weight hydrocarbons vaporize (separate) from the liquid as a gas phase causing an increase in the density of the liquid phase (i.e., change of liquid composition). Recognizing that gas as well as asphaltenes compete for solvency in the crude oil, the vaporization (loss) of light ends implies better solubility of the asphaltenes in the crude oil.

The ratio of resins to asphaltenes is more important than the absolute asphaltene content. In fact, severe asphaltene precipitation can often be encountered in reservoirs with very low total asphaltene contents. Conversely, there have been reports of high asphaltene-content oils that show no appreciable deposition. For example, the Mata-Acema crudes (in Venezuela) with asphaltene contents of only 0.4–9.8 wt% have asphaltene deposition problems; whereas, the Boscan crude with 17.2 wt% asphaltene content does not have asphaltene deposition problems. Oils containing 1:1 or greater weight ratio of resins-asphaltenes are less subject to asphaltene deposition.

Asphaltenes have an overall intrinsic charge that may be positive or negative depending on their oil source and components. If asphaltenes are placed under the influence of an external electric field, they migrate to the oppositely charged electrode. In addition, asphaltenic fluids flowing through capillaries or porous media can develop an electrical charge through the streaming potential phenomenon.

Last but not least, water, acids, and CO_2 can also contribute to the precipitation of asphaltenes. The asphaltenes concentrate at the water or acid interface and cause rigid films. CO_2 appears to cause a reduction in solubility of asphaltenes (similar to propane) and in addition can cause rigid film formation because of its effect on the pH of any water present.

Acidizing is one of the most common well treatments and can cause severe damage to a well with asphaltic crude oil. The acid causes the asphaltenes to precipitate, sludge, and form rigid film emulsions, which severely affects permeability, often cutting production by over 50%. Formation material, particularly clays, contains metals that may interact with the asphaltenes and cause the chemisorption of the asphaltenes to the clay in the reservoir.

19.5 Gas Hydrates

A gas hydrate is a crystalline solid that forms under the specific conditions of temperature and pressure, which are thermodynamically appropriate to that gas. "Hydrate" refers to the fact that water molecules surrounding a central molecule of a different kind form them. In the case of gas hydrates, this central molecule is a low molecular weight gas, such as those which commonly constitute natural gas; i.e., methane, ethane, propane, and iso-butane, SO_2, N_2, H_2S, CO_2, and others.

These molecules, which can exist in a vapor or liquid state and can be miscible or immiscible in water, are referred to as "hydrate formers." The water molecules form a repetitive geometric lattice, which is commonly referred to as a "cage."

Without the gas molecule at the center, the highly organized cage structure would be in dynamic equilibrium with free flowing water molecules, perpetually forming and collapsing. In the presence of the central gas molecule, given the right conditions of temperature and pressure, the water-molecule-based cage forms a geometric structure (usually having 12, 14, or 16 sides).

This structure is stabilized by the additional van der Waals forces acting between the gas molecule and the surrounding water molecules. On the range of a few Angstroms, solid nuclei form and begin to grow. When nuclei come in contact with other nuclei and join together to form larger particles, the process is called "agglomeration." Attraction between neighboring apolar molecules inside the cages of water molecules causes agglomeration.

Once the growing structure achieves a critical minimum size (8–30 nm), rapid growth of hydrate crystals, known as "catastrophic growth," ensues. Hydrate crystal nucleation and growth is a kinetic process, meaning that

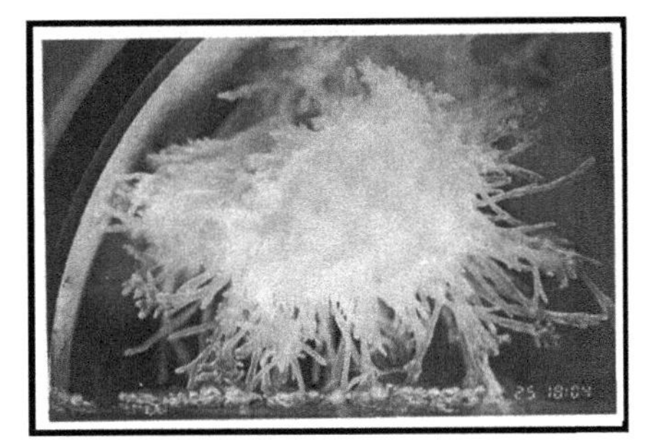

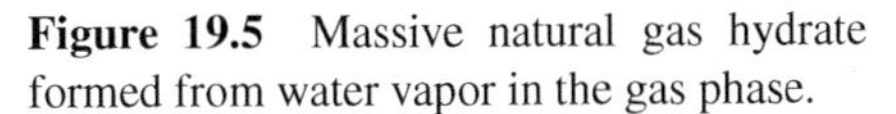

Figure 19.5 Massive natural gas hydrate formed from water vapor in the gas phase.

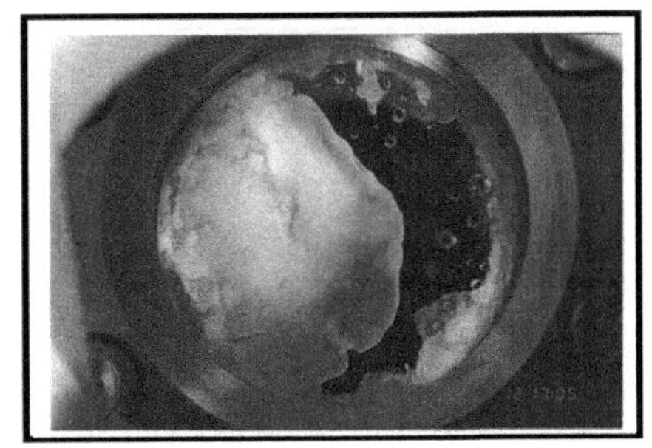

Figure 19.6 Massive methane hydrate formed in the gas phase.

hydrates do not appear instantaneously with the onset of thermodynamically favorable conditions. The lag between the time when system conditions favor hydrate formation and the appearance of hydrates is known as "induction time."

Water molecule building blocks can form a range of metastable geometric structures. Depending on the size of the cavity formed by the water molecule cages, differently sized gas molecules can enter them and stabilize them (Figures 19.5 and 19.6).

19.5.1 Conditions for gas hydrate formation

Given the time required to reach equilibrium, gas hydrates form predictably at specific thermodynamic conditions of high pressures and low temperatures. Gas composition, water chemistry, and stream turbulence are variables that affect hydrate formation as well. Gas composition determines which of the hydrate structures will develop and at what temperature and pressure the solid will form. Each hydrate is formed under specific pressure and temperature conditions according to the hydrate former. Methane hydrates have the highest formation pressure and the lowest formation temperature. When the molecular mass of the hydrate former increases, the temperature of hydrate formation increases, and pressure of hydrate formation decreases; in general, larger formers precipitate hydrates more easily than smaller ones. Water salinity is an example of a water chemistry effect: dissolved salts are known to give a hydrate inhibitor effect. Turbulence, causing efficient mixing of water and hydrocarbon phases, is a kinetic effect, producing faster hydrate formation. Thus, four elements are required for the formation of solid gas hydrates:

- Hydrocarbon phase
- Water phase

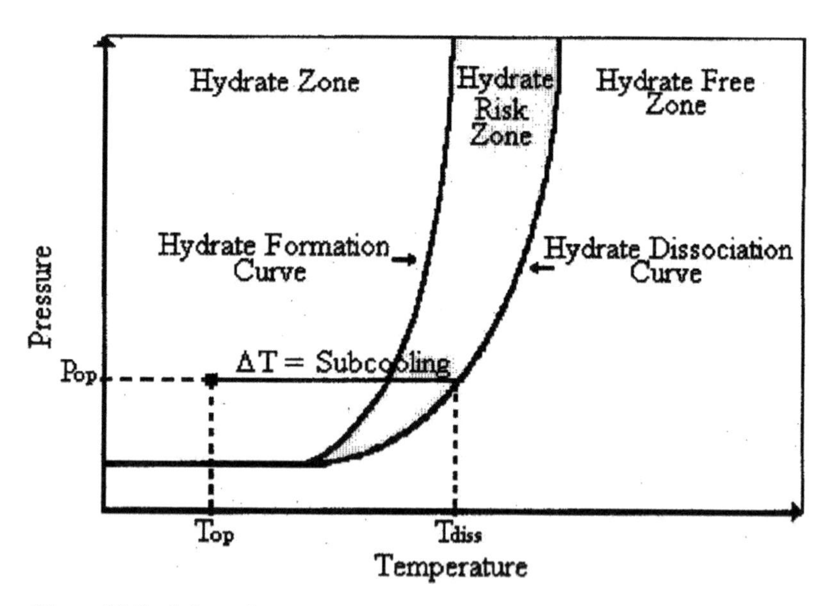

Figure 19.7 Schematic pressure vs. temperature diagram for a gas composition.

- Low temperature
- High pressure

Figure 19.7 is a generic pressure vs. temperature curve illustrating the range of system conditions including the hydrate zone (formation is favored), hydrate-free zone (formation is not thermodynamically favorable), and hydrate risk zone. Dissociation and formation curves are illustrated and a graphic explanation of sub-cooling ($T_{dissociation} - T_{operating}$) is given. Operating temperature (T_{op}) and dissociation temperature (T_{diss}) for the operating pressure (P_{op}) are illustrated.

19.5.2 Mechanism of gas hydrate formation

Hydrate and paraffin form under similar pressure, temperature, and chemical trends. Kinetics of hydrate formation is still a current research topic. Since this process is not well understood, several theories have been developed explaining the mechanisms of hydrate formation. Lederhos et al. (1996) proposed that gas hydrates form in an autocatalytic reaction mechanism, when water molecules cluster around natural gas molecules in structures similar to the ones shown in Figure 19.7. This attraction between neighboring guest molecules is termed "hydrophobic bonding," which can be described as an

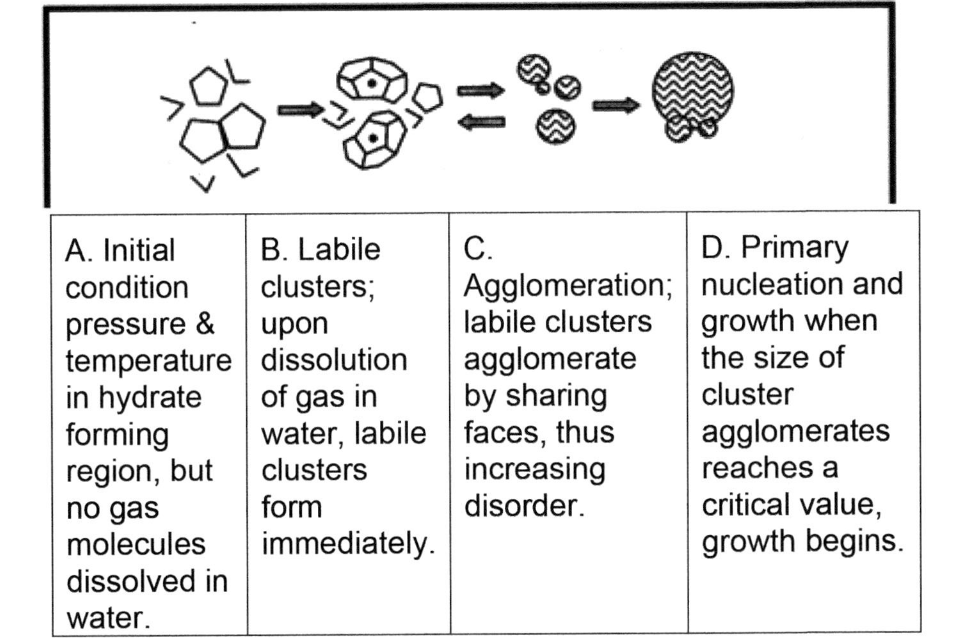

Figure 19.8 Auto catalytic reaction mechanism for hydrate formation (Lederhos et al. 1996).

attraction between the apolar molecules inside the clusters [B]. Large and small clusters forming structures I and II are termed "labiles" because they are easy to break down, but relatively long-lived. Labiles can dissipate or grow to become hydrate unit cells or agglomerations of unit cells forming what are known as "metastable nuclei" [C]. Then, growth can continue until crystals are stable, indicating the onset of secondary nucleation [D]. This process is illustrated in Figure 19.8. This theory implies reversibility of the process when the system is heated up.

19.6 Flow Assurance Management

Flow assurance generates essential information for feasibility of new areas and allows selection of the best option for minimizing the problem and allows inclusion of necessary support facilities since the design stage. The current approach in flow assurance involves fluid characterization, experimental techniques for measurement of solid deposition, and finally methods of combating deposition either through process control or chemical control. The various processes involved in providing vital input for flow assurance is given below.

19.6.1 Fluid sampling, transport, and characterization

The basic aim of a characterization method is to set up an equation of state (EOS) model for the fluid in question. This involves collection of a suitable reservoir fluid sample, which has gone through no irreversible phase changes during the sample collection, sample transfer (both in the field and the lab), and sample transportation processes. A single-phase multi-chamber (SPMC) module is normally used in collecting single-phase reservoir fluid samples.

The approach to fluid characterization consists of the following steps:

a) **Fluid compositional characterization:**
The samples are first validated for opening pressure and then characterized for the following:

- Compositional analysis
- Physical property measurements
- Phase behavior analysis
- Organic solids analysis

b) **Preliminary screening for solids:**
After completion of the fluid compositional characterization, a series of screening criteria and simulation are used to evaluate the potential of the crude oil under consideration to exhibit solid formation problems. For asphaltenes, the first screening test is the De Boer plot developed by Shell based on field observations, the second is the asphaltene to resin ratio, and the third is the colloidal instability index suggested by Baker-Petrolite. Preliminary screening for wax involves wax content of the stock tank oil (UOP procedure) and cross-polar microscopy wax appearance temperature.

19.6.2 Experimental techniques for measuring organic solid deposition

After establishing the potential for solid formation through preliminary analysis and screening criteria, the more detailed fluid behavior measurements with regard to the various techniques for asphaltene formation (1) thermodynamic conditions of wax appearance temperature (2) and the measurements of hydrate formation conditions are carried out. The measurements made on the stock tank oil (STO) are not valid for the development of model and hence it is always desirable to obtain data on both the STO and the live oil crude to improve the decision-making capability.

a) **Measurement of dead oil organic solids precipitation:**
Asphaltenes: The measurement of asphaltene precipitation from dead oils is most commonly used to establish the overall asphaltene content of a sample. The measurements are done either by a microscopic method, which involves the titration of liquid n-alkanes into crude oils, or by a microscopic visual examination for the presence of solids.

Paraffin waxes: Cross-polarized microscopy (CPM) has been the most sensitive techniques for the measurement of dead oil cloud points. The greatest advantage in the use of cross-polarized microscopy is the ability to visually detect and establish the presence of very small (and, therefore, recently formed) wax crystals. Differential scanning calorimetry (DSC) has been widely used for the determination of cloud point (and wax dissolution) temperatures in petroleum products. The DSC requires only very small samples size to initiate measurement and the entire process is quite rapid and completely automated. Cold filter test is based on the movement of dead (or live) oils through a temperature-controlled flow-loop. As the system temperature is dropped below the cloud point, crystallized paraffins collect in a filter causing an increased pressure drop across its width. This procedure seems to be more representative of true field conditions.

b) **Measurement of live oil organic solids precipitation:**
Unlike the dead oil measurements, the effects of pressure, temperature, and composition on live oil asphaltene and wax formation are critical for combating the organic solid deposition issues. Fiber optic light transmittance (FOLT) can measure the onset conditions at pressures up to 10,000 psia and temperatures up to 350 °F. High-pressure cross-polar microscopy is capable of measuring the live-oil organic precipitation under pressure, and potential exists for measuring both waxes and asphaltenes in even the darkest crude oils. High-pressure DSC can evaluate the sample at elevated pressures and temperatures for cloud point and asphaltene onset measurement. The technique developed by M/s D. B. Robinson utilizes a bulk deposition apparatus and provides accurate data on the actual quantities of solid formation.

19.6.3 Combating organic solid deposition Wax control measures – current approach and practice

There are numerous methods used to handle paraffin deposition. These can be divided into two categories:

1. Removal technique
2. Inhibition technique

Removal techniques:
The frequently used paraffin removal techniques are listed below.

Mechanical paraffin removing method:
Mechanical removal of paraffin utilizes a paraffin knife, hook, corkscrew, swab, or scratcher. The method involves working the cutting tool through the paraffin. Cutting or wire-lining is a type of downhole treatment of flowing wells to control paraffin deposition. During the cutting operations, large quantities of paraffin may be released into the flowlines with the possibility of plugging the flowline, redepositing it in separator, or as tank bottoms in tanks. The method is labor-intensive.

Pigging:
Pigging is another labor-intensive mechanical flowline/pipeline treatment method frequently used. A variety of pigging devices fit the diameter of the pipe and scrape the pipe walls as they are pushed through the pipe from one end and retrieved at the other end removing the bulk of the deposits.

Hot oil/steam well clean up:
A periodic thermal cleaning method used for removing paraffin by hot oil, water, or steam. Hot oil/water or steam is circulated through the casing to the tubing top, further heating the flowlines to avoid deposition in the flowlines. In this method, the small amount (10–40 bbl.) of treated crude is heated to 180–200 °F and is injected down the annulus and up the tubing. The basic theory here is that the oil enters the oil structure of the paraffin, breaks it, and disperses it in the flow stream.

Solvent treatment:
Solvents are added to restore solvency to the crude that may have been lost due to the escape of dissolved gases or reduction in temperature. For examples, produced condensate, gasoline, butane, xylene, toluene, etc. They are usually applied in frequent batch treatments, or continuously.

The solvent remediation technique has often been found to be the most successful method in the field but is a very costly option. However, the following aspects should be considered at the time of solvent application:

- Concentration and mol. wt. of wax determine the amount of wax that a solvent can carry.
- Normal carrying capacity of any solvent is found to be 10% of wax and an enhanced carrying capacity can be obtained by a solvent surfactant package.
- Solvent remediation methods are reserved only when hot oil/hot water application fails. If waxes and emulsion are found together, then it is desired to use a crystal modifier–demulsifier combination.
- Adequate application temperature and adequate mixing is found to be the key to the success of a solvent remediation.

Nitrogen generating system (SGN) method:
It comprises the controlled reaction between two nitrogen containing chemicals that are capable of generating large amount of nitrogen gas and heating, melting, and pushing the deposit. The method has been successfully used in more than one hundred operations in the Campos Basin area.

19.6.4 Inhibition techniques

The process of wax precipitation includes three stages:

1. Wax separationGrowing up of wax crystalsDeposition of wax

Control of any one of the three stages of wax deposition will reach the goal of paraffin inhibition. Some of the commonly used wax inhibition techniques are listed below:

1. Electric heating – control of wax at the separation stage.
2. Magnetic paraffin inhibition – control of wax deposition at the wax separation stage.
3. Paraffin inhibitors – wax control options at the crystal growing up and wax deposition stages.
4. Glass oil tube and coating oil tube – control of wax deposition at the deposition stage.

Electric heating – control of wax at the separation stage:
Pre-heating and heat tracing: The pipeline system is heated to maintain a constant temperature by a heating medium such as steam, hot water, heating oil, or electrical tracing.

Crude oil conditioning:
This process has been in use for three decades in Assam, India. It deals with static and dynamic thermal pretreatment of the crude. It involves thermal pretreatment of the crude to (80–115 °C) followed by shock chilling at a cooling rate of about 5 °C/minute to a lower temperature of 65 °C. The rate of cooling is then changed to a slow rate of about 1 °C/minute to reduce the rate of precipitation of wax crystals from the crude, since the higher rate of precipitation in this temperature range (65–128 °C) affects the morphology of the precipitated crystals in such a way so as to increase the viscosity at the limiting ambient temperature (18 °C).

This method of treatment was applied on an Egyptian residual fuel oil and three waxy crudes. The optimum conditions attained were preheating the residual fuel oil to 110 °C, followed by cooling to 85 °C at a rate of 5.5 °C/minute, then changing the rate of cooling to 0.7 °C/minute, further static cooling until 15 °C/minute, and further static cooling until 15 °C. This method of treatment lowered the maximum fluidity temperature by 18 °C for the residual fuel while reducing it by 12, 9, and 15 °C for Morgan, Ramadan, and Ras-Gharah crudes, respectively.

Magnetic paraffin – inhibition:
Strong magnetic fields have been suggested and employed in certain cases, with some reported success. The theoretical basis for such an approach suggests that forming waxes, when subjected to an intense magnetic field, will be displaced from their preferred crystal alignment. This disalignment then either prevents or interferes with the crystals continued networking.

The method requires the installation of a strong magnet on a segment of the flowline. Usually the magnet is of the permanent type.

It is reported that magnetic treatment under correct conditions (temperature, magnetic intensity, and time of exposure) could well control the crude oil wax deposition process with a reduction of 20%–25% and very obvious long lasting changes in rheological properties.

Magnetic paraffin-inhibitors are widely used in Daqin, Zhongyuan, Shenli, and Huabei oilfields of China. In the Daqin oil field, nearly 4500 wells were installed with magnetic paraffin inhibitors; the strength of the magnetic field is usually in the 150–250 mT range. Substantial improvement in the oil production and thermal washing period was observed.

Plastic coated tubular goods and fiberglass surface lines:
The initiation of a paraffin deposit has been shown to be primarily dependent upon flow surface irregularities and heat losses, and the use of plastic-coated tubular goods and fiberglass surface lines provides both a smooth surface and, to a certain extent, insulation.

Since fiberglass pipe provides a smoother surface than does steel, inhibition of paraffin deposition and oil treating to remove paraffin deposits may be better and cost effective as compared to steel pipes.

Wax control additives:
It has been found that the use of an effective paraffin inhibitor has the potential for significant saving versus other inhibition procedures. Since paraffin characteristics and content vary drastically from reservoir to reservoir, production problems and solutions also vary.

Wax control additives are known by different names as per their functions:

Flow improvers:
Ash-less polymeric material that modifies crystal growth pattern and reduces crystal size, crystal to crystal adhesion, and agglomeration. It improves flow characteristics. Flow improvers do not avoid paraffin crystallization but change crystal morphology.

Pour point depressant:
Oil soluble polymer additives, which alter the growth pattern of wax crystals and reduce pour point, viscosity, yield stress, and wax deposition on the pipe wall.

Paraffin dispersants:
Dispersants are formulated from such materials as: sulfonates, alkyl phenol derivatives, ketones, terpenes, polyamides, and naphthalenes. They work by neutralizing the attractive forces that bind the paraffin particles together and prevent the deposition of paraffins. The dispersants coat the paraffin particles moving with the fluid flow. The application can be continuous or in batches. Dispersants can be applied in water, diesel, or solvents at the location where the crude reaches the cloud point.

Paraffin crystal modifiers:
Materials that have a similar molecular structure to the wax that is precipitating. The crystal modifier co-precipitates with the wax by taking the place of a wax molecule on the crystal lattice. This prevents the paraffin crystals from adhering together and deposition is reduced or eliminated. Wax crystal modifiers are applied before the crude reaches the cloud point. They include

polyethylene, copolymer esters, ethylene/vinyl acetate copolymers, polyacrylates, polymethyl acrylates, etc.

Combating asphaltene deposition:
Actions for asphaltene precipitation control can be summarized in two categories:

Preventive measures:
After confirming that the deposition is caused by asphaltenes, the most effective approach is to regularly inject specialized organic polymers, known as asphaltene dispersants/inhibitors, into the wellbore, choke, or other surface facilities where asphaltene precipitation is most likely to occur. These polymers have a resinous nature and interact with asphaltenes to stabilize them, thereby reducing asphaltene precipitation.

If the zone of maximum probability is located on the perforations or near the wellbore, a hydraulic fracturing treatment should be considered, so the pressure profile can be moved in a manner such that the bubble point pressure can be reached as near as possible to the surface since asphaltene flocculation is predominantly a pressure phenomenon.

The choke size is reduced with the purpose of moving the pressure profile upward in the production tubing so that the bubble point pressure will be located nearest to the wellhead as possible. Pressure maintenance through injection fluids also helps minimize the asphaltene precipitation in the reservoir.

In cases where the zone of maximum probability of precipitation is in the vicinity of the perforations, injection of asphaltene dispersants/inhibitors into the formation will prevent asphaltene flocculation and deposition in the reservoir and in the near wellbore.

Corrective measures:
Once the precipitation of asphaltenes has occurred within the system, the following corrective action should be taken to solve the problem:

Mechanical methods can be used to periodically remove asphaltene deposits from flowlines, producing tubing, and pipelines; however, asphaltenes can be more brittle and harder to remove than typical wax deposits.

These methods include rod scrapers, wire-line scrapers, flowline scrapers, free-floating piston scrapers, "pigging" flowlines, and wire-lining tubing. A disk or a cup pig is used since they can apply much more force on the pipe wall. A bypass pig (one that allows part of the fluid stream to go through the pig) allows the removed solids to be dispersed into the crude oil ahead of the pig. This prevents a solid buildup in front of the pig and decreases the likelihood of sticking of the pig.

When the asphaltene deposition is in the near wellbore area, hydraulic fracturing should be considered to reduce the pressure differential at the face of the formation and displace the pressure profile so that the bubble point pressure is reached nearest to the wellhead and then the problem can be handled with the dispersant/inhibitor injection at the surface.

Unlike wax, asphaltenes are very soluble in aromatic solvents such as benzene and xylene even at seabed temperatures. For downhole, near wellbore and flowline treatment solvent soaks with aromatic solvents and/or aromatic solvents blended with dispersants are the most common and effective remedial measures currently used for removing the asphaltene deposits.

19.6.5 Hydrate control measures

Hydrate formation can be prevented by several methods including controlling temperature, pressure, removing water, and shifting thermodynamic equilibrium with chemical inhibitors such as methanol or mono-ethylene glycol. Measures employed to prevent the formation of gas hydrates are:

Insulation:
Insulation is used to retain heat in the produced fluids to stay above hydrate formation temperatures, or in other words, to the right of the hydrate formation region in the phase envelope. For wellbores, vacuum-insulated tubing (VIT) or gelled completion fluids can be used. In the Gulf of Mexico, VIT has been used to minimize well restart times and to reduce the minimum operating/restarting flow rate of individual wells. For flowlines and risers, there are a number of insulation options like externally insulated rigid pipe, insulated flexible pipe, pipe-in-pipe and bundle, etc.

Heating:
It can be used to warm equipment and/or produced fluids to stay on the right of the hydrate formation region. Examples include hot oil circulation and electrical heating. Active heating is almost always combined with insulation. Hot fluid/hot oil circulation can be applied in continuous operations, shutdown, and restarts. Electrical heating can eliminate flowline depressurization and displacement, providing ability to quickly remediate hydrate blockages where other techniques might not be effective.

Water removal:
If the water can be removed from the produced fluid, then hydrate formation is not likely to occur. This is the common hydrate prevention technique applied to transmission pipelines. For subsea production systems, there are two potential

techniques to remove water. The first is downhole water separation and disposal, and the second is subsea (i.e., mudline) water separation and disposal.

Chemical inhibition:
Chemical inhibitors are routinely injected into gas pipelines to protect against gas hydrate formation in the event of an unscheduled shutdown or flow interruption. Chemical inhibitors are classified by their mode of action, viz. thermodynamic inhibitors, kinetic inhibitors, or crystal modifiers/ anti-agglomerates.

Thermodynamic inhibitors:
Thermodynamic inhibitors using chemical compounds as inhibitors are added in high concentrations (10–60 wt%), to alter the hydrate formation conditions, allowing hydrates of the new mixtures form at lower temperatures or higher pressures. Methanol, mono-ethylene glycol, and salt solutions like potassium formate are some of the main chemicals used as hydrate inhibitors. Thermodynamic inhibition is still the widest method used worldwide and continues to be the industry standard, but its associated costs, environmental concerns, and operational complexity have made researchers look for a different approach to the problem.

Kinetic inhibitors:
Kinetic inhibitors delay the nucleation and growth of hydrate crystals from several hours to days, which may exceed the residence time of fluids in process flowlines. These chemicals can be effective at very low concentrations (<1 wt%). Its effect appears to be independent of the amount of water in a system and the thermodynamic conditions of hydrate formation; thus, it gives them an advantage over thermodynamic inhibitors.

Anti-agglomerants:
These chemicals are polymers and surfactants, which are also added at low concentrations (1 wt%). They allow hydrate formation but work by preventing agglomeration of hydrates so that the hydrate crystals do not grow large enough to plug flowlines, but are transportable as a slurry. Anti-agglomerants only work in the presence of both water and liquid hydrocarbon phases.

Hydrate remediation techniques:
Thermodynamic inhibitors: These inhibitors can essentially melt blockages when they come in direct contact with the hydrates. In the case of a riser or wellbore, it may be possible to gravity feed an inhibitor to the blockage. MEG is preferred due to its high density. If the blockage can be accessed with coiled tubing, then methanol can be pumped down the coiled tubing

to the hydrate blockage. Some field applications have reached as far as 7 miles (11.3 km).

Depressurization:
Depressurization appears to be the most common technique used to remediate hydrate blockages. From both safety and technical standpoints, the preferred method to dissociate hydrate blockages is to depressurize from both sides of the blockage. Rapid depressurization should be avoided because it can result in Joule–Thomson cooling, which can worsen the hydrate problem and form ice blockages as well. Two-sided depressurization is partially recommended because of safety concerns. If only one side of a blockage is depressurized, then there will be a large pressure differential across the plug, which can potentially turn the plug into a projectile.

Active heating:
Active heating can remediate hydrate plugs by increasing the temperature and heat flow to the blockage. It is important to note that there are safety concerns with applying heat to a hydrate blockage. During the dissociation process, gas will be released from the plug. If the gas is trapped within the plug, then the pressure can build and potentially rupture the flowline. However, with evenly applied heating to a flowline, as is the case with electrical heating and heated bundles, remediation can be performed safely and effectively. The advantage of active heating is that it can remediate a blockage in hours, whereas depressurization alone can take days or weeks. The ability to quickly remediate hydrate blockages can enable less conservative designs for hydrate prevention.

19.7 Summary

The earlier approach to combating solid deposition involved techniques that were remedial or mitigative in nature. The current approach for combating solid deposition has evolved in the form of flow assurance, which is used to identify and prevent potential fluid-related problems impacting production throughout the asset life. As oil and gas exploration moves into subsea and deepwater, solid deposition challenges become more prevalent and the system design must address the issues from a fresh perspective. It is necessary to investigate a wide range of possible operating conditions to effectively bound the system design and select a suitable system that is capable of meeting all the flow assurance challenges for a given field. Since small changes in fluid properties or reservoir properties can significantly impact the flow assurance results, it is recommended that studies encompassing a wide range of operating parameters be undertaken as early in the design phase as possible to provide the operator with the information that is needed to make an educated choice about how best to develop the asset. To meet these challenges, one needs to develop new predictive tools and different production practices. Cost effective solutions will not just rely upon ongoing chemical cures but must be linked to design.

20

Oil Field Development

20.1 Introduction

The development activities of oil and gas fields start from the point of striking a field while undertaking exploratory drilling. The second phase starts proving its commerciality by interpreting its exploratory results and seismic information. The results have to necessarily indicate the volume of oil and gas in surface conditions, normally in the units of million metric tons (MMT) of oil, and million standard cubic meter (MMSCM) of gas.

The Exploration and Development Directorate (E&D) releases the (Reserve Estimate Committee) REC figures, i.e., the total in-place reserve of oil and gas in the explored field. This information and all the exploration information is passed on to reservoir group in the basin and also to the Institute of Reservoir Studies (IRS) to firm up a development plan. Normally a computer model of the reservoir is generated with all seismic, logging, fluid properties, and existing database and knowledge. A different exploitation plan is made to optimize the rate of fluid withdrawal from the reservoir with minimum investment. The development items include development wells, water injection wells (if required), and any artificial lift (if required) pipelines, production facility and delivery to customers through pipeline/tankers. Existing developed infrastructure and combination with adjacent potential reservoir development are also looked into. The final scheme is selected based on the best return on investment.

The study thus made is released on the final scheme with the following information (by Basin or IRS):

- Recoverable reserve
- Annual oil and gas rate
- Annual water injection quantity
- Period of production/field life

- Number of production and injection wells
- Well locations
- Reservoir fluid and physical parameters

The above information and data are taken into account while working on development and evacuation options. The study is to identify the following basic requirement of development:

- Location of the production platform
- Optimizing the number of platforms
- Optimize facility and space on platforms
- Well completion strategy
- Identify the required flow lines
- Identify the production facility
- Identify processing equipment requirement
- Identify the evacuation methodology

Technologically viable solutions are identified. Different combinations are worked out and a set of solutions are created. All cost elements are identified based on:

- Earlier project experience
- Existing database
- Personal experience
- Cost software

Capital cost of facility, pipeline, and wells are identified. A typical capital cost shall include the following heads:

- Well drilling and completion
- Well platform topside equipment and jacket
- Well flow lines
- Processing facility and pipeline/modification of existing facility cost
- Installation cost (platform and pipeline)
- Taxes and duties

Table 20.1 Operating cost for various items in oil and gas fields.

Sl. no.	Items	Unit rate	MMUS$/ year	Rs Cr/ year
1	R&M	3% CAPEX	6.558	30.168
2	Insurance	0.26% of capex	0.568	2.615
3	Well maintenance	6% of well cost	3.778	17.380
4	OSV charges	1.5 OSV/process complex	1.190	5.475
5	Helicopter charges	Rs. 7.2 crores/helicopter (1/2 helicopter)	0.783	3.600
6	Manpower	5.0 lakh/person/year for 60 × 2 persons	1.304	6.000
7	Catering	'@Rs. 140/ = for 60 persons	0.067	0.307
8	Consumables and stores	Lump sum	0.435	2.000
9	**Subtotal**		**14.684**	**67.544**
10	Overheads	6% of Subtotal	0.881	4.053
11	Interest on working capital	25% of 17% on 75% of WC	0.468	2.153
12	**Total:**		**16.033**	**73.750**

The operating cost is also worked out based on software, operating experience, or empirical formulas. Following is a typical operating cost (Table 20.1).

Economic analysis of all viable alternatives is made. The prevalent oil and gas prices, the tax regime (including cess and royalty) is considered. Internal rate of return (IRR) and net positive values (NPVs) are worked out. A typical set of assumptions for the analysis is reproduced below (Table 20.2):

The most economically viable development scheme that is the investment scenario with highest rate of return is recommended through the conceptual study report (CSR).

A feasibility report is prepared with all technical and cost information recommending for an investment. Sensitivity analysis on the investment is checked on varied oil price, fluctuation of capital, and operating expenditures to assess robustness of the proposal.

View of ONGC board is sought on the proposal. Necessary checks corrections and views are incorporated. Financial vetting is done through financial institutes like SBI caps, IDBI Bank, etc., and the project is presented before the Project Appraisal Committee; a board consists of some ONGC directors, the Sectary of Ministry of petroleum, and external members. This is the final body to approve the investment proposal.

Table 20.2 Assumed various parameter values for internal rate of return and net positive values.

Basis:	
Database year:	Dec' 2003
Year of investment:	2005–06
Year of first oil:	2007–08
Oil price:	**US$ 18/BBL**
Gas price :	Rs: 4742/1000 M^3 against US$ = 46/ = Rs. & 'oil @ US$18/bbl
Cess on oil:	(Deducted) 1800 Rs/ = per metric ton
Royalty on oil:	(Deducted) 1/11th of 90% oil gross price
Royalty on gas:	(Deducted) 1/11th of 90% gas gross price
Capex:	Escalated @6%/year
Opex:	Escalated @8%/year
Wells:	Escalated @6%/year
Rupee:	Devaluation considered @ 4%/year
Abandonment:	Deducted from revenue @ 10% of capex spread over the field life in proportion to production
Discount factor:	14%

A project coordinator is identified, preferably of GM level by CMD or ONGC board, and a team of engineers for steering the proposal to execution. Support of engineering services and materials management is obtained to identify an engineering consultant to create a basic engineering and bid document for the proposed construction activity. The process is generally done through an International Competitive Bidding (ICB).

Drilling of wells can however be undertaken through the existing capability of drilling services or can be through a drilling consultant and a drilling contractor. The engineering contractor is to ensure the availability of required deck or subsea hardware required to start drilling.

The basic engineering and bid document together with contractual clauses constitute the tender document for the work. Tendering process is initiated for the detailed engineering and construction of the project.

Pre-bid discussion once or twice is organized with the interested vendors; clarifications are provided on technical issues (by engineering services and engineering consultants) and commercial issues (by the related MM section) and their suggestions are also looked into for a better control of cost and time.

Work is awarded on competitive bidding. The project coordinator with the association of the engineering consultant does monitoring of the project. The performance run is done after commissioning and the production facility is handed to the asset operation group.

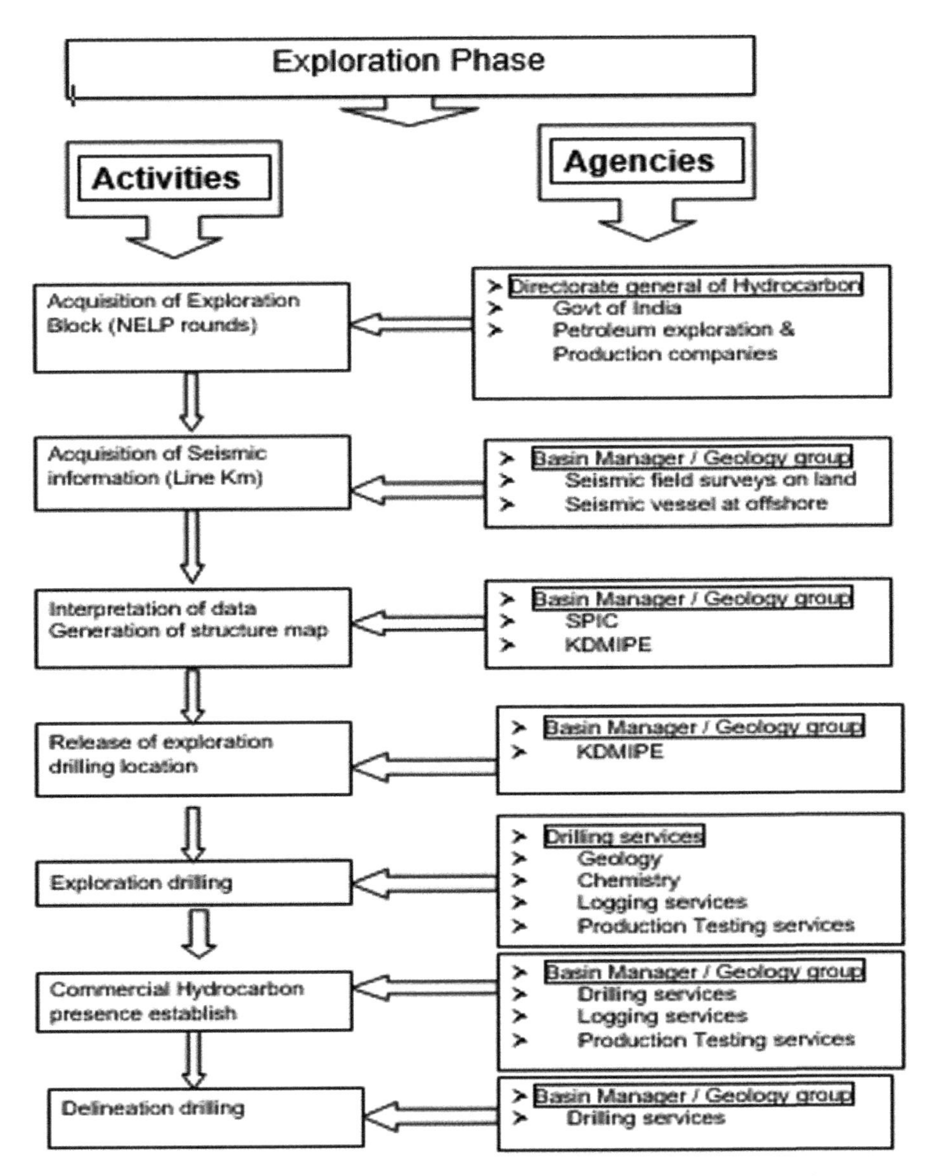

Figure 20.1 Different activities and agencies that come under exploration phase.

20.2 New Approach: Data to Delivery

A new methodology is being used in the process of development of South Bassein East oil and gas field development. Where the exploration data (Figure 20.1) like seismic (Figure 20.2), production testing, well logging,

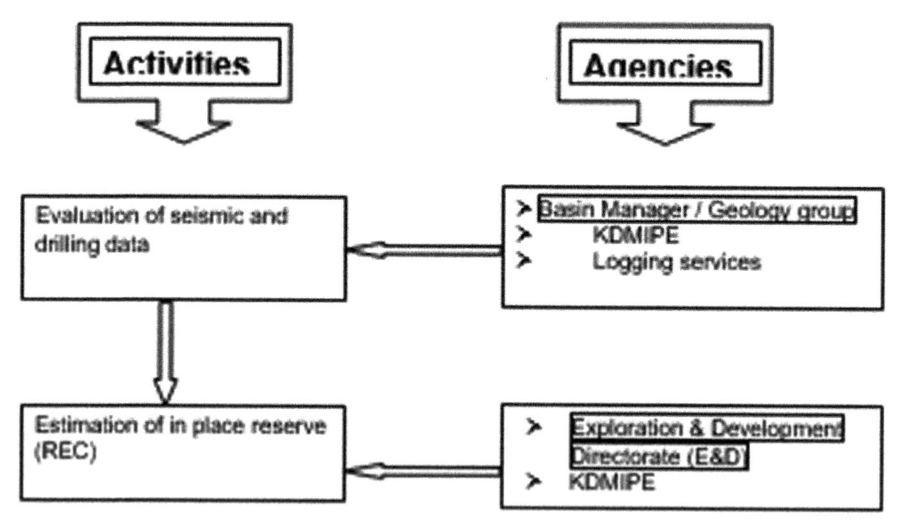

Figure 20.2 Different activities and agencies that come under geophysical method implementation.

and metocean data are offered as the basic information to the prospective bidder, who must in turn estimate reserves and recovery as well as the development proposal (Figure 20.3) as a lump sum contract value. It is believed that no single company can offer all the services required in the process; hence, a consortium of companies is expected to come forward (Figure 20.4). The evaluation of the bidders and linking ONGC's and contractor's interests (Figure 20.5) for the total life of the project is being worked out. Consultants having a global idea of handling contracts is being hired to work out the modalities.

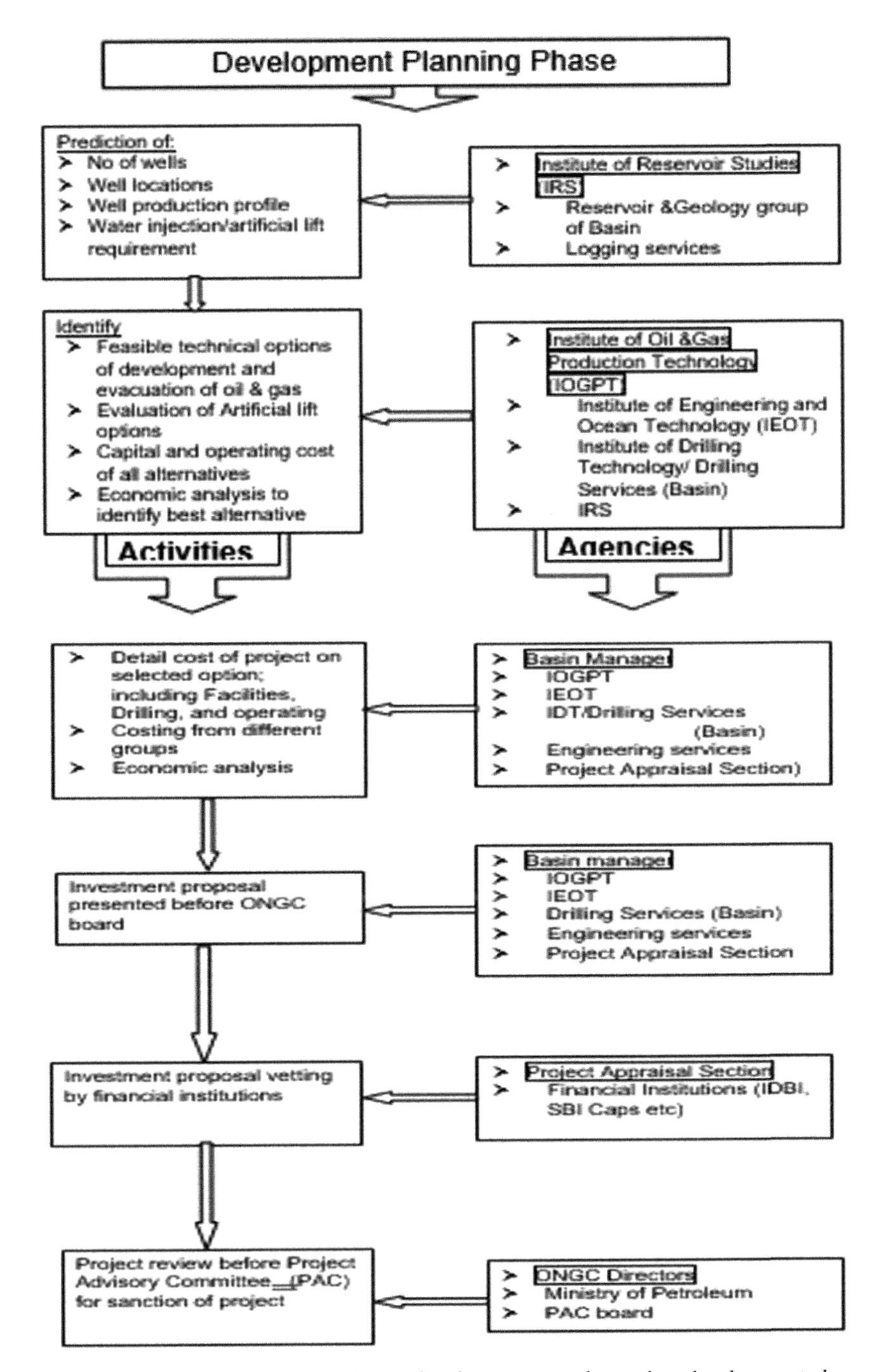

Figure 20.3 Various activities and agencies that comes under project development phase.

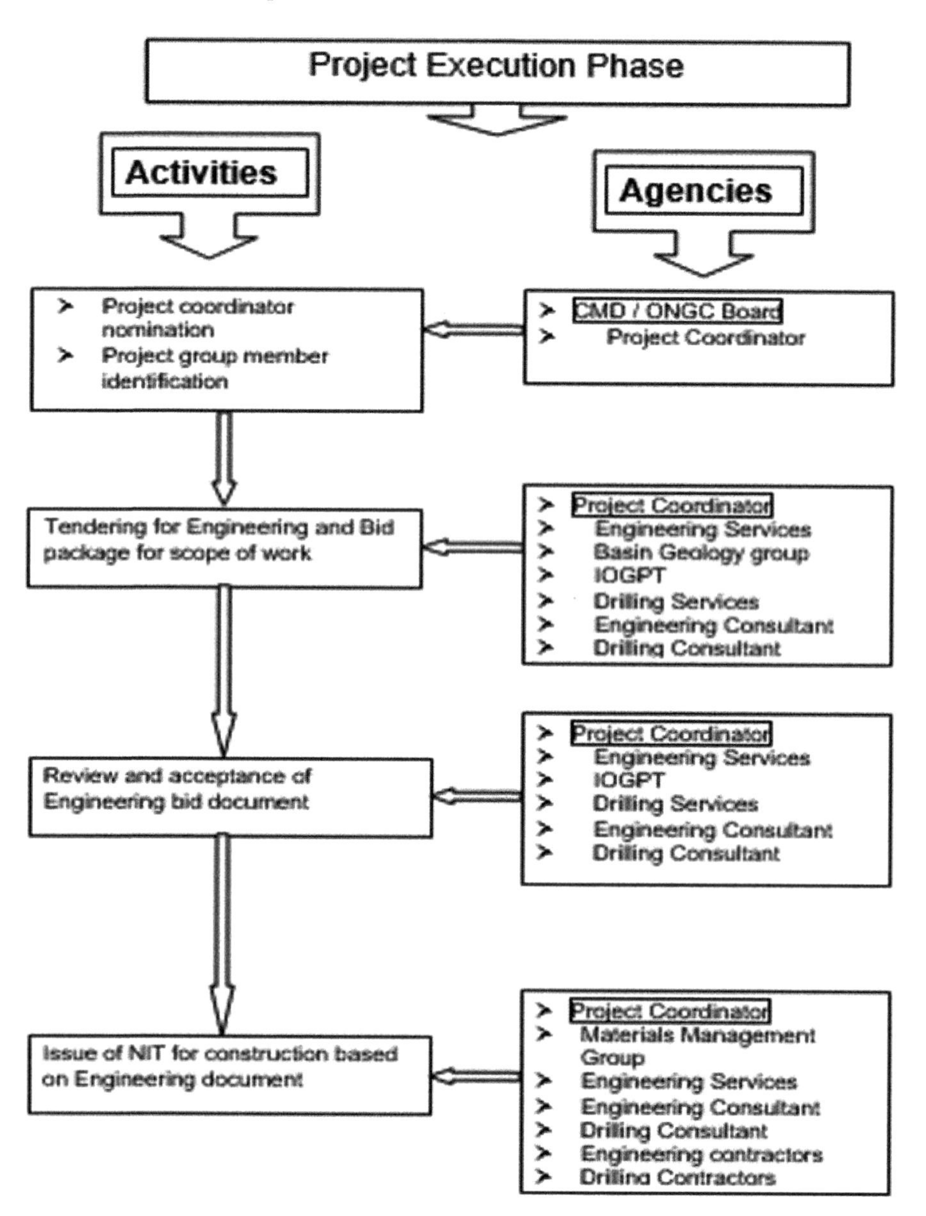

Figure 20.4 Various activities and agencies that come under the project execution phase.

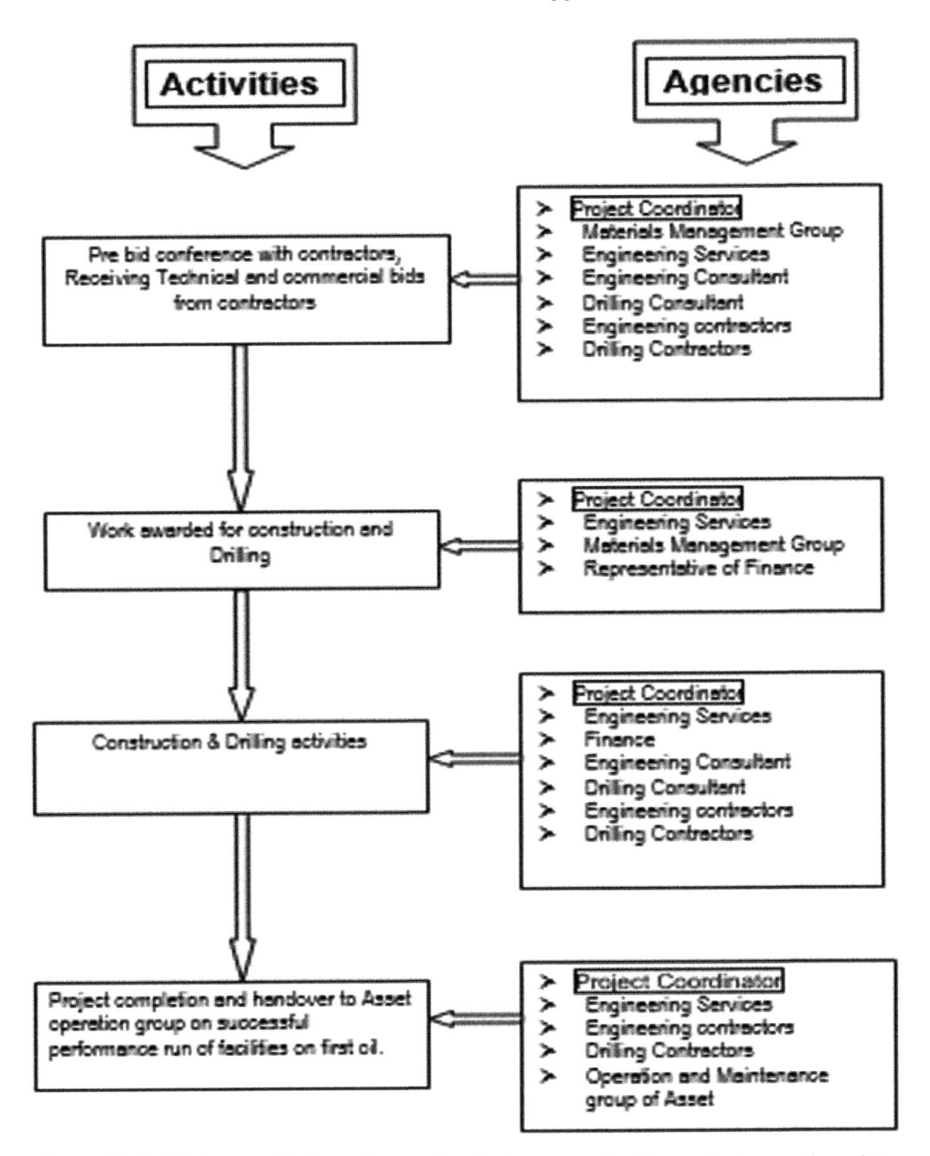

Figure 20.5 Various activities and agencies that come under the project execution phase with contractors.

Index

About the Authors

Dr. Mohammed Ismail Iqbal is a techno-managerial professional, who has hands-on experience from the classroom to the corporate boardroom, and everything in between. He is an engineer by training and a management enthusiast by choice. From being a team player in the oil field to a management consultant amongst the business leaders, he brings in the best of both worlds, be it related to education or corporate.

He has been a part of many initiatives that started with a concept on paper and took the shape of full-fledged business operations on the ground. His consolidated work experience is more than a decade but by the virtue of the mentors he worked under, he was able to utilize time in multiplicity to gain the depth and quality of experience to add value to whatever he came across.

His research papers are published in Elsevier, Scopus (Q4) Journal, peer-reviewed journals in area of general management, oil and gas, and IOT.

Dr. Vamsi Krishna Kudapa is Associate Professor, working in the Department of Chemical and Petroleum Engineering, UPES, Dehradun since 2013. Before UPES, he worked as an Assistant Professor and Head of the Department of Petroleum Engineering, at Sri Aditya Engineering College, Andhra Pradesh. He pursued his Ph.D. in oil and gas – modelling and simulation at UPES, Dehradun in 2018. In his 10 years of experience, labs like the Drilling Fluids & Cementation Lab and Petroleum Product Testing Lab were developed. In addition, a unique prototype model "Digital Oil Field" was developed at UPES in 2015, which is a combination of gas lift and oil-spill control mechanism apparatus. Presently, he has 35 publications, 2 patents, 4 book chapters, and 2 books to his credit. He is presently working on the modelling and simulation of unconventional gas reservoirs, solar energy applications in agriculture, nanoparticle applications in drilling fluids and cementation, and enhanced oil recovery.

9788770042031